COMET HALLEY
Investigations, Results, Interpretations
Volume 1: Organization, Plasma, Gas

THE ELLIS HORWOOD LIBRARY OF SPACE SCIENCE AND SPACE TECHNOLOGY

SERIES IN ASTRONOMY

Series Editor: JOHN MASON
Consultant Editor: PATRICK MOORE

This series aims to coordinate a team of international authors of the highest reputation, integrity and expertise in all aspects of astronomy. It makes valuable contributions to the existing literature, encompassing all areas of astronomical research. The titles will be illustrated with both black and white and colour photographs, and will include many line drawings and diagrams, with tabular data and extensive bibliographies. Aimed at a wide readership, the books will appeal to the professional astronomer, undergraduate students, the high-flying 'A' level student, and the non-scientist with a keen interest in astronomy.

PLANETARY VOLCANISM: A Study of Volcanic Activity in the Solar System
Peter Cattermole, formerly Lecturer in Geology, Department of Geology, Sheffield University, UK, now Freelance Writer and Consultant and Principal Investigator with NASA's Planetary Geology and Geophysics Programme

DIVIDING THE CIRCLE: The development of critical angular measurement in astronomy 1500–1850
Allan Chapman, Centre for Medieval and Renaissance Studies, Oxford, UK

SATELLITE ASTRONOMY: The Principles and Practice of Astronomy from Space
John K. Davies, Royal Observatory, Edinburgh, UK

THE ORIGIN OF THE SOLAR SYSTEM: The Capture Theory
John R. Dormand, Department of Mathematics and Statistics, Teesside Polytechnic, Middlesbrough, UK, and Michael M. Woolfson, Department of Physics, University of York, UK

THE DUSTY UNIVERSE
Aneurin Evans, Department of Physics, University of Keele, UK

SPACE-TIME AND THEORETICAL COSMOLOGY
Michel Heller, Department of Philosophy, University of Cracow, Poland

ASTEROIDS: Their Nature and Utilization
Charles T. Kowal, Space Telescope Institute, Baltimore, Maryland, USA

COMET HALLEY — Investigations, Results, Interpretations
Volume 1: Organization, Plasma, Gas
Volume 2: Dust, Nucleus, Evolution
Editor: J. W. Mason, B.Sc., Ph.D.

ELECTRONIC AND COMPUTER-AIDED ASTRONOMY: From Eyes to Electronic Sensors
Ian S. McLean, Joint Astronomy Centre, Hilo, Hawaii, USA

URANUS: The Planet, Rings and Satellites
Ellis D. Miner, Jet Propulsion Laboratory, Pasadena, California, USA

THE PLANET NEPTUNE
Patrick Moore, CBE

ACTIVE GALACTIC NUCLEI
Ian Robson, Director of Observatories, Lancashire Polytechnic, Preston, UK

ASTRONOMICAL OBSERVATIONS FROM THE ANCIENT ORIENT
Richard F. Stephenson, Department of Physics, Durham University, Durham, UK

EXPLORATION OF TERRESTRIAL PLANETS FROM SPACECRAFT: Instrumentation, Investigation, Interpretation
Yuri A. Surkov, Chief of the Laboratory of Geochemistry of Planets, Vernandsky Institute of Geochemistry, USSR Academy of Sciences, Moscow, USSR

THE HIDDEN UNIVERSE
Roger J. Taylor, Astronomy Centre, University of Sussex, Brighton, UK

AT THE EDGE OF THE UNIVERSE
Alan Wright, Australian National Radio Astronomy Observatory, Parkes, New South Wales, Australia, and Hilary Wright

COMET HALLEY
Investigations, Results, Interpretations
Volume 1:
Organization, Plasma, Gas

Editor

J. W. MASON B.Sc., Ph.D.

ELLIS HORWOOD

NEW YORK LONDON TORONTO SYDNEY TOKYO SINGAPORE

First published in 1990 by
ELLIS HORWOOD LIMITED
Market Cross House, Cooper Street,
Chichester, West Sussex, PO19 1EB, England

A division of
Simon & Schuster International Group
A Paramount Communications Company

Printed and bound in Great Britain
by Hartnolls, Bodmin, Cornwall

British Library Cataloguing in Publication Data

Comet, Halley: Investigations, results, interpretations.
1. Halley's Comet
I. Mason, J. W.
523.642
ISBN 0–13–171075–3

Library of Congress Cataloging-in-Publication Data

Comet Halley: Vol. 1. Investigations, results, interpretations / editor. J. W. Mason
p. cm. — (The Ellis Horwood library of space science and space technology. Series in
astronomy)
Includes bibliographical references and index.
Contents: v. 1. Organization, plasma, gas
ISBN 0–13–171075–3
1. Halley's comet. I. Mason, John (John W.) II. Series.
QB723.H2C63   1990
523.6'42–dc20

90–4789
CIP

Directory of Contributors

Abe, T.
Institute of Space and Astronautical Science, 3-1-1 Yoshinodai, Sagamihara, Kanagawa, 229 Japan.

Axford, W.I.
Max-Planck-Institut für Aeronomie, D-3411 Katlenburg-Lindau, Federal Republic of Germany.

Balsiger, H.
Physikalishes Institut, University of Bern, Sidlerstrasse 5, CH-3012 Bern, Switzerland.

Brandt, J.C.
Laboratory for Atmospheric and Space Physics, University of Colorado at Boulder, Campus Box 92, Boulder, Colorado 80309-0392, USA.

Eberhardt, P.
Physikalishes Institut, University of Bern, Sidlerstrasse 5, CH-3012 Bern, Switzerland.

Erdös, G.
Central Research Institute for Physics, PO Box 49, H-1525 Budapest, Hungary.

Feldman, P.D.
Dept. of Physics and Astronomy, Johns Hopkins University, Baltimore, Maryland, MD 21218, USA

Festou, M.C.
Observatoire de Besançon, 41 bis Avenue de l'Observatoire, F-25044 Besançon-Cedex, France.

Gringauz, K.I.
Space Research Institute, USSR Academy of Sciences, Profsoyuznaya 84/32, SU-117810, Moscow GSP-7, USSR.

Hirao, K.
Dept. of Aeronautics and Astronautics, Tokai University, 1117, Kitakaname, Hiratsuka, 259-12 Japan.

Ip, W.-H.
Max-Planck-Institut für Aeronomie, D-3411 Katlenburg-Lindau, Federal Republic of Germany.

Johnstone, A.D.
Mullard Space Science Laboratory, University College London, Holmbury St. Mary, Dorking, Surrey, England.

Kitayama, M.
Institute of Space and Astronautical Science, 3-1-1 Yoshinodai, Sagamihara, Kanagawa, 229 Japan.

Kömle, N.I.
Space Research Institute, Austrian Academy of Sciences, Lustbühel Observatory, A-8042 Graz, Austria.

Krankowsky, D.
Max-Planck-Institut für Kernphysik, Postfach 103980, D-6900 Heidelberg 1, Federal Republic of Germany.

McKenna-Lawlor, S.M.P. St. Patrick's College, Maynooth, Co. Kildare, Republic of Ireland.

Miyake, W. Communications Research Laboratory, 4-2-1 Nukuikitamachi, Koganei, Tokyo, 184 Japan.

Mukai, T. Institute of Space and Astronautical Science, 3-1-1 Yoshinodai, Sagamihara, Kanagawa, 229 London.

Neubauer, F.M. Institut für Geophysik und Meteorologie der Universität zu Köln, Albertus-Magnus-Platz, D-5000 Köln 41, Federal Republic of Germany.

Newburn, R.L., Jr. International Halley Watch, Jet Propulsion Laboratory, California Institute of Technology, 4800 Oak Grove Drive, Pasadena, California, CA 91109, USA.

Oyama, K-i. Institute of Space and Astronautical Science, 3-1-1 Yoshinodai, Sagamihara, Kanagawa, 229 Japan.

Rahe, J. Dr.-Remeis-Sternwarte, Sternwartstrasse 7, D-8600, Bamberg, Federal Republic of Germany.

Reinhard, R. Space Science Dept. of ESA, ESTEC, Noordwijk, The Netherlands.

Reme, H. Centre d'Etude Spatiale des Rayonnements, 9 Avenue due Colonel-Roche, BP 4346, 31029 Toulouse Cedex, France.

Saito, T. Onagawa Magnetic Observatory and Geophysical Institute, Tohoku University, Aramaki, Aoba, Sendai, 980 Japan.

Shapiro, V.D. Space Research Institute, USSR Academy of Sciences, Profsoyuznaya 84/32, SU-117810, Moscow GSP-7, USSR.

Shevchenko, V.I. Space Research Institute, USSR Academy of Sciences, Profsoyuznaya 84/32, SU-117810, Moscow GSP-7, USSR.

Somogyi, A.J. Central Research Institute for Physics, PO Box 49, H-1525 Budapest, Hungary.

Teresawa, T. Geophysical Institute, Kyoto University, Sakyo, Kyoto, 606 Japan.

Verigin, M.I. Space Research Institute, USSR Academy of Sciences, Profsoyuznaya 84/32, SU-117810, Moscow GSP-7, USSR.

Table of Contents

List of names, acronyms and abbreviations

AIAA	American Institute of Aeronautics and Astronautics
AMPTE	Active Magnetospheric Particle Trace Experiment
amu	Atomic mass unit
APV-N	Wave and plasma analyzer (Vega)
APV-V	Wave and plasma analyzer (Vega)
ASP-G	Automatic stabilized platform (Vega)
ASTRO-1	Three UV telescopes carried in Shuttle payload bay
ASTRON	Franco-Soviet UV space observatory
AU	Astronomical unit (150 million km)
BD	Bonner Durchmersterung
BRL	High data rate telemetry channel (Vega)
BS	Bow shock
BTM	Low data rate telemetry channel (Vega)
CA	Closest approach
CCD	Charge-coupled device
CD	Compact disk
CHON	Carbon-hydrogen-oxygen-nitrogen
CIS	Capacitive impact sensor (Giotto)
CNES	Centre National des Etudes Spatiales
CNRS	Centre National de la Recherche Scientifique
C/P	Coronagraph/polarimeter (SMM)
CRA	Cometary ram analyzer (Vega)
CRAF	Comet Rendezvous and Asteroid Flyby
CSE	Cometocentric solar ecliptic (coordinate system)
CSIRO	Commonwealth Scientific and Industrial Research Organization
CSG	Centre Spatial Guyanais (Kourou, French Guyana)

DE	Disconnection event
DE-1	Dynamics Explorer-1
D/H	Deuterium to hydrogen isotope ratio
DID (DIDSY)	Dust impact detection (system) (Giotto)
DSN	Deep Space Network
DUCMA	Dust particle detector (Vega)
EA	Electrostatic analyzer (Vega)
EESA	Electron electrostatic analyzer (Giotto)
EPA	Energetic particle analyzer (Giotto)
EPONA	Energetic particle experiment (Giotto)
ESA	European Space Agency
ESO	European Southern Observatory
ESP	Energy spectrum of particles plasma experiment (Suisei)
EUV	Extreme ultraviolet
FES	Fine error sensor (IUE)
FFS	Fast forward shock
FIS	Fast ion sensor (Giotto)
FITS	Flexible image transport system
FLD	Field line draping (coordinate system)
FRS	Fast reverse shock
FTS	Fourier transform spectrometer
GRE	Radio science experiment (Giotto)
GRT	Ground received time
GSFC	Goddard Space Flight Center
GSRT	Ground station received time
HERS	High-energy range spectrometer (Giotto)
HIS	High intensity spectrometer (Giotto)
HMC	Halley multicolour camera (Giotto)
HSE	Halley-centred solar ecliptic (coordinate system)
IACG	Inter-Agency Consultative Group
IAU	International Astronomical Union
ICE	International Cometary Explorer (formerly ISEE-3)
IGY	International Geophysical Year
IHW	International Halley Watch
IIS	Implanted ion sensor (Giotto)
IKS	Infrared spectrometer (Vega)
IMC	Interstellar molecular cloud
IMF	Interplanetary magnetic field (also experiment on Sakigake)

IMP-8	Magnetometer experiment on ICE
IMS	Ion mass spectrometer (Giotto)
ING	Neutral gas mass spectrometer (Vega)
IPM-M, -P	Impact momentum and plasma sensor (Giotto)
IPS	Interplanetary scintillation
IR	Infrared
IRAM	30-metre millimetre wave telescope
IRAS	Infrared Astronomy Satellite
ISAS	Institute of Space and Astronautical Sciences
ISEE-3	International Sun-Earth Explorer-3
ISTP	International solar-terrestrial phenomena
IUE	International Ultraviolet Explorer
JPA	Johnstone plasma analyzer
JPL	Jet Propulsion Laboratory
KAO	Kuiper Airborne Observatory
KSC	Kagoshima Space Center
LHS	Left-hand side (of equation)
LJO	Lear-Jet Observatory
L-SPN	Large-scale phenomena network (of IHW)
LWP	Long-wavelength primary camera (IUE)
Ly-α	Lyman-α line of hydrogen
MAG	Magnetometer experiment (Giotto)
MHD	Magnetohydrodynamic
MISCHA	Magnetometer experiment (Vega)
MIT	Massachusetts Institute of Technology
MPAE	Max-Planck-Institut für Aeronomie
MPB	Magnetic pile-up boundary
MSM/RSM	Meteoroid shield momentum sensor (Giotto)
MS-T5	Name for Sakigake before launch
MVA	Minimum variance analysis
NASA	National Aeronautics and Space Administration
NMS	Neutral mass spectrometer (Giotto)
N-NN	Near-nucleus network (of IHW)
NS	Neutral (current) sheet
NSF	National Science Foundation
OAO	Orbiting Astronomical Observatory
OB	Outer boundary
OGO	Orbiting Geophysical Observatory

OPE	Optical probe experiment (Giotto)
PB	Pile-up boundary (cometopause)
PDS	Planetary data system
PHOTON	Shield penetrator detector (Vega)
PIA	Dust mass spectrometer (Vega)
PICCA	Positive ion cluster composition analyzer (Giotto)
Planet-A	Name of Suisei before launch
PLASMAG	Plasma energy analyzer (Vega)
POM	Polyoxymethylene
PUMA	Dust mass spectrometer (Vega)
PVO	Pioneer Venus Orbiter
PWP	Plasma wave probe (Sakigake)
RFC	Ram faraday cup (Vega)
RHS	Right-hand side (of equation)
rms	Root mean square
RPA	Plasma analyzer (Vega)
SCET	Spacecraft event time
SDA	Solar direction analyzer (Vega)
SDFC	Solar direction faraday cup (Vega)
SFS	Slow forward shock
SMM	Solar Maximum Mission
SOFA	Second-order fermi acceleration
SOW	Solar wind experiment (Sakigake)
SP-1, -2	Dust particle detector (Vega)
SPARTAN	University of Colorado experiment on *Challenger*
SR	Sub-region
SRS	Slow reverse shock
TKS	Three-channel spectrometer (Vega)
TÜNDE-M	Energetic particle analyzer (Vega)
TVS	Television camera system (Vega)
UT	Universal time
UV	Ultraviolet
UVI	Ultraviolet imager (Suisei)
VLA	Very Large Array
VLBI	Very-long baseline interferometry
WG	Working group (of IHW or IACG)
WG$^+$	Water group ions

Editor's Preface

In 1985/86, thousands of scientists around the world turned their attention to perhaps the most famous of all cosmic visitors — Halley's comet. Returning to the inner Solar System once every 76 years or so, Halley's comet is the brightest periodical comet, and the return of 1985/86 was the thirtieth consecutive apparition to be recorded since that of 240 BC. The comet is named after the second Astronomer Royal for England, Edmond Halley, not because he discovered it but because he was the first person to calculate its path around the Sun.

Halley saw the comet which now bears his name in 1682, five apparitions ago and, believing it to be the same comet that had been seen previously in 1531 and 1607, predicted it would return in 1758. The comet was duly recovered by an amateur astronomer Johann Palitzsch on Christmas Night 1758, and it passed perihelion in March of the following year. Halley's correct prediction and in particular his use of Isaac Newton's laws of motion incorporating the newly formulated theory of universal gravitation were major scientific achievements — all the more so when one considers the superstitious fears associated with comets in the 17th century.

Studies of Halley's comet at successive returns have mirrored the rapid progress in technological developments during the last 150 years. In 1835, scientific studies were based on telescopic drawings made of the comet: in 1910 the newly discovered tools of photography and spectroscopy were used to secure a permanent record of the comet's appearance at visual wavelengths and determine its chemical composition.

In 1986, a flotilla of six spacecraft from four space agencies — nicknamed the 'Halley Armada' — encountered the comet and conducted *in situ* measurements. Few people alive in 1910 (less than seven years after the Wright brothers had made their first flight) would have believed that such a feat would be possible just 76 years later.

The 1985/86 return has turned out to be the most significant apparition ever of Halley's comet. A wealth of new and important data were obtained by the spacecraft which examined the comet at close range, including the first ever images of a cometary nucleus — the tiny dust-and ice-ball at the heart of the comet. In addition, many exciting and intriguing discoveries were made by observers using spacecraft in orbit around the Earth and Venus, and some of the world's largest ground-based telescopes in conjunction with the most sensitive electronic detectors, working in the infrared and ultraviolet wavebands as well as at visual wavelengths. When combined with the other remote observations carried out from rockets and high-flying aircraft, this was, at the time, the largest and most intensive observing campaign ever mounted on any astronomical object. Only the attention given to supernova SN1987A in the Large Magellanic Cloud, since February 1987, is in any way comparable.

Coordinating the vast comet Halley observing campaign — which began with the comet's recovery in October 1982 and continues even as this book goes to press, nearly eight years later — presented enormous difficulties. It was vital to standardize observing techniques and instrumentation, and help ensure that all data and results were carefully collated and properly documented and archived for future reference. This was especially important in view of some of the problems that were encountered in 1910. At that time, although the project had well-defined aims, many observatories did not cooperate with the central committee and, unfortunately, sufficient money and manpower were not available to utilize efficiently the enormous quantity of data which were collected. Indeed, the only comprehensive study dealing with the 1910 apparition of Halley's comet appeared in 1931 — some 21 years after the comet's perihelion passage — and the most detailed series of photographs, taken at the Lick Observatory, could not be included.

For the 1985/86 return, several thousand professional astronomers from more than 50 countries and tens of thousands of amateur astronomers worldwide, all coordinated by a body called the International Halley Watch (IHW), combined their efforts to monitor the comet's activity almost continuously at a wide range of wavelengths. The activities of the nations involved in spacecraft studies of the comet were coordinated by a separate body called the Inter-Agency Consultative Group (IACG). It is a testimony to the success of this organizational structure that such a wealth of scientific data has been obtained, collated and archived from the comet's most recent apparition.

As usual, when an important, worldwide scientific investigation, such as that of Halley's comet, is carried out, major conferences and symposia around the world provide a forum for the many scientists involved to present their results, identify areas of agreement and conflicting opinion, and discuss the conclusions which can be drawn from them. Conference proceedings containing the multitude of contributed oral and poster papers presented at these meetings are one way in which this information is published and disseminated. Indeed, several of these have already appeared, the most important being those from the conferences held in Heidelberg (October 1986), Brussels (April 1987) and Bamberg, FRG (April 1989), Discussion of observations of Halley's comet was also a major item on the agenda at the 20th General Assembly of the International Astronomical Union (IAU), held in Baltimore, USA in August 1988.

However, with an observational programme as complex, diverse and of such importance as that conducted during the 1985/86 return of Halley's comet, it was felt that something more than just conference publications and papers in the learned journals was needed to review the wealth of data obtained, and summarize in a shortened, more manageable form the most important results of the research carried out. It is with this in mind that the present work has been produced.

Comet Halley — Investigations, Results, Interpretations, has been published in two volumes, each dealing with three specific aspects of cometary research conducted during the recent Halley apparition. Volume 1 deals with Organization, Plasma and Gas and Volume 2 with Dust, Nucleus and Evolution. Here, the wide range of investigative techniques involved in the comprehensive studies of Halley's comet and the results obtained are carefully reviewed and discussed by 54 of the world's top cometary scientists, representing 17 different countries. This is not a collection of scientific papers, but a selection of review articles, each constituting a separate chapter in one of the two volumes. In general, the reviews contained in the present work are based on the published material available prior to October 1989 and include relevant results from IAU Colloqium No. 116, 'Comets in the Post-Halley Era', held in Bamberg, FRG during April 1989.

Although every effort has been made to minimize the degree of overlap between the various chapters in each section, some repetition has been unavoidable. This is particularly so where investigators involved in each of the five principal spacecraft missions, Giotto, Sakigake, Suisei, Vega-1 and Vega-2 have reviewed the results obtained by the instruments on that spacecraft within a particular subject area and compared them with those obtained elsewhere.

Finally a note on consistency. I have attempted to impose some uniformity of style between the chapters, while retaining the flavour of the original contributions. The meanings of all symbols are explained in the chapters as they appear, and, although some standard symbols appear throughout, there are minor differences in style from one chapter to another. No attempt has been made to settle differences of opinion between expert contributors, when to do so would be to fly in the face of common sense. Such differences demonstrate only too well that even after an investigation as intensive as that conducted during the recent apparition of comet Halley, a great many questions remain unanswered. Some of these may remain so until the comet next returns in 2060/61. Other problems may be resolved sooner, when planned future spacecraft missions to comets take place.

This project would not have succeeded without the enormous help that I received from the contributing authors. The task of drawing together all the material for the two volumes took considerably longer than originally anticipated, and I would like to thank all of the contributors for their patience and enthusiasm, and for responding promptly and helpfully to requests for information.

I am also pleased to acknowledge the assistance provided by the staff of Ellis Horwood; special thanks are due to Felicity Horwood for her guidance and encouragement, and to Beverley Ford for dealing with correspondence and secretarial matters.

John Mason
Chichester, England, 1990

Editor's Introduction

The extensive observations of comet Halley made throughout the recent apparition, since its recovery in October 1982 and continuing through to the present, have greatly expanded our knowledge of comets in general. The ICE spacecraft fly-by of Giacobini-Zinner in September 1985 and the encounters of the 'Halley Armada' with comet Halley in March 1986 have provided the first ever *in situ* measurements of the cometary environment. In particular, great advances have been made in our understanding of the nature of the complex interaction between the solar wind and the cometary plasma, and of the neutral gas and ion species present in the head and plasma tail of the comet. Most of the species identified at the time of the 1910 apparition of Halley's comet have been detected, together with a number of new species.

This book is divided into three sections, entitled Organization, Plasma and Gas. Part I, Organization, consists of just two chapters. In chapter 1, R. Reinhard describes the spacecraft missions to Halley's comet and the scientific payloads carried. Satellite, rocket and airborne observations are also covered. The structure and activities of the Inter-Agency Consultative Group (IACG) are described. Chapter 2 by R.L. Newburn and J. Rahe deals with the organization of the International Halley Watch (IHW), with a brief history of its conception, birth and growth. Its successes and failures are identified, and the problems of logging and archiving the enormous volume of scientific data are presented.

Part II, Plasma, is the largest section of the book and contains 11 chapters altogether. In chapter 3, J.C. Brandt describes the large-scale structure of the comet's plasma tail. The circumstances of some important Disconnection Events (or DEs) are also discussed. Chapter 4 by A.D. Johnstone describes some of the Giotto plasma analyzer measurements, which have helped to unravel the complex interactions between the solar wind plasma and its intrinsic magnetic field with the plasma of the comet. In chapter 5, F.M. Neubauer reviews observations made by the Giotto magnetometer experiment. The spatial structure of the cometary magnetoplasma was characterized by four major spatial regions with definite magnetic signatures. These are the far-upstream wave region, the magnetosheath, the magnetic pile-up region and the magnetic cavity region, separated by the bow shock, pile-up boundary/cometopause and ionopause, respectively. Chapter 6 by H. Rème describes plasma analysis measurements made by the two detectors of the RPA-COPERNIC experiment on Giotto. The positions of the main transitions at Halley (shock and ionopause) were consistent with the total gas production rate of the comet and the associated mass loading and neutral drag on the inflowing solar wind. Several other sharp discontinuities and regions were observed in the cometosheath. In chapter 7, T. Saito describes how the main plasma tail disturbances (DEs) identified from more than 500 ground-based photographs of the comet may be tabulated and classified. The observed disturbances are explained further by possible interaction models which utilize IMF observations made by Sakigake. Chapter 8 by H. Balsiger discusses Giotto measurements of the composition and dynamics of ion species within the coma of comet Halley. These have contributed to our knowledge of the composition of the volatile components of the nucleus, and of the mass-loading of the solar wind with cometary ions by the pick-up process. Chapter 9 by K.I. Gringauz and M.I. Verigin summarizes measurements made by the PLASMAG-1 plasma scientific packages aboard Vega-1 and -2, which incorporated neutral particle flux sensors and ion and electron energy spectrometers. The results generally confirmed the MHD description of the solar wind interaction with cometary plasma, but led to the discovery of a number of phenomena that were not envisaged. In chapter 10, K.-i. Oyama and T. Abe discuss the detection of various ion species, including water group ions, in the coma of comet Halley from the observations of Sakigake. Fluctuations in the solar wind velocity were related to the influence of the comet and the *in situ* gyrofrequencies of several ion species were measured. In chapter 11, S. McKenna-Lawlor reviews spacecraft measurements of very energetic pick-up ions in the environment of comet Halley. Various mechanisms whereby further acceleration and energization of the pick-up particles could take place are also discussed. In chapter 12, W. Miyake *et al.* describe plasma observations from Suisei, which reveal a shell structure for both protons and water group ions inside the cometosheath and in the upstream region. The spatial distribution of cometary ions up to 10 million km from the nucleus is summarized. In the final chapter of the Plasma section, chapter 13, A.J. Somogyi *et al.* comprehensively review the various ion acceleration mechanisms

operating in the vicinity of comets. The energy range and spatial distribution of high-energy ions observed by the spacecraft was far greater than expected, and many features of the high-energy processes around comet Halley are unexplained.

Part III, Gas, consists of five chapters. In chapter 14, K. Hirao describes observations by Suisei's ultraviolet imager of Lyman-alpha radiation from the hydrogen coma of comet Halley. The coma was found to be very active and variable, which was unexpected. From the strong 'breathing' of the hydrogen coma, the spin period of the comet's nucleus was determined as 2.2 days. Chapter 15 by N. Kömle describes the theoretical modelling of gas jets on the nucleus surface and observations of jet and shell structures within the coma. The activity of near-nucleus jets, the structure of large-scale jets, anisotropies and expanding haloes in the coma of comet Halley, and the physical nature of the CN-haloes are all discussed. M.C. Festou describes variations in the gaseous output of the nucleus of comet Halley in chapter 16. The factors governing the onset of gas emissions from the nucleus, long-term activity of the nucleus, visual brightness estimates and the light curve of the comet, possible seasonal effects and short-term fluctuations and outbursts in the brightness of the comet are all discussed. In chapter 17, P.D. Feldman describes how monitoring of the production rate of the dominant cometary species, water, in comet Halley was achieved by remote rocket and satellite observations of the ultraviolet emissions of H, OH and O. These observations showed the evolution of the gaseous coma, short-term variability of the water production rate, and the higher water production rates noted post-perihelion compared with preperihelion. Finally, in chapter 18, D. Krankowsky and P. Eberhardt use observations of the chemical species in the coma to deduce the composition of the ices in the comet's nucleus. Water constitutes about 80% by volume of the gases in the coma. An inventory of probable parent molecules other than water detected in the coma is presented and their likely origins discussed. Measurements of the D/H ratio in Halley were not compatible with the D/H in gaseous hydrogen of the protosolar nebula or the Jovian and Saturnian atmospheres.

Clearly, until we have new spacecraft data on comets, the results obtained in the investigations of Giacobini-Zinner and Halley in 1985 and 1986 will have to serve as our best and at present only guide as to what typical comets are really like. Fortunately, there is a good chance that several of the spacecraft involved in the encounters with comet Halley will visit other comets well within the next decade. Giotto is on course to rendezvous with comet Grigg-Skjellerup in July 1992, an old, relatively inactive comet that would provide an interesting contrast to Halley. Sakigake could be redirected to flyby comet Honda-Mrkos-Pajdusakova in 1996 and Giacobini-Zinner in 1998, while Suisei could be sent on to encounter comet Tempel-Tuttle in 1998 and Giacobini-Zinner later the same year.

Hopefully, if these missions continue as planned, then before too long we should have a wealth of new data on the plasma environment of comets, their complex interaction with the solar wind, and of the neutral gas and ion species present in their coma and plasma tails. Much of this data should be directly comparable with that obtained in the neighbourhood of comet Halley. This may go some way to answering some of the old questions which remain unanswered at the present time, as well as the new ones which have arisen as a consequence of the spacecraft encounters.

Part I
Organization

1

Space missions to Halley's comet and international cooperation

R. Reinhard

1 INTRODUCTION

For several reasons Halley's comet is the most important of about 750 known comets, and it comes therefore as no surprise that during the 1985/86 apparition not one but six spacecraft from four space agencies were sent to this comet, and that ground-based astronomers mounted one of the largest observational campaigns ever on an astronomical object. With its roughly 76-year period, Halley's comet is truly a 'once-in-a-lifetime' opportunity.

Halley's comet is the only highly active comet with a well-known orbit. Its orbit is inclined at 18° with respect to the ecliptic. During the 1985/86 apparition it crossed the ecliptic on 9 November 1985 (ascending node), and again on 10 March 1986 (descending node). The time of the second crossing was only 4 weeks after perihelion (9 February 1986), and Halley's comet was only 0.84 AU away from the Sun and still highly active. To send a spacecraft to this point required one of the lowest launch energies of all possible cometary missions, so that even with a moderate launch vehicle a fairly comprehensive scientific payload could be sent to Halley. Finally, Halley was observable from the Earth during the flyby, which was important for correlating the *in situ* observations with remote ground-based observations.

Halley's comet is also the most famous comet. Because of its high activity it has been recorded at each apparition since 240 BC, and was certainly observed much earlier. Five apparitions ago, the English astronomer Edmond Halley discovered in 1705, the periodicity of what later became known as 'his' comet and correctly predicted its return in 1758, a triumph for science best appreciated in the context of contemporary views, or rather fears, about comets at that time.

A mission to Halley's comet, however, also has a disadvantage: the comet's orbit is retrograde, i.e. it orbits the Sun in the opposite direction to the Earth and any spacecraft launched from the Earth. This has the consequence that the relative flyby velocity is very high (~ 70 km s^{-1}). A spacecraft traverses the whole coma, the extent of which is comparable to the Earth–Moon distance, in only 1.5 h. To achieve a reasonable spatial resolution, scientific data must be transmitted at a very high rate. Secondly, during the coma passage, dust particles impact on the spacecraft at that high velocity, and could cause serious damage. So, a special dust protection shield is required for spacecraft that are going to penetrate deeply into the cometary coma.

2 THE VEGA-1 AND VEGA-2 MISSIONS

An unique opportunity to combine missions to Venus and comet Halley was available in 1985/86 by employing a two-element space vehicle consisting of a Venus lander and a Halley flyby probe. The mission was called Vega, a contraction of the Russian words Venera (Venus) and Gallei (Halley), and was conducted by the USSR in cooperation with a number of other countries.

Two spacecraft, Vega-1 and Vega-2, were launched by Proton rockets from the Baikonur Cosmodrome on 15 and 21 December 1984, respectively. The two spacecraft were identical, and the redundancy was aimed at increasing the overall reliability of the mission. The basic design of the spacecraft chosen for this mission is the same as has been used many times to deliver Soviet landers and orbiters to Venus. On 11 and 15 June 1985, the two Vega spacecraft successfully delivered the first balloons into the Venusian atmosphere (each carrying four scientific experiments) as well as delivering landers with nine experiments to the surface of the planet. The results of this part of the mission are published in Sagdeev & Moroz (1986), Sagdeev *et al.* (1986a, b). The flyby modules made a gravitational manoeuvre near Venus, and finally Vega-1 and Vega-2 encountered comet Halley on 6 and 9 March 1986.

Each Vega spacecraft was composed of the Halley flyby probe and a Venus descent module; the whole system weighed about 4.5 t at launch. The Halley probe is shown in Fig. 1 in its nominal Halley flyby configuration. On its trajectory the probe was still surmounted by the Venus descent module (not shown in Fig. 1) which was a spherical structure with a diameter of 2.5 m and a mass of approximately 2 t. The spacecraft is 3-axis stabilized with a wingspan of the order of 10 m. It carries 14 experiments (Table 1), among them a TV system for imaging the inner coma and cometary nucleus.

Successful imaging of the comet from the spacecraft required a steerable platform which could be automatically pointed with great accuracy at the cometary nucleus. The most difficult problem was to locate the nucleus with its very low albedo against the background of the very bright coma with gas and dust jets. The accurate pointing achieved allowed the examination of the innermost parts of the coma and near-nucleus region by means of television cameras and infrared and optical spectrometers placed on the steerable platform. The platform is shown in the lower part of Fig. 1. It has a mass of 82 kg and carries 64 kg of instrumentation. It could scan an angular sector of 110° in the ecliptic plane and 60° in a plane perpendicular to the ecliptic. The pointing platform could track the nucleus with an accuracy of the order of 2.5 arc min.

Table 1 — VeGa scientific payload

Acronym	Experiment	Mass [kg]	Power [W]	Direct telemetry [bit s^{-1}]
TVS	Television system	32	50	32 768
IKS	Infrared spectrometer	18	18	2048
TKS	Three-channel spectrometer	14	30	12 288
PHOTON	Shield penetration detector	2	4	108
DUCMA	Dust particle detector	3	2	100
SP-2	Dust particle counter	4	4	1024
SP-1	Dust particle counter	2	1	150
PUMA	Dust mass spectrometer	19	31	10 240
ING	Neutral gas mass spectrometer	7	8	1024
PLASMAG	Plasma energy analyzer	9	8	2048
TÜNDE	Energetic particle analyzer	5	6	512
MISCHA	Magnetometer	4	6	512
APV-N	Wave and plasma analyzer	5	8	2048
APV-V	Wave and plasma analyzer	3	2	512

Recorded telemetry [bit/20 min]	Goal and instrument parameters	Collaborating institutes (Principal Investigators)
–	Inner coma and nucleus imaging. Two CCD cameras (fields of view, $0°43 \times 0°57$ and $3°5 \times 5°3$)	LAS, Marseille, France (P. Cruvellier) Central Research Institute for Physics, Budapest, Hungary (L. Szabo) IKI Moscow, USSR (G. Avanesov)
4320	Detection of infrared emissions of coma and thermal radiation of nucleus ($2.5 < \lambda < 12\ \mu$m)	Observatoire de Meudon, France (M. Combes) IKI, Moscow, USSR
–	Spectral mapping of coma emissions in the range $0.12 < \lambda < 19\ \mu$m	Observatoire de Besançon, France (G. Moreels) IKI, Moscow, USSR (V. Krasnopolskii) Bulgaria (M. Gogoshev)
–	Large dust particle detection (under anti-dust shield)	USSR
100	Dust particle flux and mass spectrum ($m > 1.5 \cdot 10^{-13}$ g)	University of Chicago, USA (J. Simpson) MPI, Lindau, W. Germany IKI, Moscow, USSR Central Research Institute for Physics, Budapest, Hungary
2160	Dust particle flux and mass spectrum ($m > 10^{-16}$ g)	IKI, Leningrad, USSR (E. Mazets)
2160	Dust particle flux and mass spectrum ($m > 10^{-16}$ g)	IKI, Moscow, USSR (O. Vaisberg)
–	Dust particle elemental composition	MPI, Heidelberg, W. Germany (J. Kissel) Service d'Aéronomie, Verrières, France (J.-L. Bertaux) IKI, Moscow, USSR (R. Sagdeev)
1080	Neutral gas composition	MPI, Lindau, W. Germany (E. Keppler) Central Research Institute for Physics, Budapest, Hungary IKI, Moscow, USSR University of Arizona, USA
15 120	Ion flux composition, energy spectra of ions and electrons	IKI, Moscow, USSR (K. Gringauz) Central Research Institute for Physics, Budapest Hungary MPI, Lindau, W. Germany ESA Space Science Dept., ESTEC, Netherlands
6480	Energy and flux of accelerated cometary ions	Central Research Institute for Physics, Budapest, Hungary (A. Somogyi) IKI, Moscow, USSR MPI, Lindau, W. Germany ESA Space Science Dept., ESTEC, Netherlands Nuclear Research Institute, Moscow, USSR
2160	Magnetic field	Space Research Institute, Graz, Austria (W. Riedler) Izmiran, Troitsk, USSR
28 080	Plasma waves, 0.01–1000 Hz, plasma ion flux fluctuations	IKI, Moscow, USSR (S. Klimov) Aviation Institute, Warsaw, Poland Geophysical Scientific Institute, Prague, Czechoslovakia
15 120	Plasma waves, 0–300 kHz, plasma density and temperature	ESA Space Science Dept., ESTEC (R. Grard) LPCE, Orléans, France Izmiran, Troitsk, USSR

The Vega spacecraft is derived from the Venera series of spacecraft. A number of modifications improve the reliability of the probe, for example, 5 m² of shield have been added to protect the most essential subsystems against the bombardment of dust particles with masses 0.01 g. A dual-sheet bumper shield has been adopted, composed of a thin metallic front sheet (0.4 mm) and a thicker rear sheet, separated by several centimetres.

The spacecraft structure resembles a cylindrical body connected to two conical skirts. The lower skirt houses a motor for orbital manoeuvres and a toroidal

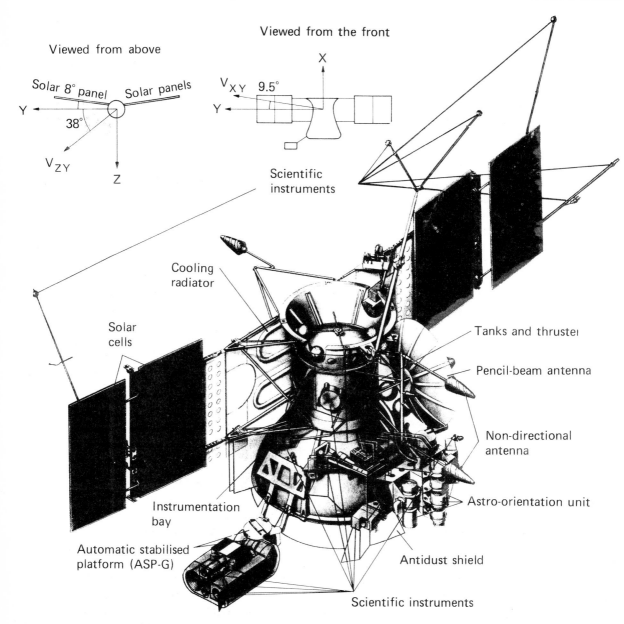

Fig. 1 — The Halley probe of the Vega spacecraft (Vega-1 and Vega-2 were identical), showing the locations of scientific instruments. Also shown is the orientation of the Vega spacecraft during the Halley flyby in the spacecraft coordinate system.

pressurized utility instrument bay. The cylindrical compartment contains the fuel tanks, and the upper skirt is the interface that held the Venus lander. Two pairs of deployable solar panels are mounted on each side of the cylindrical section; the solar array has a total area of nearly 10 m². Three-axis stabilisation with 1° accuracy in flight is achieved by a gyroscopic system and a number of gas nozzles, most of which are mounted on the solar panels.

The telemetry system consists of a high-data-rate channel (BRL) and a low-data-rate channel (BTM). The BRL channel is used for real-time transmission only. Its telemetry rate of 65 536 bit s⁻¹ could be reduced by half if required by propagation conditions. The telemetry rate of the BTM channel is 3072 bit s⁻¹. Scientific data could also be stored by onboard magnetic tape recorders with 5 Mbit capacity and subsequently telemetered through the BTM channel. This was done every 20 days during the interplanetary cruise phase, and once every 20 minutes during the cometary flyby. The spacecraft orientation must be such that the high gain antenna is pointed at the Earth during high data rate transmission periods.

The spacecraft attitude during the encounter (Fig. 1, inset) was determined by the requirements to point the high-gain antenna at the Earth and to obtain maximum power from the solar panels. Consequently, the solar power panels could not be aligned with the relative velocity vector, and damage to the panels was inevitable. After the encounters, the power from the solar cells decreased by almost 50%.

The flight operations centre was located in Evpatoria (Crimea), but most investigators were in Moscow at the Space Research Institute, where they were able to obtain all data in real time. Deep-space antennae in Evpatoria (70 m) and Medvezy Ozera (64 m), near Moscow, received the telemetry.

All experiments were switched on two days before closest approach. The direct high-speed telemetry was transmitted from −48 to −46 hours, from −24 to −22 hours, from −3 to +1 hour, from +21 to +24 hours, and from +45 to +48 hours during the cometary flyby. A limited amount of data were also stored on the onboard tape recorder and transmitted every 20 minutes to cover the 2-hour gaps between high-speed telemetry intervals.

The Vega mission, in particular the experiments, is described in a comprehensive document entitled *Venus–Halley mission, Experiment description and scientific objectives of the international project Vega (1984–1986)*, published by Imprimerie Louis-Jean, 05002 Gap, France, in May 1985. A shortened version can be found in Grard *et al.* (1986).

3 THE SUISEI AND SAKIGAKE MISSIONS

Japan's Institute of Space and Astronautical Science (ISAS) decided in 1979 to send a spacecraft to comet Halley. This decision was supported both by Japanese scientists who were anxious to carry out an interplanetary mission and by those Institute engineers who wanted to demonstrate the capabilities of a new launch vehicle.

The Japanese mission to comet Halley used two spacecraft, a test spacecraft called 'Sakigake' (Japanese for 'forerunner', this spacecraft was called MS-T5 until launch) and a dedicated comet Halley encounter spacecraft called 'Suisei' (Japanese for 'comet', this spacecraft was called Planet-A until launch).

Sakigake and Suisei are almost identical spacecraft (Fig. 2) except for the scientific experiments they carry. The spin-stabilized spacecraft are cylindrical with a diameter of 140 cm and a height of 70 cm. An offset despun parabolic antenna of 70 cm diameter is mounted on top of the cylinder. At the bottom there is a low-gain, cross dipole antenna and a medium-gain antenna of the colinear type. The total height of the spacecraft, including antennae, is ∼2.5m. Around 70% of the outside of the cylinder is covered with solar cells which produce about 100 W during cruise, when the spin axis of the spacecraft is kept perpendicular to the ecliptic plane. The Sakigake and Suisei spacecraft weigh 138.1 and 139.5 kg, respectively, including 10 kg of hydrazine propellant in each. The attitude and orbit control system is composed of a two-thruster complex mounted diagonally at the top of the cylindrical body, the thrusting force of each thruster being 3 N. The spacecraft is usually spin-stabilized at 6 rev min⁻¹ which may be reduced to 0.2 rev min⁻¹ in the case of Suisei by means of a momentum wheel when the ultraviolet imager of Suisei is in operation. Information on the spacecraft attitude is provided by a Sun sensor and a star scanner.

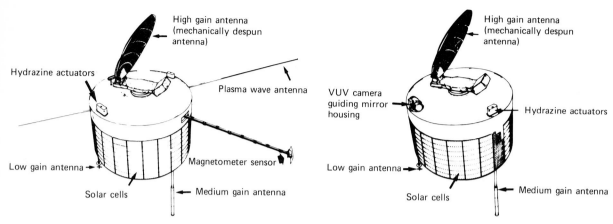

Fig. 2 — The Sakigake (left) and Suisei (right) spacecraft.

The communications subsystem operates in S-band, using a 5 W transmitter. The spacecraft possess two data rates, a 'high bit rate' mode (2048 bit s^{-1}) and a 'low bit rate' mode (64 bit s^{-1}). The bit rate used depended on the communication distance. The ground station for the mission was built at Usuda, about 170 km northwest of Tokyo. The station's 64 m dish antenna has a gain of approximately 63 dB. During the Halley encounter data were received at Usuda at a rate of 64 bit s^{-1}. A 1 Mbit magnetic bubble memory onboard the spacecraft recorded data obtained during periods of non-contact with the Usuda Deep Space Station.

Suisei (Fig. 3*) carried two scientific experiments, a

Table 2 — The Suisei (top) and Sakigaki (bottom) scientific payloads

Suisei Experiment	Mass (kg)	Power (W)	Data rate (bit s^{-1}) Format 1/2	Principal Investigator
UVI	7.5	8.9	H 1568/0 L 98/0	E. Kaneda, Geophys. Res. Lab., Faculty of Science, Univ. of Tokyo
ESP	4.7	4.9	H 0/1408 L 0/88	T. Mukai, Res. Div. Planetary Science, ISAS, Tokyo

Format 1	UVI and UVI check	H = High bit rate
2	ESP	L = Low bit rate

Sakigate Experiment	Mass (kg)	Power (W)	Data rate (bit s^{-1}) Format 1/2/3/4	Principal Investigator
PWP	5.2†	1.7	H 512/768/768/0 L 32/48/48/0	H. Oya, Faculty of Science, Tohoku Univ., Sendai
SOW	2.0	1.1	H 512/768/0/768 L 32/48/0/48	K. Oyama, Res. Div. Planetary Science, ISAS, Tokyo
IMF	5.4††	2.0	H 400/0/768/768 L 25/0/48/48	T. Saito, Faculty of Science, Tohoku Univ., Sendai

Format 1	PWP	SOW	IMF	H = High bit rate
2	PWP	SOW		L = Low bit rate
3	PWP	IMF		
3	SOW	IMF		

† Includes dipole atenna, search coil, and electronics
†† Includes boom

* to be found in the colour illustration section in the centre of the book.

UV imager (UVI) and a plasma experiment (ESP) for the observation of the solar wind plasma and cometary ions. Sakigake carried three scientific experiments, a plasma wave probe (PWP), a solar wind experiment (SOW), and, on a 2 m boom, a magnetometer (IMF). Some key experiment parameters are given in Table 2. The Suisei and Sakigake missions are described in more detail in Hirao (1986).

Both spacecraft were launched by M-3SII launchers. This launcher is a three-stage solid propellant rocket with a solid kick-stage motor attached. Both spacecraft were injected directly into a heliocentric orbit.

All experiments onboard Sakigake were switched on at the end of February 1985, and both experiments onboard Suisei were switched on in September 1985. UV images of the comet's hydrogen coma were recorded almost continuously from 26 November 1985 until 15 April 1986 except for the period 11 January to 9 February when the elongation angle of the comet precluded observation. At specific times during and a few days after the encounter, data were obtained in the photometry mode.

Operation of the UV imager had to alternate with operation of the plasma experiment because together they needed a higher data rate than the capacity of the spacecraft telemetry allowed, and also because the experiments required different spacecraft spin rates. UV observations of the comet in the photometry mode were made until 1220 UT on 8 March. Data from the plasma experiment are available from 1232 UT on 8 March until 0402 UT on 9 March.

Suisei passed the comet's nucleus at a distance of 151 000 km on the sunward side (flyby parameters in Table 6). It crossed the bow shock inbound presumably around 1100 UT and outbound between 1443 and 1449 UT. Closest approach was at 1305:49 UT. Thus the inbound crossing was missed, but good data are available for the outbound crossing except for the data gap of 6 minutes. At 1254 and at 1326 UT Suisei was hit by cometary dust particles, which were of sufficient mass (a few milligrammes and tens of microgrammes, respectively) to cause a noticeable change of the spacecraft attitude and spin period. Nevertheless, Suisei survived the encounter undamaged and without any loss of data due to dust-impact induced nutation.

Sakigake passed the comet's nucleus at a distance of 6.99×10^6 km on the sunward side. The time of closest approach was 0417:51 UT on 11 March. All experiments worked flawlessly but could not be operated continuously because of the limited visibility from the Usuda Deep Space Station. During the Halley encounter, scientific data from the experiments onboard Sakigake are available on 10 March from 0247 to 0347 and on 10, 11, and 12 March from 1940 to 0345. Data are also available in the storage mode on 11 March from 0345 to 0837 UT.

Both spacecraft are now orbiting the Sun. After an orbit correction manoeuvre on 25–28 January 1987 ($\Delta V = 37$ m s^{-1}), Sakigake is now on an Earth return trajectory, due to arrive at the Earth on 8 January 1992. If still functional it could be redirected, using a series of three Earth-swingby manoeuvres, to flyby comet Honda–Mrkos–Pajdusakova on 4 February 1996 at a distance of 10 000 km. Almost 3 years later, it would also be in the vicinity of comet Giacobini–Zinner, passing its nucleus on 29 November 1998 at a distance of several million km.

Suisei's orbit was corrected in April 1987 to return to the Earth. Using an Earth swingby in August 1992 it could be redirected to flyby comet Tempel–Tuttle on 28 February 1998 at several million kilometres distance. Almost at the same time as Sakigake, on 24 November 1998, it would also make a very close approach to comet Giacobini–Zinner.

4 THE GIOTTO MISSION

As was the case for the Japanese Suisei and Sakigake missions, the Giotto mission to Halley's comet was also ESA's first interplanetary mission. The ESA mission was named 'Giotto' after the Italian painter Giotto di Bondone who in 1304 depicted comet Halley as the 'Star of Bethlehem' in one of the frescoes in the Scrovegni chapel in Padua.

The Giotto spacecraft (Fig. 4) is spin-stabilized, nominally at 15 rev min^{-1}. Its diameter is 1.86 m and the overall height is 2.85 m. At launch Giotto weighed 960 kg, reducing to 573.7 kg during the Halley flyby, after the solid-propellant kick motor, used to inject the spacecraft from a geostationary transfer orbit into a heliocentric orbit, had burnt out, and 9 kg of the available 69 kg of hydrazine had been used for three

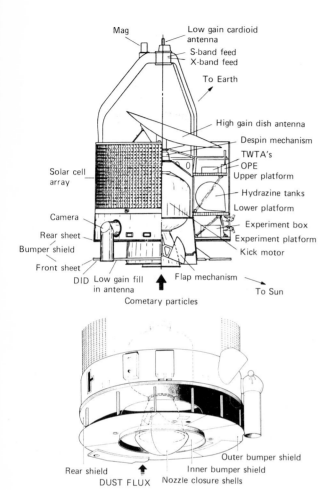

Fig. 4 — left: Cross-section through the Giotto spacecraft. The dust protection system is shown below. right: The Giotto spacecraft mounted on top of the Ariane 1 launch vehicle.

orbit corrections and a large number of attitude corrections.

As it was intended to target Giotto to within a few hundred kilometres of the nucleus where the dust fluxes are high, protection of the spacecraft from high-velocity dust particle impacts was essential. Giotto carries a dual-sheet bumper shield, the face of which was forward during the Halley flyby (downward in Fig. 4). This shield consists of a thin front sheet (1 mm of aluminium) and a thick rear sheet (Kevlar sandwich), separated by 23 cm. The principle of the bumper shield is as follows. A dust particle striking the thin front sheet is completely vaporized. The vapour cloud then expands into the empty space between the two sheets and strikes the rear sheet, where its energy is distributed over a large area. It had been calculated that the shield which weighs about 50

kg can withstand 70 km s^{-1} impacts of dust particles up to 0.1 g in mass or even higher.

At the other end, well protected from dust particle impacts, Giotto carries a high-gain antenna with an effective reflector diameter of 1.47 m. This antenna can be operated in either the S-band (receiving at 2.1 GHz, transmitting at 2.3 GHz) or the X-band (transmitting at 8.4 GHz). The antenna beam is inclined 44°.3 with respect to the spacecraft spin axis, and the antenna itself is despun, so that the beam could point continuously at the Earth during the encounter. The pointing requirements in the X-band are rather stringent: if the antenna beam were to be misaligned by more than 1° the telemetry link to the ground receiving station would be lost. It was calculated before the mission that the impact of a 0.1 g dust particle on the rim of the bumper shield could change the spacecraft attitude and hence the antenna pointing direction by more than 1°.

The solar cell array provided 196 W of power during the encounter, which was not quite sufficient to power the spacecraft and all experiments. Also, the possibility of damage to the solar cell array by dust impacts during the flyby had to be accounted for. Giotto, therefore, carries in addition four Si–Cd batteries.

Giotto's scientific payload comprises 10 hardware experiments (Table 3). All experiments are mounted on the 'experiment platform' of the spacecraft which is on top of the rear sheet of the bumper shield. The experiment sensors protrude up to 17 cm (this limit was imposed by the fairing of the Ariane 1 launch vehicle) from the spacecraft side wall to allow measurements in the undisturbed flow of cometary particles. The sensors of the Dust Impact Detection System (DID) are mounted on the outer face of the front bumper shield; the two magnetometer sensors (MAG) are mounted on the carbon fibre tripod as far away from the spacecraft's magnetic field sources as possible; and the Optical Probe Experiment (OPE) is mounted on the upper platform inside the spacecraft, looking rearward. All other experiments look in the forward direction (downward in Fig. 4), apart from some plasma experiments, which look to the side to observe the changing characteristics of the solar wind plasma. The camera can be rotated by 180°. This, together with the spacecraft spin, allows viewing in all directions.

Giotto has no onboard data storage capability; all data are transmitted in real time. Two experiments (MAG and EPA: acronyms are defined in Table 3) have small memories to bridge gaps in ground-station contact during the cruise phase.

As the spacecraft traversed the coma it was decelerated by dust and gas impacts (atmospheric drag); a radio science experiment was designed to measure this deceleration by observing the Doppler frequency shift of the Giotto radio signal, and to derive from it the dust column density along the Giotto trajectory. Detailed experiment, spacecraft, and mission descriptions can be found in the ESA Special Publication No. 1077 (1986), in Lo Galbo (1984), and Reinhard (1987a).

Giotto was launched on 2 July 1985 by an Ariane 1 launch vehicle from Centre Spatial Guyanais (CSG) in Kourou. The interplanetary trajectory (Fig. 8*) was in the ecliptic, with the closest approach to the Sun of 0.72 AU.

For routine operation of the spacecraft in the S-band, ESA's 15 m ground station at Carnarvon, Australia, was used, while the 64 m ground station at Parkes, Australia, was used for high-rate transmission of scientific data (Format 1 or 2, Table 3). The Parkes antenna is owned and operated by the Commonwealth Scientific and Industrial Research Organisation (CSIRO), and is normally used for radioastronomy. During the cruise phase, scientific data were transmitted a few times per week to Carnarvon and the 30 m station at Weilheim, W. Germany at low rate (Format 3). During the encounter, continuous data coverage for ∼70 h before and ∼30 h after closest approach in high-data rate mode was provided by the Parkes ground station together with the NASA 64 m Deep Space Network (ground stations at Goldstone, Madrid, Canberra). To save power, not all the experiments were operated throughout that period. Also, during orbit and attitude manoeuvres, the payload was switched off (except for MAG and EPA which stayed on continuously). The periods for which encounter science data are available are given for each experiment in Table 4.

The geometry of the Giotto encounter with comet Halley is shown in Fig. 9*, together with the encounter geometries of the other five Halley flyby spacecraft. The phase angles and flyby velocities were all very similar. Giotto was the last spacecraft to encounter comet Halley, and made the closest approach.

Giotto was targeted to fly past the nucleus at a closest distance of 540 km on the sunward side. This distance was chosen as a compromise between the partly conflicting requirements of three groups of experiments. The camera experimenters wanted to fly past ideally at 1000 km but no closer than 500 km because at distances <500 km the camera could not rotate fast enough to follow the apparent motion of the nucleus during the flyby. A second group of experimenters (NMS, IMS, OPE, MAG) wanted to approach as close as possible to the nucleus, even if the spacecraft would not survive, and a third group (PIA, DID, JPA, RPA) also wanted to approach as close as possible, while maintaining a high survival probability. The distance of 500 km was acceptable to all three

---

* to be found in the colour illustration section in the centre of the book.

Table 3 — Giotto scientific payload

Experiment (acronym)	Mass (kg)	Data rate (bit s^{-1}) F1	F2	F3	Goal and instrument parameters
Camera (HMC)	13.51	20058	20058	723	Inner coma and nucleus imaging, CCD narrow-angle camera, 11 m resolution from 500 km
Neutral mass spectrometer (NMS) Mass analyzer Energy analyzer	12.70	4156	4156		Neutral gas composition Mass analyzer: 1–36 amu Energy analyzer: 1–57, 9–89 AMU
Ion mass spectrometer (IMS) High-energy-range spectrometer (HERS) High-intensity-range spectrometer (HIS)	9.00	3253	3253	1084	Ion composition HERS: 1–35 AMU/q HIS: 12–57 AMU/q
Dust mass spectrometer (PIA)	9.89	2891	5782		Dust particle flux and composition, 1–110 AMU, $3 \times 10^{-16} – 5 \times 10^{-11}$ g
Dust impact detection system (DID) Meteoroid shield momentum sensor (MSM/RSM) Impact plasma and momentum sensor (IPM) Capacitor impact sensor (CIS)	2.26	361	903		Dust particle flux and mass distribution IPM: $6 \times 10^{-17} – 6 \times 10^{-11}$ g CIS: $> 10^{-10}$ g MSM/RSM: $10^{-10} – 10^{-1}$ g
Plasma analysis 1 (JPA) Fast ion sensor (FIS) Implanted ion sensor (IIS)	4.70	3975	1265	1355	FIS: 3-dimensional ion velocity distributions, 10 eV–20 keV IIS: ion flux, mass and velocity distributions, 90 eV–90 keV, 1–45 AMU/q
Plasma analysis 2 (RPA) Electron electrostatic analyzer (EESA) Positive ion cluster composition analyser (PICCA)	3.21	2530	1807	904	EESA: 3-dimensional electron velocity distributions, 10 eV–30 keV PICCA: composition of cold ions, 10–50 AMU, 50–203 AMU
Energetic particle analyser (PEA)	0.95	181	181	181	Energy and flux of electrons and accelerated ions, $\geqslant 20$ keV
Magnetometer (MAG)	1.36	1265	1265	407	Magnetic field, 0.004–65 536 nT, up to 25.4 vector s^{-1}
Optical probe experiment (OPE)	1.32	723	723		Coma brightness in 4 continuum bands (dust), 4 discrete emissions (OH, CN, CO^+, C_2)
Radio-science experiment (GRE)					Dust and gas column densities in the coma
Total	58.90	39 393	39 393	4654	

groups; however, the combined 1 σ uncertainty of the spacecraft and Halley nucleus positions of 40 km had to be added to this distance.

The flyby distance that was actually achieved was 596 ± 2 km, as derived by Curdt *et al.* (1988) from camera observations (variation of the camera offset angle with time). All relevant encounter parameters are summarized in Table 4. The time of closest approach was determined as 0003:01.84 (± 0.20 s) UT on 14 March. This time is given as 'spacecraft event time' (SCET); however, some experimenters prefer to use 'ground station received time' (GSRT).

Table 4 — Periods of encounter science data availability after the last orbit correction manoeuvre (completed 12 Mar, 05:00) (Intervals of intermittent data transmission due to spacecraft nutation around closest approach are omitted)

Exp. acronym		Data available		Damage assessment after the encounter
		from (SCET)*	until (SCET)*	
HMC		12 Mar, 2038	13 Mar, 0033	mirror degradation
		13 Mar, 0619	13 Mar, 0641	baffle loss
		13 Mar, 0830	13 Mar, 0946	
		13 Mar, 1944	14 Mar, 0002:52.8	
NMS		13 Mar, 2018	14 Mar, 0002:53.4	CCDs of both detectors dead
IMS	HERS	12 Mar, 0515	13 Mar, 0357	HV damage
		13 Mar, 0500	14 Mar, 0002:41	
	HIS	13 Mar, 0507	14 Mar, 0221	no damage
PIA		13 Mar, 1923	14 Mar, 0349:42	Mass spectrum channel 30% damage
DID		13 Mar, 1925:17	14 Mar, 0350:11	MSM/RSM: no damage CIS: indeterminate IPM-P: anomalous behaviour – damaged? IPM-M: no damage
JPA	IIS	12 Mar, 0516:52	13 Mar, 0400:40	no damage
		13 Mar, 0454:14	14 Mar, 0002:59	
		14 Mar, 0033:03	15 Mar, 0233:53	
	FIS	12 Mar, 0515:48	13 Mar, 0400:59	HV drop of 400 V
		13 Mar, 0454:06	14 Mar, 0115:52	
RPA	EESA	12 Mar, 0535	13 Mar, 0351	partial damage
		13 Mar, 0453	14 Mar, 0002:29	
	PICCA	13 Mar, 1903:09	14 Mar, 0002:41	HV damage
EPA		<12 Mar, 0500 (continuous data available)	15 Mar, 0247:39	no damage
MAG		<12 Mar, 0500 (continuous data available)	15 Mar, 0251:26	no damage
OPE		13 Mar, 2006	14 Mar, 0331	no damage

Format 2 from 13 Mar, 2238:30 until 14 Mar, 0104:00; other times Format 1
* SCET — spacecraft event time

The two are related by: GSRT = SCET + 8 min 0.1 s.

During its Halley flyby the Giotto spacecraft was struck several times by relatively 'large' (> 1 mg) dust particles as is evident from discrete jumps in the downlink frequency and spacecraft attitude and spin period.

At 0010:54.2 UT (GSRT), i.e. 7.6 seconds before closest approach, Giotto was struck by a particularly 'large' dust particle. This impact caused the spacecraft angular momentum vector to shift by 1°, and the spacecraft performed a nutation around the new axis with a period of 3.2 s and an amplitude of 1°. Thus the maximum deviation from the desired attitude was 2° and, as had been expected in such an event, the telecommunications link to Earth could no longer be maintained.

The effect of this dust impact is clearly seen in the strength of the Giotto radio signal received at the Parkes ground station around the time of closest approach (Fig. 5). The signal strength dropped by almost 40 dB at 0010:54.2 and recovered within 0.4 s. This short-duration loss of signal is due to the switch-over from travelling wave tube 1 to travelling wave tube 2, which is thought to be caused by a large electrical discharge resulting from the formation of a plasma cloud after the impact of the 'large' dust particle. About one second later the signal strength dropped again, this time caused by the change in spacecraft attitude due to the dust impact induced nutation.

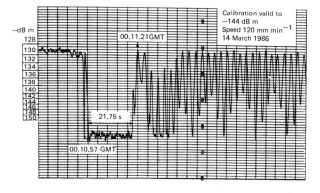

Fig. 5 — Giotto radio signal strength recording received at the Parkes ground station from 0010:32 to 0012:47 UT (GSRT). Closest approach was at 0011:01.94. The vertical lines are 5 s apart.

Thereafter, the signal level however did not fluctuate with the spacecraft inertial nutation period of 3.2 s but stayed low ('blackout') for the next 21.75 s. Possibly a counting error in the despin control system of the high gain antenna drove the antenna pointing direction away from the Earth direction. After 21.75 s the antenna despin motor was adjusted to the correct phase and spin rate and the signal strength fluctuated with a period of 3.2 s. Superimposed on the 3.2 s period in the received radio signal strength is a 16.4 s period which is the onboard nutation period.

For the next 32 minutes scientific data were received only intermittently. By that time the onboard passive nutation dampers (two 60 cm-long tubes filled with a viscous liquid) had reduced the maximum deviation from the Earth-pointing direction to <1°, and continuous data were again received.

A few hours after closest approach the camera was rotated twice to 150°. In this position the camera baffle should have cast a shadow on the solar cell array, causing a decrease in the solar cell array current. In contrast to similar pre-encounter operations such a decrease was, however, not observed, and it must be concluded that the camera baffle was sheared off by impacting dust particles. Analysis of the effects of post-encounter camera rotations on the dynamic of the spacecraft body independently also led to the conclusion that the camera lost its baffle during the Halley flyby.

Analysis of spacecraft housekeeping, radio telemetry and camera data (Bird *et al*, 1988; Fertig *et al.*, 1988; Curdt and Keller, 1988; Nye, 1988) revealed that the mass of the impacting single large dust particle which impacted on the spacecraft 7.6 s before closest approach was in the range 0.1–0.2 g and that the most probable impact location was close to the rim of the dust shield and approximately opposite to the camera. The single large dust particle could therefore not have sheared off the camera baffle.

In the few tens of seconds around closest approach a large number of dust particles impacted on the spacecraft and the baffle. During that time the baffle was inclined at 40° to the stream of dust particles and thus very vulnerable to destruction. Most likely the baffle was eroded by numerous dust impacts during this time interval. This is consistent with the observed increase of the spacecraft spin period from 3.998 to

4.010 s during this interval as the baffle was at that time the only large oblique area in the forward direction which was inclined opposite to the spin direction.

Due to the loss of the camera baffle (330 g) the spacecraft is no longer properly balanced and now has a permanent wobble ($0.4°$ for a camera rotation angle of $40°$). The camera baffle is not the only mass loss. From the observed change in the spacecraft intertia ratio before/after the encounter it is concluded that a few hundred grammes of mass were lost in addition, presumably eroded away by dust impacts on the outer face of the dust bumper shield.

During the flyby Giotto was decelerated by 23.05 cm s^{-1}, as determined independently from both two-way Doppler and ranging data. This deceleration corresponds to a total mass of 1.7 g for all impacting dust particles.

About half of the experiments worked flawlessly throughout the encounter, while the other half suffered damage due to dust impacts (Table 4). The last experiment was switched off $26\frac{1}{2}$ hours after closest approach.

The spacecraft also suffered some damage during the encounter, but it was possible to redirect it to the Earth before it was put in a hibernation configuration on 2 April 1986. This manoeuvre was carried out in three steps on 19/20 March, using ~ 35 kg of hydrazine. At present 25.45 kg (corresponding to ~ 85 m s^{-1} in ΔV) of hydrazine are left. Giotto will swingby the Earth on 2 July 1990 at a distance of 22 000 km. Using the Earth's gravity assist it would be possible to encounter another comet. The short-period comet Grigg–Skjellerup has been identified as the most suitable target. It is an 'old', relatively inactive short period comet which is favourable for imaging and would also provide an interesting contrast to Halley. Its orbit is very well known; the flyby would occur on 10 July 1992. Details of the Giotto extended mission are described in Reinhard (1987b).

It is planned to reactivate and checkout the spacecraft and the experiments in early 1990. The Giotto mission extension will then be decided upon, weighing spacecraft and experiment performance against mission cost.†

† On 2 July 1990, Giotto was given a new lease of life by the first ever gravity-assist manoeuvre to use the Earth. ESA has given its tentative approval for the Grigg-Skjellerup fly-by, but extra funding of around £10 million will be required.

5 THE INTERNATIONAL COMETARY EXPLORER (ICE) AND PIONEER-7 MISSIONS

The ICE spacecraft (Fig. 6), formerly called ISEE-3, is a NASA spacecraft which was launched on 12 August 1978 as part of a three-spacecraft mission called International Sun-Earth Explorer to study the solar wind interaction with the Earth's magnetosphere. The spacecraft monitored the solar wind input from a halo orbit around the sunward libration point L_1, at 0.01 AU geocentric distance.

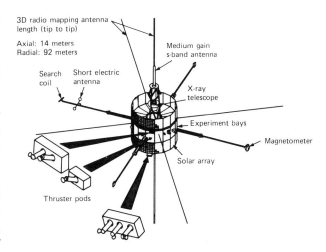

Fig. 6 — The ICE spacecraft.

At launch the spacecraft carried a large amount of mono-propellant hydrazine as a safeguard against large injection errors. The launch was nominal, however, so that after 4 years, at the end of the nominal mission in the halo orbit, 62 of the original 88.57 kg of hydrazine were still available, giving the spacecraft a ΔV capability of 311.6 m s^{-1}. It was therefore decided to first send the spacecraft to the geotail and then to comets Giacobini–Zinner and Halley. The spacecraft left the halo orbit on 10 June 1982. After several passes through the geotail involving five lunar swingby manoeuvres, ISEE-3 was 'launched' on its interplanetary trajectory toward the comets on 22 December 1983. During this last lunar swingby the spacecraft passed only 120 km above the Moon's surface. After this manoeuvre the spacecraft

Table 5 — The ICE scientific payload

Experiment	Goal & instrument parameters	Principal investigator
Solar wind plasma	speed, density, temperature of solar wind electrons, 5–1500 eV (ion portion failed)	S. Bame, Los Alamos National laboratory Los Alamos, NM, USA
Magnetometer	magnetic field in 8 ranges from 4 nT to 1.4×10^5 nT sensitivity: 1/256 of each full range	E. Smith Jet Propulsion Laboratory Pasadena, CA, USA
Radio waves	samples the range 30 kHz–2 MHz at 24 frequencies	J.-L. Steinberg Paris Observatory Meudon, France
Plasma waves	magnetic fields 20 kHz–1 kHz electric fields 20 Hz–100 kHz	F. Scarf TRW Systems Redondo Beach, CA, USA
Plasma composition	flow speeds from α-particles and ion composition (14–34 amu)	K. Ogilvie NASA/Goddard Space Flight Center Greenbelt, MD, USA
Low-energy cosmic-rays	ions in the range 2 keV/charge to 80 MeV/nuc electrons $3–560$ keV/q	D. Hovestadt, Max-Planck-Institut für Physik und Astrophysik Garching, W. Germany
Energetic protons	ion fluxes and angular distributions in the range 35–1600 keV	R. Hynds, Imperial College London, UK

Note: The total ICE payload consists of 13 experiments. Only those experiments are listed which produced useful data during a cometary encounter.

name was changed to International Cometary Explorer (ICE). The primary ICE scientific objective was to study the interaction between the solar wind and the cometary atmosphere. Clearly the instrumentation onboard the spacecraft was well-suited to this objective. Of the 13 experiments onboard the spacecraft, seven were thought capable of producing useful data during a cometary encounter (Table 5).

The spacecraft is a 16-sided cylinder, 1.74 m in diameter and 1.6 m high, weighing 478 kg at launch. Solar cell arrays mounted on the faces of the cylinder provide 160 W of primary power. The spacecraft body supports a total of 10 appendages. Two equatorial experiment booms (3 m long) support the magnetometer and plasma-wave sensors. Four wire antennae (49 m each) are deployed in the spin plane as part of the radio wave and plasma-wave investigations. Two

axial antennae (7 m each) extend above and below the spacecraft parallel to the spin axis to render the radio-wave measurements three-dimensional. Finally, two short inertial booms provide stability.

The radio system consists of two redundant S-band transponders operating at 2217 and 2270 MHz. The spacecraft has two low-gain antennae and a medium-gain telemetry antenna. The telemetry antenna is mounted on the 'tower' above the spacecraft body. The standard telemetry rate is the halo orbit was 2048 bit s^{-1}.

Attitude information and control are provided by two fine Sun sensors and a panoramic attitude sensor. The spacecraft is spin-stabilized at 19.75 rev min^{-1}. The spin axis is maintained perpendicular to the ecliptic plane to within $<1°$. The spacecraft is described in more detail in Brandt (1986a).

After $3\frac{1}{2}$ years of interplanetary cruise (Fig. 8*) ICE encountered comet Giacobini–Zinner on 11 September 1985. It passed the nucleus at a distance of 7800 km on the anti-sunward side and crossed the centre of the comet ion tail at 1102 UT. In the comet frame of reference ICE traversed the cometary tail essentially from below, crossing all major boundaries and the current sheet in the plasma tail. The scientific results of the Giacobini–Zinner encounter are summarized in a special issue of *Science* (**232** 297–428, 1986).

Half a year later, at the end of March 1986, ICE passed comet Halley at a distance of 28 million km on the sunward side (time of closest approach 25 March, 1030 UT). The geocentric distance during the Halley encounter of 0.47 AU greatly exceeded the distance of 0.01 AU while in the halo orbit. Hence a lower bit rate (1024 bit s^{-1}) and tracking by NASA's 64 m DSN were required. Outfitting antennae in the DSN and at Arecibo, Puerto Rico to operate at the ICE frequencies was a major effort of the mission. At the time of the encounter, data were collected by Arecibo, Madrid, and Goldstone, and later by Canberra and Usuda, Japan.

For several days around the time of closest approach, bursts of heavy ions were observed by the ICE energetic ion experiment. The most likely source for these ions is comet Halley. However, as these bursts were observed while ICE was in a recurrent high-speed solar wind stream it is also possible that the observed heavy ions are solar wind iron. The latter possibility is supported by the absence of ion cyclotron waves in high-resolution magnetic field data. The level of plasma wave intensity is enhanced during the distant Halley flyby, similar to the enhancement observed during the Giacobini–Zinner flyby, but the observations are inconclusive as to whether this is due to Halley ions or not.

After the Halley flyby on 7 April 1986, ICE was retargeted to the Earth. Since that time spacecraft attitude manoeuvres and tracking have continued. The current bit rate is 256 bit s^{-1}. On 10 February 1988 the spacecraft still had 28.56 kg of hydrazine onboard, corresponding to a ΔV capability of 149.3 m s^{-1}. In its current orbit the spacecraft will return to the Earth–Moon system in 2014. If the spacecraft were subsequently brought back to Earth it would make a splendid addition to the Smithsonian Air and Space Museum in Washington, DC.

Even closer to Halley's comet than ICE came the Pioneer-7 spacecraft. Pioneer-7 was launched in August 1966 into a heliocentric orbit which is very near the ecliptic and between 1 and 1.126 AU from the Sun. On 20 March 1986, Pioneer-7 reached a closest distance to the comet of 12.1 million km on the anti-sunward side. From 18 to 21 March, Pioneer-7 was in the vicinity of the plasma tail, between 8 and 10 million Km to the north of it (Brunk 1986). At the time of the Halley encounter, only the University of Chicago Cosmic Ray Telescope and the Ames Research Center Plasma Analyzer were functioning, and only in a power-sharing mode. The time profile of 0.6–13 MeV ions observed by the Cosmic Ray Telescope during a period of several days around closest approach was analysed, but it was difficult to draw any conclusions on the origin of the particles observed (cometary ions or interplanetary protons). The Plasma Analyser data, on the other hand, show a significant increase of He^+ ions that could arise from charge exchange of solar wind He^{++} with cometary gas. This increase is about 50% higher than peak values of solar wind minor ions, and is observed for an interval of about five hours after closest approach (Mihalov *et al.* 1987).

6 REMOTE SENSING OBSERVATIONS FROM SPACE

Remote sensing observations in the UV and in the IR (at several wavelengths between 1 and 20 %Gmm, and above 20 %Gmm) are difficult from the ground because the Earth's atmosphere absorbs most of the radiation. Instruments onboard various spacecraft, rockets, and high-flying airplanes operating at these wavelengths complemented the ground-based observations.

Also, ground-based images of comet Halley could not be obtained during an interval of some six weeks near perihelion when the comet was in angular proximity to the Sun. This gap was filled by instruments onboard the Pioneer–Venus Orbiter and the Solar Maximum Mission.

---

* to be found in the colour illustration section in the centre of the book.

Pioneer–Venus Orbiter (PVO)

The PVO (Fig. 7*) is a NASA spacecraft carrying 21 experiments designed to conduct a comprehensive investigation of the atmosphere of Venus. The 517 kg spacecraft is a solar-powered cylinder, about 250 cm in diameter, with its spin axis spin-stabilized perpendicular to the ecliptic plane. PVO was launched on 20 May 1978 into an orbit around Venus (periapsis: 200 km, apoapsis: 66 000 km). comet Halley passed within 0.27 AU of Venus on 3 February and 0.4 AU of Venus on 9 February 1986, and the PVO was thus well placed to observe the comet during perihelion passage when observation from Earth was difficult. The Ultraviolet Spectrometer on the PVO made systematic measurements of the comet from 28 December 1985 through 6 January, 1986, and from 31 January through 7 March 1986. The spectrometer was used as a spin scan photometer to build up a Ly-α image of the entire comet. The images show an extensive hydrogen cloud around the nucleus, well over 10^7 km in extent. The water molecule production rates determined from Ly-α measurements range from 3.3×10^{29} to 1.2×10^{30} s^{-1} and are consistent with other measurements in both amplitude and periodicity (Stewart 1987).

Solar Maximum Mission (SMM)

The SMM was a NASA spacecraft carrying seven instruments designed for the study of solar flares over a wide range of wavelengths. SMM was launched on 14 February 1980 into a circular orbit with an altitude of 572.5 km and an inclination of 28.5°. On 23 November 1980, a power failure crippled its ability to point precisely at the Sun. During a Space Shuttle flight in April 1984 the SMM was retrieved and repaired and all instruments except for the Hard X-ray Imaging Spectrometer were working again. Two of them, the Coronagraph/Polarimeter and the UV Spectrometer and Polarimeter had the capability, in principle, to provide unique observations of comet Halley during the time interval when it was difficult to observe from the Earth (Niedner 1986).

Comet Halley observations with the Coronagraph-/Polarimeter began on 26 January and continued until 28 February 1986. Because of its design as a solar instrument, the C/P's strong rejection of stray light permitted observations of the comet down to solar elongation of 10°, resulting in a gap of observations from 1–12 February only. Images taken through the wideband blue filter show an extended coma, elongated approximately in the anti-solar direction. In the images taken on 28 January there is definite evidence for both a plasma and a dust tail (Brandt 1986b). The images taken during the last two weeks of February complement the ground observation beginning again in March in providing a good record of the structure and evolution of the comet.

The second potentially useful instrument on SMM was limited in its performance because the wavelength drive was stuck, and only single (Ly-α) wavelength measurements were still possible. (SMM re-entered the Earth's atmosphere on 2 December 1989.)

International Ultraviolet Explorer (IUE)

IUE was launched on 26 January 1978 into an eccentric quasi-geosynchronous orbit (apogee 44 000 km, perigee 27 000 km). IUE is a collaborative programme operated by NASA, ESA, and the UK Science and Engineering Research Council.

IUE's primary mirror is 45 cm in diameter. It is equipped with echelle spectrographs capable of operating in the UV. The short wavelength primary (SWP) camera operates in the 115–190 nm region, the long-wavelength primary (LWP) and redundant cameras in the range 190–320 nm. Observations of faint sources, such as comets, are normally performed in low resolution (0.6–0.7 nm) mode. The pointing accuracy of IUE is 1–2 arcsec. The IUE is described in more detail in Kondo & Brandt (1986).

The UV spectrum of comet Halley was monitored with the IUE between 12 September 1985 and 8 July 1986 at regular time intervals except for a two-month period around the time of perihelion when Halley was in angular proximity to the Sun. The first IUE star-tracker image of comet Halley was obtained on 11 September, and the UV spectrum was recorded on 12 September. These are believed to be the first observations of Halley from space. Of particular interest are the observations made during six 4–8 h observing sessions in the period 9–16 March 1986. The IUE fine error sensor, used for target acquisition

---

* to be found in the colour illustration section in the centre of the book.

and tracking, was used in a photometer mode to provide nearly continuous light curves over an 8-hour observing shift, a unique capability. It showed brightness variations over timescales of the order of 1–2 h. At the time of the Giotto encounter, this activity was seen to be near a minimum. From the brightness of the lines in the spectra recorded by the SWP and LWP cameras production rates for various species can be calculated (Feldman *et al.* 1987). In the period 9–16 March the water production rate fluctuated between 5 and 9×10^{29} molecules s^{-1}, while atomic carbon was a few percent and $CO \sim 10$–20% of these values.

Dynamics Explorer 1 (DE-1)

The NASA DE-1 spacecraft was launched on 3 August 1981 into an eccentric polar orbit with initial perigee and apogee altitudes of 570 and 23 000 km, respectively. The spacecraft spin axis is normally orientated perpendicular to the orbit plane. Observations of comet Halley in the post-perihelion period are available owing to significant reorientations of the spacecraft spin vector. For the period of observations discussed here, the apogee of DE-1 was located at high southern latitudes which provide a nearly ideal viewing geometry for imaging comet Halley.

The observations with the VUV-wavelength photometer on DE-1 comprise fully two-dimensional imaging of the comet at the 121.6 nm wavelength of resonantly-scattered solar Ly-α radiation. From the measurement of the two-dimensional projected brightness distribution the hydrogen production rate can be calculated (Craven & Frank, 1987). The hydrogen production rates vary with heliocentric distance according to $Q_0 \times r^{-n}$ with $Q_0 = 9.1 \times 10^{29}$ atoms s^{-1} and $n = 2.3$ for heliocentric distances 1.5–0.68 AU pre-perihelion, and $Q_0 = 1.1 \times 10^{30}$ and $n = 1.62$ for heliocentric distances 0.63–1.2 AU post-perihelion. These values are consistent with water production rates derived for the Giotto and Vega encounters from *in situ* measurements.

Rocket observations

A sounding rocket experiment specifically designed for UV spectroscopy of comets and other extended astronomical sources was successfully flown to observe Halley from White Sands Missile Range on 26 February, 1202 UT, recovered, refurbished, and reflown on 13 March 1986, 1120 UT (Woods *et al.* 1987). The payload consisted of a 40 cm diameter Dall–Kirkham telescope and Roland circle spectrograph with an ellipsoidal grating. The two-dimensional photon counting detector provided long-slit spectra of the comet with a projected slit length of 5×10^5 km and a resolution of 10^4 km and a spectral range of 120–175 nm. By using the Earth's limb to occult the Sun, the rocket observation was possible near perihelion.

On two different rocket flights an experiment was flown to obtain images and objective spectra of comet Halley (Opal *et al.* 1987). One electronographic camera obtained images at Ly-α with a 20° field of view and 3 arc min resolution. The second camera pointed at a grating and obtained objective spectra with 12° field of view and 1 arc min resolution. This experiment was successfully flown from White Sands Missile Range on 24 February, 1153 UT and on 13 March 1986, 1054 UT.

Airborne observations

Infrared observations of comet Halley were made from the NASA Kuiper Airborne Observatory (KAO) which consists of a 91 cm telescope carried onboard a C-141 aircraft and from NASA's Lear Jet Observatory (LJO) carrying a 30 cm telescope. The altitudes reached by these planes (up to 13.5 km) are above 97% or more of the water vapour in the Earth's atmosphere.

A total of 22 flights were made with the KAO to study comet Halley, eight flights in the period 10–24 December 1985 from California and 14 flights in the period 15 March to 6 May 1986 from New Zealand (Haughney & Mumma 1986). These observation periods were determined by the viewing constraints of the KAO: the comet had to be at an elevation angle higher than 35° and more than 30° from the Sun.

The LJO made 12 flights in March and April 1986 to observe Halley's comet from the Kwajalein atoll 9° north latitude in the eastern Pacific Ocean (Russel *et al.* 1986). Typical flight duration on the LJO was 2.5 h and 6.5 h on the KAO. The observations made during

Table 6 — Key data on missions to Halley's comet

Mission	Agency	Launch date	Flyby date/time (UTC)	Heliocentric distance (AU)	Flyby distance (km)	Flyby speed (km s^{-1})	Phase angle (deg)
Vega-1	Intercosmos (IKI)	15 December 1984	6 March 1986/0720	0.79	8890	79.2	111.2
Vega-2	Intercosmos (IKI)	21 December 1984	9 March 1986/0720	0.83	8030	76.8	113.4
Sakigake (formerly MS-T5)	ISAS	8 January 1985	11 March 1986/0418	0.86	6.99×10^6	75.3	109.4
Giotto	ESA	2 July 1985	14 March 1986/0003	0.89	596	68.4	107.2
Suisei (formerly Planet-A)	ISAS	19 August 1985	8 March 1986/1306	0.82	151×10^3	73.0	104.2
ICE	NASA	12 August 1978	25 March 1986/1044	1.07	28.1×10^6	64.9	111.2

these flights constitute the most comprehensive study of infrared emissions yet made for any comet.

The investigations concentrated mainly on the detection of water vapour (v_3 band near 2.65 μm), the examination of dust particles in the wavelength range 5–160 μm, and the search for molecular species in the far infrared and submillimetre regions. The spectra obtained in the 5–160 μm range yield information on the dust densities, size distribution, colour temperature, and composition. Of particular interest was the water investigation. The first positive detection of H_2O in any comet was obtained from the KAO on 22 December 1985.

From the high-resolution spectra obtained of Halley in December 1985 and March 1986 the H_2O absolute production rate, its spatial distribution, temporal variability, outflow velocity, and the H_2O ortho/para ratio were determined.

7 INTERNATIONAL COOPERATION IN THE IACG

Four space agencies — Intercosmos representing ten Eastern Bloc countries, the Japanese Institute of Space and Astronautical Science (ISAS), NASA, and ESA — sent spacecraft to Halley's comet and made Halley observations from space during the comet's 1985/86 apparition. Intercosmos sent Vega-1 and Vega-2; ISAS sent Sakigake and Suisei; ESA sent Giotto; and NASA sent ICE and was able to mount a significant space-based observation programme, using existing spacecraft and sounding rockets.

The spacecraft that encountered Halley complemented each other in flyby distance, which ranged from 600 km to 28 million km, all spacecraft passing on the sunward side (Fig. 9*). The ICE spacecraft had earlier explored the anti-sunward side of comet Giacobini–Zinner.

The flybys occurred between 6 and 14 March 1986, covering more than three comet nucleus rotations and the corresponding variations in activity. (See also Table 6) The scientific experiments on the various spacecraft together provided the full complement of experiments that can be flown on flyby spacecraft. Some experiments on the various spacecraft were also very similar, which provided a stimulating basis for data comparison after the encounters.

Realizing that many aspects of mission planning, spacecraft and experiment design, and data evaluation were common to all missions, and that the overall scientific return could be increased through cooperation, the four agencies agreed in 1981 to form the Inter-Agency Consultative Group for Space Science (IACG). The IACG had the task of informally coordinating all matters related to the space missions to Halley's comet and the observations of the comet from space.

During its first five years, the IACG demonstrated an ever-growing usefulness for the various Halley flight projects, as a focal point for the exchange of information, discussion of common problems, and mutual support to enhance the overall scientific return. Since its formation the IACG has met annually with the task of organizing the meeting, and consequently the meeting place, rotating within the four agencies.

At the IACG's first meeting (13–15 September

* to be found in the colour illustration section in the centre of the book.

1981, Padua, Italy) three Working Groups were formed in which many of the problems common to all space missions to Halley's comet were discussed, resulting in recommendations to the flight projects or actions on the Working Group members to carry out specific tasks. The three Working Groups formed were:

- the Halley Environment Working Group (WG-1)
- the Plasma Science Working Group (WG-2)
- the Spacecraft Navigation and Mission Optimization Working Group (WG-3)

The Halley Environment Working Group had the task of providing the scientific community and the spacecraft engineers of the Halley flight projects with the best possible estimates for the key parameters of the Halley nucleus and with dust and gas models for the Halley coma.

The Plasma Science Working Group had the task to:

- study the various plasma physical processes resulting from the solar wind/comet interaction,
- optimize the scientific return from the cruise phase by making use of the exceptional close co-existence of seven spacecraft (Vega-1, Vega-2, Giotto, Suisei, Sakigake, ICE, Pioneer-7) in interplanetary space, and characterizing the interplanetary medium during the various flybys,
- stimulate discussion between the plasma experimenters on the various spacecraft,
- investigate the effects of the impact-generated plasma around the spacecraft.

The Spacecraft Navigation and Mission Optimization Working Group had several tasks, among them data dissemination and spacecraft targeting. Targeting the spacecraft with respect to the comet nucleus was a major problem as the nucleus, being masked by the gas and dust in the coma, could not be observed from Earth. The Astrometry Net of the International Halley Watch (IHW)† had the task of providing the cometary ephemeris to the scientific community, i.e. to all the astronomers who wanted to observe Halley's comet and to the flight projects that were flying to it.

In March 1986 the expected 1 σ uncertainty in the target plane was 370×100 km, achieved by ground-based observations over a long period. The reduced comet positions were used as input for a model that included the effects of the nongravitational forces. These are due to the emission of gas and dust from the heated side of the nucleus, i.e. towards the Sun ('rocket effect'). This uncertainty was more than sufficient for all observations from the ground and near-Earth space, and also for the Sakigake and Suisei spacecraft flying by at distances > 100 000 km. It was even sufficient for Vega-1 and Vega-2 flying by some 8000–9000 km away. For Giotto, which was intended to be targeted a few hundred kilometres on the sunward side of the nucleus, a smaller targeting uncertainty was highly desirable. Fortunately, Giotto was the last of the spacecraft to encounter the comet, and the information on the nucleus position, obtained by the cameras onboard the earlier arriving Vega-1 and Vega-2 spacecraft, could be passed on to the Giotto Project.

This was the 'Pathfinder Concept', the principle of which is illustrated in Fig. 10*. The uncertainty in spacecraft position (30–50 km) is shown in Fig. 10* as a small circle around the Giotto path, at the time when the last orbit-correction manoeuvre was made. The large circle around the path of the comet reflects the relatively large 1σ uncertainty that could be achieved via the Astrometry Net. Vega located the comet nucleus during its flyby on 6 March 1986 with an uncertainty given by the angular uncertainties in spacecraft attitude and pointing direction of the platform on which the camera was mounted. The Vega position uncertainty also contributed to the error.

Using only conventional (6 GHz) ranging and Doppler, Soviet experts expected a geocentric Vega position accuracy of several hundred kilometres. It was estimated that the Vega position uncertainty could be reduced to ∼40 km by using Very Long Baseline Interferometry (VLBI) techniques. This was achieved by NASA, using the widely separated tracking stations of its Deep Space Network (DSN) and precise L-band VLBI. After processing all the

---

† See Chapter 2: The International Halley Watch.

* to be found in the colour illustration section in the centre of the book.

data, the position of the comet nucleus was known with much better accuracy, represented by the small circle around the Halley path. Giotto was targeted to the centre of the small circle projected to the intersection of the comet and the spacecraft paths. Between the time of nucleus detection by Vega and the Giotto encounter, the uncertainty grew slightly owing to the nongravitational forces, which cannot be modelled precisely. In fact, both Vega-1 and Vega-2 were used as pathfinders (Fig. 10* shows one Vega for simplicity). The final Giotto orbit correction manoeuvre was made 48 hours before Giotto's closest approach. Giotto was aimed to flyby at 540 km on the sunward side, and actually achieved a flyby distance of 596 km. The Pathfinder Concept reduced the targeting uncertainty to 40 km (1σ).

While the improvement of the Giotto targeting accuracy through the Pathfinder Concept was undoubtedly the most visible achievement of the IACG, it is only one in a long list:

- exchange of information on spacecraft (in particular the dust bumper shield) and experiment design;
- intercalibration of experiment sensors;
- implementation of the Pathfinder Concept;
- development and distribution of Halley gas and dust models, prediction of dust jets, determination of parameters for the nucleus;
- exchange of results from models on spacecraft charging due to the dust and gas impacts on the spacecraft during the flybys;
- exchange of Halley spacecraft orbital data;
- exchange of information on spacecraft and experiment performance and mission planning and, in particular, on experiment operation times;
- use of common coordinate systems;
- definition of special periods for cruise science;
- near-real-time exchange of scientific data, i.e. within days of the encounters;
- exchange of all encounter science data at a later time, approximately three years after the encounters;
- joint publication of 'first results' just two months after the encounters in a special issue of *Nature* (**321**, 15 May 1986).

- support of a major Halley Symposium (Heidelberg, Germany, October 1986) and of Topical Workshops,
- exchange of information on post-Halley targets for the flyby spacecraft.

The *in situ* observations by the various experiments onboard the six flyby spacecraft were complemented by a number of remote observations from space and a very large number of observations from the ground, the latter being coordinated through the International Halley Watch (IHW). The remote observations from space were particularly useful when Halley was in solar conjunction (near its perihelion passage) and difficult or impossible to observe from the ground. The ground-based observations began with the comet's recovery in October 1982, and are expected to continue until about 1990, covering the whole apparition. The IHW, although not a space agency, has participated in all IACG meetings, just as the space agencies have been represented at all general meetings of the IHW. The communication and coordination achieved through this cross-representation has turned out to be highly effective.

The IACG and its computerpart on the ground, the IHW, have formed the cornerstones of a global effort to explore Halley's comet as completely as possible during its present apparition. By the end of the 1980s, when Halley will disappear again into the outer Solar System, it will be the most thoroughly studied comet ever, with more data having been collected on it than on all other comets put together.

REFERENCES

Bird, M.K., Pätzold, M., Volland, H., Edenhofer, P., Buschert, H., & Porsche, H. (1988) Giotto spacecraft dynamics during the encounter with comet Halley, *ESA J.* **12** 149.

Brandt, J.C. (1986a) The International Cometary Explorer (ICE) mission to comets Giacobini–Zinner and Halley, *ESA SP-1066* 99.

Brandt, J.C. (1986b) Space observations of comet Halley, *Nature* **321** 391.

Brunk, W. (1986) NASA-supported smaller programmes to observe comet Halley, *ESA SP-1066* 151.

Craven, J.D., & Frank, L.A. (1987) Atomic hydrogen production rates for comet P/Halley from observations with Dynamics Explorer 1, *Astron. Astrophys.* **187** 351.

Curdt, W., & Keller, H.U. (1988) Collisions with cometary dust recorded by the Giotto HMC camera, *ESA J.* **12** 189.

Curdt, W., Wilhelm, K., Craubner, A., Krahn, E., & Keller, H.U. (1988) Position of comet P/Halley at the Giotto encounter, Astron. Astrophys. **191** L1.

Feldman, P.D., Festou, M.C., A'Hearn, M.F., Arpigny, C., Butterworth. P.S., Cosmovici, C.B., Danks, A.C., Gilmozzi, R. Jackson, W.M., McFadden, L.A., Patriarchi, P., Schleicher, D.G., Tozzi, G.P., Wallis, M.K., Weaver, H.A., & Woods, T.N. (1987) IUE observations of comet P/Halley: evolution of the ultraviolet spectrum between September 1985 and July 1986, *Astron. Astrophys.* **187** 325.

Fertig, J., Marc, X., & Schoenmaekers, J. (1988) Analysis of Giotto encounter dynamics and post-encounter status based on AOCS data, *ESA J.* **12** 171.

Grard, R., Gombosi, T.I., & Sagdeev, R.Z. (1986) The Vega missions, *ESA SP-1066* 49.

Haughney, L.C., & Mumma, M.J. (1986) NASA C-141/KAO comet Halley observations, *Proc. 20th ESLAB Symposium, ESA SP-250* **III** 215.

Hirao, K. (1986) The Suisei/Sakigake (Planet-A/MS-T5) missions, *ESA SP-1066* 71.

Kondo, Y., & Brandt, J.C. (1986) International Ultraviolet Explorer (IUE) observation of Halley's comet, *ESA-SP-1066* 129.

Lo Galbo, P. (1984) The Giotto spacecraft system and subsystem design, *ESA J.* **8** 215.

Mihalov, J.D., Collard, H.R. Intrilligator, D.S., & Barnes, A. (1987) Observation by Pioneer 7 of He + in the distant coma of Halley's comet, *Icarus* **71** 192.

Niedner, M.B. (1986) Observations of Halley's comet by the Solar Maximum Mission (SMM), *ESA SP-1066* 135.

Nye, H.R. (1988) Giotto: Post-encounter status assessment, *ESA J.* **12** 209.

Opal, C.B., McCoy, R.P., & Carruthers, G.R. (1987) Far-ultraviolet objective spectra of comet P/Halley from sounding rockets, *Astron. Astrophys.* **187** 320.

Reinhard, R. (1987a) The Giotto mission to comet Halley *J. Phys. E: Sci. Instrum.* **20** 700.

Reinhard, R. (1987b) The Giotto extended mission, *Proc. of the Symp. on the Diversity and Similarity of comets, ESA SP-278* 523.

Russel, R.W., Lynch, D.K., Rudy, R.J., Rossano, G.S., Hackwell, J.A., & Campins, H.C. (1986) Multiple aperture airborne infrared measurements of comet Halley, *Proc. 20th ESLAB Symp., ESA SP-250* **II** 125.

Sagdeev, R.Z., & Moroz, V.I. (1986). *Astr. Zh. Lett.* **12** 5–9

Sagdeev, R.Z. et al. (1986a) *Astr. Zh. Lett.* **12** 10–16.

Sagdeev R.Z. Linkin, V.M., Blamont, J.E., & Preston, R.A. (1986b) *Science* **231** 1407–1408.

Stewart, A.I.F. (1987) Pioneer Venus measurements of H, O, and C production in comet P/Halley near perihelion, *Astron. Astrophys.* **187** 369.

Woods, T.N., Feldman, P.D., & Dymond, K.F. (1987) The atomic carbon distribution in the coma of comet P/Halley, *Astron. Astrophys.* **187** 380.

2

The International Halley Watch

R. Newburn and J. Rahe

1 HISTORICAL ANSCHAUUNG

1.1 The 1910 effort

In November of 1909 a circular was mailed to astronomers all over the world by the 'Committee on comets' of the Astronomical and Astrophysical Society of America (Comstock *et al.* 1909). The circular was primarily a ten-page letter, over the signature of George C. Comstock, Chairman, on behalf of the committee, seeking cooperation in observations of Halley's comet during its 1910 apparition. The committee consisted of E.E. Barnard, E.B. Frost, C.D. Perrine, and E.C. Pickering in addition to Comstock, a most distinguished group of astronomers. Consideration was given in the circular to astrometry, photometry, and spectroscopy, but the most attention focused on photography. Comstock particularly emphasized the importance of photographs in the study of the rapidly changing tail morphology and the development of envelopes and jets in the head. He noted the difficulties of obtaining such photographs of an object having small angular separation from the Sun, and the need for cooperation of observatories worldwide in obtaining them. He mentioned plans to send an observer to the Hawaiian Islands to fill the huge gap in coverage caused by the Pacific Ocean. Finally he stated that 'The Committee will be pleased to receive from every astronomer who may cooperate in the matter, copies (glass positives) of his negatives of Halley's comet, and it will undertake the comparison and discussion of the material thus collected.'

Unfortunately the 1910 effort was not notably successful. A late start, slow communications, and modest funding undoubtedly contributed to a somewhat disappointing result. The Committee itself formally reported in 1915 (Comstock *et al.* 1915). They noted that 'Subsequent developments have made it seem inexpedient to carry out the [intended] programme.' The subsequent developments were the fact that a few observatories obtained most of the photographs, with Lick Observatory and the National Observatory of Argentina (Cordoba) obtaining almost half of the total and wishing to publish their own monographs. Those monographs were finally published, but not for 21 and 24 years respectively after Halley's perihelion (Bobrovnikoff 1931, Perrine 1934). The committee published only the photographs of Ellerman (Ellerman & Barnard 1915), whose trip to Hawaii they sponsored with National Academy funds, and a catalogue of all images about which they could obtain information (Comstock *et al.* 1915). Many of the photographs were never published or studied, and there is a current effort to rectify this deficiency (Donn *et al.* 1986).

The 1910 Halley effort provided advocacy, suggested some standardization, offered the very limited coordination possible with slow communication and little lead time, and volunteered to research and publish the collected results. It has widely been considered a failure because it was unsuccessful in the last of these. It failed because it asked for a considerable effort by a limited number of people and really offered *only* the satisfaction of contributing to a worthwhile scientific undertaking.

1.2 Conception of the IHW

The original drive for some form of Halley coordination for the 1986 apparition came from an engineer, Louis Friedman of the Jet Propulsion Laboratory (JPL). Friedman was on leave of absence from JPL in 1979 as an AIAA (American Institute of Aeronautics and Astronautics) fellow detailed to the United States Senate Committee on Science and Technology. Recognizing an enormous fascination with Halley on the part of both scientists and public, and knowing the great synergism possible with cooperation, he convinced the NASA (National Aeronautics and Space Administration) Division of Solar System Exploration to set up a small study at JPL in the autumn of 1979 to investigate what might be accomplished. Friedman recruited three JPL scientists to work on the study: celestial mechanics specialist and historian D. Yeomans to provide ephemerides and the general circumstances of apparition as well as historical background; astronomer and comet expert R. Newburn to outline a ground-based programme, and astronomer J. Bergstrahl to study research possibilities from existing or planned Earth satellites and sounding rockets. Friedman himself looked at the space probes being planned in other countries as well as the probe possibilities still under study in the USA at that time.

Comets are very dynamic objects, showing perceptible change in hours or even minutes. Simple advocacy and coordination sufficient to provide appropriate time coverage and to make possible correlation of data acquired nearly simultaneously using differing observing techniques would be a useful scientific contribution. A truly unique contribution to the understanding of comets could be made, if a real Halley database were created, available to scientists everywhere. A formal statement of these goals developed during the JPL study became of the goals of the International Halley Watch (IHW) as expressed on the frontispiece of the first *IHW Newsletter* (Edberg 1982). These were:

Advocacy
The IHW will encourage and support any scientifically valid means of studying comet Halley.

Coordination
The IHW will coordinate activities among the ground-based disciplines and with flight projects in order to maximize the scientific value of the entire body of observations.

Standardization
The scientists in each IHW discipline will set useful standards for observations in that discipline. There is no absolute requirement that such standards be followed for IHW participation, however.

Archiving
All properly documented data (reduced data, not interpretations) will be published in a Halley Archive in 1989. This Archive will complement the usual interpretive papers in the open literature, not supplant them.

Newburn suggested that the keys to having adequate data submitted to the Archive were several. Most important, in most areas of research astronomers would publish their own data, as always, in the technical journals, but they would in addition submit the reduced data to the Archive. The second key would be to offer a free copy of the Archive to each astronomer submitting data to it. A third key would be to have a large enough group of participants to avoid domination of the total data by a few institutions, and to make up for the inevitable losses caused by impossible weather conditions. A final key would be to make the IHW so big and successful that everyone would want to be a part of it.

All of these general ideas became a part of the IHW, although variations were necessary in some types of research. Among the astronomers consulted during the JPL study phase were J.C. Brandt of NASA's Goddard Space Flight Center (GSFC) and J. Rahe, on sabbatical leave at GSFC from the Astronomical Institute of the University of Erlangen–Nürnberg. Brandt, Rahe, and collaborators already had contacted numerous astronomers worldwide, asking their cooperation in a photographic study of

Halley ion tail morphology. Such a study requires as near continuous coverage as possible, when the comet is near Earth, and could not possibly be carried out from a single site. They proposed, as did the comet Committee in 1910, that copies of all plates be sent to them for collective analysis.

Looking for a way to organize hundreds of astronomers and to collect and archive gigabytes of data without an enormous administrative apparatus and cost at the Lead Center, was another major problem. The organization proposed (and eventually instituted) is shown in Fig. 1. The Discipline Specialists would be experts in a given observing technique who would organize an observing net, suggest standards, and collect the data from the net for inclusion in the Archive. The Steering Group would offer advice and criticism while detached from day to day operation. They also would select the Discipline Specialists. Contact would be made and maintained with flight projects through Project Representatives chosen by the projects. The Lead Center would coordinate everything, supply communications, and prepare the Archive. Finally, enthusiastic amateurs worldwide would be made an official part of the organization.

IHW ORGANIZATION

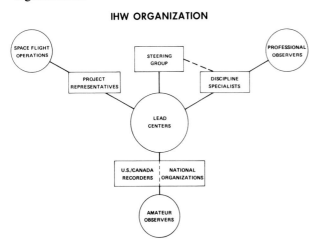

Fig. 1 — IHW organisation and communications flow chart.

A formal presentation of these ideas and of progress in all phases of the study was given at NASA headquarters on 23 January 1980. It resulted in the creation of a NASA Science Working Group, chaired by Brandt, with Newburn as Vice-chairman, and

Yeomans as Executive Secretary. Two-day meetings of the 20 Working Group members were held in March and May of 1980 at which the scientific goals, proposed administrative structure, and budget as developed by the study team were critiqued and modified by the Working Group. Opinions also were sought from many professional and amateur cometary astronomers not formally part of the Working Group. The fully developed plans for the International Halley Watch were endorsed and reported in print in July (Brandt et al. 1980), followed by a final review of the conclusions at NASA headquarters on 1 August. Then we all waited.

1.3 Birth of the IHW

NASA decided very soon to support the IHW concept, but they had two centres with sufficient expertise to act as Lead Center, JPL and GSFC. After two months of discussion the job was given to JPL, and the IHW (in principle) was under way. The rationale for this decision included the facts that the concept and first year of planning had been carried out at JPL, that Brandt's laboratory at GSFC supported sending the work to JPL, and that some elements of the operation such as travel and communications would be easier to conduct from a non-government laboratory. JPL is operated by the California Institute of Technology under contract to NASA.

It was clear from the beginning that a successful effort would have to be international in scope, and no name was ever seriously considered for the activity but 'The International Halley Watch.' Giving it the name was rather easier than making the name factual. Friedman had informed the European Space Agency (ESA) of the IHW study on 17 October 1979, and they sent a representative to the study meetings. He gave formal papers on the concept to the space community beginning in January 1980. Friedman's original idea had been to create an umbrella organization to coordinate both space and ground-based research, but in fact NASA and world support were given initially only to coordinate the ground-based work. The international cooperation in space missions to Halley did not begin to be seriously considered until a year later (see section 1.4). The first formal presentation to cometary astronomers came in a paper at the GSFC conference 'Modern Observational Techniques for Comets' in October 1980 (Friedman & Newburn 1980). This was followed by a briefing and question and answer session conducted by Newburn at the major international comet conference held in Tucson in March 1981.

Louis Friedman was instigator and leader of the IHW study at JPL. In a letter to NASA associate administrator Thomas Mutch dated 24 June 1980 he expressed the pleasure he had derived from the programme and initiative and stated that 'I would think it appropriate to phase out my involvement following completion of the report so that the cometary scientists can run it.' Over the next few months he did just that, and Newburn became interim Study Leader and, later, when JPL became the US Lead Center, interim IHW Leader. Friedman soon left JPL to aid the birth and become Executive Director of 'The Planetary Society,' which he has helped to grow into the world's largest space enthusiast group, with over 100 000 members.

To help assure the international character of the IHW, a second Lead Center was established in Bamberg, the Federal Republic of Germany, under the leadership of Jürgen Rahe. This eastern hemisphere Lead Center is supported by the government of the Federal Republic of Germany. Split responsibility may not seem ideal organizationally, but there were many advantages. The nine hour time difference between Europe and California would have made real time communication difficult without a European office that could speak for the IHW. Travel expenses were reduced. Contacts with scientists in Eastern Bloc countries were rather easier from the German Federal Republic than from the USA in 1981.

It was obvious from the beginning in 1979 that the IHW could never be the broadly based international organization needed to carry out the ground-based observing without International Astronomical Union (IAU) endorsement. A substantial organization had to be presented to the IAU General Assembly in Patras, Greece, in August 1982. The IAU Commission on comets, Minor Planets and Meteorites had to submit a resolution for such endorsement to the IAU executive committee some months earlier, demonstrating that the IHW was a reality and was proceeding on schedule with a well developed plan.

Time was already short five years before Halley's perihelion!

In early 1981 NASA began to assemble a Steering Group membership list. The Steering Group would advise the Lead Centers on all activities of the IHW. By early April the 21 well known international scientists selected had all been notified (see Appendix 1a), although not all were able to be present on short notice for the first meeting on 30 April and 1 May 1981. (Later, the Steering Group became an independent entity with their own chairman, Ian Halliday, and they selected their own members, Appendix 1b.) It had been decided by NASA that Discipline Specialist Centers should be established as the result of an open competition among scientists in response to a NASA letter announcing the opportunity. The letter, dated 20 April 1981, was sent to 4500 addresses worldwide. The primary purposes of the first meeting of the Steering Group were to acquaint members with the prior 18 months' work on the IHW concept and the NASA letter, and to set up the mechanisms by which the resulting proposals would be evaluated.

Bertram Donn, then President of Commission 15 of the IAU, had been a member of the NASA Science Working Group and also accepted membership on the Steering Group. He handled the initial IAU relations and provided a note on IHW activities for the *IAU Bulletin* that summer. Rahe was then secretary of Commission 15, and several Steering Group members were on the Commission 15 organizing committee. This made it a simple matter to gain the support of the commission that had to propose a resolution of support to the IAU General Assembly.

The NASA letter announcing the Discipline Specialist competition allowed only three months in which to respond. While NASA would fund Discipline Specialist Centers in the USA, each non-US proposer had to get a promise of support from his or her own government or institution. At the second Steering Group meeting in November Murray Geller, recently appointed western deputy leader, became executive secretary for all general meetings. Seven Discipline Specialists were selected, all from the USA. It was soon decided that non-US scientists had to be added in each speciality. NASA would pay for operation of the US Discipline Centers, but each team had to be international. An eighth discipline, Meteor

Studies, was created in 1984 as a result of Steering Group recommendation, with its centre in Ottawa, Canada. The Discipline Specialists and their disciplines are listed in Appendix 2.

The entire administrative apparatus for the ground-based programme was in place a few weeks before the Patras, Greece, General Assembly of the IAU in August 1982. Copies of the first *IHW Newsletter* (actually a 30-page booklet) describing the plans and structure of the IHW were distributed to every astronomer at the meeting. Two resolutions were passed by the 18th General Assembly, one asking that observatories give high priority to Halley observations, and the other endorsing the IHW 'as the international coordinating agency for comet Halley observations.' (The complete text of these will be found in Appendix 3.) With this, after almost three years of work, IHW had been successfully instituted and the word international in the title fully justified. Newburn and Rahe became IHW Leaders for the western and eastern hemispheres respectively (see Appendix 4). It was now less than $3\frac{1}{2}$ years until perihelion, Halley could be recovered any time, and the IHW had to go to work.

1.4 The space connection

It had been Friedman's hope that the IHW would be an umbrella organization overseeing cooperation both on the ground and in space. As it began to appear that the USA would not have a space mission to Halley, while the Soviet Union, Japan, and Europe might all have them, this possibility faded. It was clear, however, that there could be a synergism in cooperation, if several missions went to Halley. The late Giuseppi Colombo of the University of Padua especially pushed this effort. The ESA therefore invited a few senior people from each space agency to meet in Padua, Italy, in September 1981, with Colombo and his university as host. NASA brought along the IHW as 'technical experts' in the persons of Newburn, Rahe, and Yeomans. It soon became clear that the immediate and chief concern of the flight projects was navigation, and that the IHW could provide essential support. The Soviets questioned our international bona fides rather closely, since we seemed to be present totally as a part of NASA, and at that time we had only Commission 15 support. Patras

was still eleven months away. Nevertheless, they agreed to work with us and soon gave us a major contact in the Soviet Union, Yaroslav Yatskiv. Yatskiv began setting up a Halley observing organization within the USSR as part of the IHW. This Padua meeting was the first of what came to be known as the IACG (Inter-Agency Consultative Group).

The second IACG meeting took place in Dobogokö, Hungary, in November 1982. At that time Newburn reviewed the Patras resolutions and pointed out the inconsistency of treating an independent, international organization as if it were part of NASA, although admittedly still very dependent upon NASA funding. Thereafter the IHW was invited to all IACG meetings as a non-voting 'member.' 21 November 1982 was widely toasted at the IACG banquet that night in Dobogokö as 'IHW Independence Day.'

An important agreement at the third IACG meeting in Kagoshima, Japan, in December 1983 was that all Halley space missions would send their data to the IHW for inclusion in the Halley Archive two years after the encounters. The IHW in turn would supply copies of the Archive to all space scientist contributors and would supply urgently needed data such as astrometric positions as soon as available. The IHW continued to participate in this space connection in Tallinn, USSR, in November 1984, in Washington, DC, USA during the comet Giacobini–Zinner encounter in September 1985, in Moscow, USSR, and Darmstadt, FRG, during the Vega and Giotto Halley encounters in March 1986, and finally back in Padua (and Rome), Italy, once more in November 1986. The IACG and the space agencies have moved on to new projects, and the IHW is not involved in the IACG any more except for the data link.

1.5 Growth of the IHW

The Patras IAU meeting not only gave the IHW its imprimatur but also provided a forum from which to publicize its goals and to recruit observers. The *Newsletter* distributed to everyone had application forms attached in the back which could be removed and mailed to the appropriate Discipline Specialist to join an observing net. It also had a form for a free subscription to the *Newsletter*. The subscription list finally settled at 1900 names. A separate *IHW Amateur Bulletin* was begun in December 1982. The recovery of Halley on 16 October 1982, with its attendant publicity, helped enormously in recruiting observers, both professional and amateur. By December 1983 over 700 professional astronomers had joined one or more nets.

In mid-1981 Stephen Edberg, an astronomer with close amateur ties, was appointed Coordinator for Amateur Observations. He produced a two-volume *International Halley Watch Amateur Observers' Manual for Scientific* comet *Studies* (Edberg 1983) which detailed the observing procedures expected of amateurs, if they were to produce results potentially useful to science. About 10 000 English language copies were eventually sold by the USA Government Printing Office and two private publishers. In addition it was translated into nine other languages, and an unknown number of additional copies were produced. By the end of 1983 the *IHW Bulletin* was going to 2800 amateurs, and this distribution eventually exceeded 4000. About 1250 amateurs eventually filled out the detailed forms about their equipment and plans, information required to officially become a part of the amateur net. Of these, 800 have actually submitted data.

In 1981 Robert Farquhar of the NASA Goddard Space Flight Center discovered a complex series of Earth and lunar swingbys that would permit an existing US solar–terrestrial physics probe, ISEE-3 (International Sun Earth Explorer No. 3), to escape on a trajectory to comet Giacobini–Zinner from its position near the Earth–Moon libration point. The ISEE-3 spacecraft was well instrumented for cometary tail studies, which none of the Halley probes would perform, so ISEE became ICE (International cometary Explorer) and was sent off to Giacobini–Zinner. The IHW added Giacobini–Zinner observations to the Halley plans.

It was decided quite early that everyone should be encouraged to use FITS (Flexible Image Transport System) for digital data transmission, as this system was rapidly becoming the international astronomical standard. Discipline Specialists prepared FITS header records. Software was created at the Lead Center for transforming FITS records into an agreed-upon printed format. A trial run was needed to check our ability to transport data from the observer to the Discipline Specialist, to the Lead Center, and finally

to the printed page. This trial was performed on comet Crommelin in March 1984, and a Crommelin Archive was printed in 1985 (Sekanina & Aronsson 1985). The trial run was a successful test and produced useful scientific data. Use of the Crommelin Archive also showed how it could be improved in many ways for Halley and Giacobini–Zinner.

Images are a type of data that can be copied in print only as halftones or some similar form of reproduction. An image that might originally have hundreds or thousands of grey levels is reproduced with perhaps 10 or 12 levels in print. Clearly some form of digital archive would also be needed, if much of the scientific value of the Halley data was not to be lost. Magnetic tape would be bulky, inordinantly expensive for widespread distribution, and does not have archival durability unless copied every few years. The new digital disk technology clearly was the best possibility, and its growth was watched carefully during the growth of the IHW. Compact disks (CDs) holding about 680 Mbytes per disk now have an international standard format, and readers for them are widely available. The disks are indeed compact, 12 cm in diameter, moderately priced, and offering long life. It now seems certain that the primary archive will take this form.

The prime Halley observing window occurred from August 1985 through June 1986. As the window opened we had just over 1000 professional astronomers in 51 countries who had expressed an intent to observe the comet.

2 THE INEVITABLE PROBLEMS

2.1 'Political' problems

'Political' problems come in all shapes and sizes when attempting to create an international organization, but most are easily overcome in a purely scientifically motivated organization. It proved to be extremely important to work within the framework of existing international entities, in this case especially the IAU. It was equally important to have the support of the space agencies. Many countries, the United Kingdom, France, the USSR, Brazil, and the Peoples Republic of China among them, created national Halley organizations to coordinate observing time and funding as part of the international IHW. The

IHW always worked very closely with these national organizations.

No attempt was ever made to 'control' the work of individuals or of any laboratory, university, or national or international organization. However, in some cases suggestions were made to observatory scheduling committees when requested, since it was important that strong observing teams be at key sites during certain critical times.

The most difficult problems to deal with are usually raised by individuals. At the beginning many individuals asked, 'Why should I give you my data?' The answer was always, 'We want your data only after you have sent it off for publication.' The IHW always planned to give more than received, an Archive of everyone's measurements in exchange for the inclusion of the individual's data. In spite of these intentions, one should realize that not all scientists are prepared to submit their data to the Archive. One can only hope that there are enough data in that discipline to make up for this loss.

Occasionally a complaint was heard from an individual that the USA was over-represented or their country in some way under-represented in the IHW. The USA has more astronomers, more astronomers in the IAU, and more astronomers in the IHW than any other country. It is the financial support of NASA which made the IHW and the IHW Archive possible and the reason it is being prepared largely in US Discipline Centers and at JPL.

2.2 Time and money

It was the rule from the beginning that each nation would support its own scientists in work on Halley. The IHW supported no individual researchers, although NASA and the NSF (National Science Foundation) jointly purchased 75 sets of standard comet filters which were distributed by the IHW photometry net for worldwide use. NASA supplied six portable 8-inch Schmidt cameras used by the large-scale phenomena net to fill gaps in geographical coverage (Niedner & Liller 1987) and transferred another Schmidt from Kitt Peak to Hawaii. Two CCD cameras were purchased to extend necessary near nucleus study net coverage. Four sets of special narrow bandpass, infrared, interference filters were procured for the infrared net. On the other hand the

IHW in its advocacy role tried to create a climate of excitement and enthusiasm which made it easier for scientists everywhere to gain the support of their normal funding agencies for new equipment with which to observe comets.

Within the IHW administrative structure all travel expenses and other forms of support have been born by agencies within each country. An equal number of administrative meetings have been held in Europe and the USA to keep travel costs at an equal minimum for those on either side of the Atlantic.

Publication costs of the Archive are being borne by NASA, and support has been promised into 1990. This means that all data must be at the Lead Center by early 1989 in order to put together a printed Archive later in 1989 and have the compact disks mastered and pressed for distribution in 1990. This is a real problem, since many scientists like to work with their data for several years before releasing it. The IHW is trying very hard to encourage data release as early as possible, but inevitably some data will not be included. As a volunteer organization the IHW cannot and should not be tightly controlled. Our only recourse is to friendly persuasion.

2.3 The public

Without public interest and support there could be no space programme, nor would astronomers be likely to have the fine observatories and equipment they sometimes take for granted. Without anticipated public enthusiasm it is unlikely there would have been a Halley Watch. Nevertheless, requests of IHW astronomers for information and interviews began to build as soon as Halley was recovered in 1982, and became overwhelming in 1985. The IHW tried to satisfy all requests by the media. This was a means of reaching millions of people simultaneously and making sure that they received the best available information. Tens of thousands of brochures and fact sheets were mailed and handed out about the IHW and about Halley, especially on where and how to see the comet. Halley 'Hotline' systems were set up quite widely, providing phone numbers for the public to call to receive a recorded message of the latest information on the comet. A comet information centre was set up at the Pasadena Lead Center with models and graphics displays suitable for touring school classes.

Comprehensive slide sets were distributed free of charge to lecturers.

Major media, television, radio, newspapers, and magazines did a very creditable job of informing the public about Halley. For the most part they avoided sensationalism and properly informed the public that, while the current apparition of Halley was of tremendous scientific importance, it was not going to be spectacular. Halley would be nearly impossible to see from the city. Inevitably some fringe publications did their usual irresponsible job, printing all sorts of nonsense. Entrepreneurs offered pins, shirts, hats, mugs, anything the public would buy, but usually gave reasonable value for money received. Unfortunately many people who knew nothing about astronomy bought expensive telescopes when a pair of 7×50 binoculars would have served them much better and longer. The IHW refused to be a part of any commercial venture but allowed our logo to be used for non-commercial purposes such as some postage stamps. A copyrighted 'official' Halley logo that had nothing official about it was *sold* by an entrepreneur to some 50 governments for use on their postage stamps. They could have had the legitimate one for free!

3 ON BUILDING INTERNATIONAL ORGANIZATIONS

3.1 Communications

A proper system of communications is an absolute necessity for any organized undertaking. Direct IHW communication with *all* members takes place only via our *Newsletter*, which is printed twice per year. Each of the eight observing nets also has its own newsletter for more technically specialized communications, and these have also appeared roughly twice a year. During 1985 and 1986 we had an electronic mail system set up, using GTE Telenet, subsidized by GTE so that its use was free within the USA or on the US side of the foreign entry node. In Europe such systems are generally controlled by the governments of each country, and special permits are required for connection and use. Even use of the regular telephone connections to the USA and then entry into telenet was of sufficient difficulty that the system was not widely used outside North America.

Primary administrative communications of the IHW used a separate electronic mail system with the postal system as back-up. Necessary rapid communication used telex and telephone. High-speed transfer of critical technical data for the flight projects was made computer to computer via satellite.

Although we did what we could, it seems likely that a greater sense of 'Halley Watchness,' of belonging, of *esprit de corps*, of participation, could have been built with even more frequent communications.

3.2 General requirements

To build worldwide enthusiasm and voluntary cooperation for a special effort you need to have something fairly unusual like Halley or a bright supernova outburst. Connection with a space mission definitely helps, both as a potential source of funds and to build the sense of participation in something big and important. Of course you can always gain a large following, if you are in a position to give financial support to participants. Smaller groups of specialists can always be assembled to address a limited effort of great scientific importance, but major efforts such as the IHW and the IGY (International Geophysical Year, 1957) succeeded in the first case because of the magic attraction of Halley and in the second because of worldwide governmental funding and excellent organization. Both took place within the framework of existing, respected international organizations.

Timing is extremely important. Begin as early as possible and have planning and organization well in hand before seeking endorsements. Don't overlap another major effort involving the same people, if you can possibly avoid it. Don't ask too much of individuals without offering them something in return. Seek the cooperation of observatory scheduling committees, and build on enthusiasm. Make sure your own organization is built up from respected scientists. You can't demand cooperation from volunteers, but you may well be able to wheedle cooperation, given a competent and trusted staff. All of this is obvious, but it is surprising how often these simple precepts are violated.

4 FINAL COMMENT

As this work goes to press (March, 1990) the IHW anticipates successful completion of its goals in mid-

1990. A test CD containing all Giacobini-Zinner data was produced in April 1989, and it seemed to meet all necessary criteria that had been established for ready access to the desired data. It's use was demonstrated each day during the international comet conference 'Comets in the Post-Halley Era' held in Bamberg, FRG, April 24–28, 1989. The final deadline for acceptance of ground-based data has now passed. Each observing network met or exceeded its data acquisition goals, although we know of data sets not submitted or, in a few cases, submitted so late that Discipline Center staffs already had been dissolved and were unavailable to format the data. The space data is due next month, and we anticipate having a large fraction of it in the Halley Archive. The archive will contain in excess of 20 Gbytes of data, to be delivered on only about 25 CDs because data compression is being used for the images. Each participating scientist will have the entire archive in one small box on his or her desk. We certainly have improved upon the results obtained in 1910. We hope our experiences will help lead to an even better conclusion in 2061.

REFERENCES

Bobrovnikoff, N.T. (1931) Halley's comet in its apparition of 1909–1911, *Pub. Lick Obs.* **XVII** 305–482.

Brandt, J.C., *et al.* (1980) The International Halley Watch, Report of the Science Working Group, NASA TM 82181, pp. 72 + x.

Comstock, G.C., Barnard, E.E., Frost, E.B., Perrine, C.D., & Pickering, E.C. (1909) Circular respecting observation of Halley's comet 1910, *Astron. & Astrophys. Soc. America*, pp. 12 + i.

Comstock, G.C., Barnard, E.E., Frost, E.B., & Pickering, E.C., 1915, Report comet Committee 1909–1913, *Astron. & Astrophys. Soc. America Pubs.* **2** 177–218.

Donn, B., Rahe, J., & Brandt, J.C., 1987, Atlas of comet Halley 1910 II U.S. Government Printing Office, Washington, DC., NASA SP-488.

Edberg, S.J., ed., 1982, *International Halley Watch Newsletter* No. 1, JPL 410–5, pp. 25 + iv.

Edberg, S.J. 1983, International Halley Watch Amateur Observers' Manual for Scientific comet Studies, JPL Pub. **83–16**, Parts I and II.

Ellerman, F., & Barnard, E.E. 1915, Photographs of Halley's comet and Notes on These Photographs, *Astron. & Astrophys. Soc. America Pubs.*, **2**, 219–227.

Friedman, L.D., & Newburn, R.L., 1980, The International Halley Watch: a Program of Coordination, Cooperation, And Advocacy, pp. 313–314 in *Modern Observational Techniques for comets*, J.C. Brandt, B. Donn, J.M. Greenberg, and J. Rahe eds., JPL Pub. **81–68**.

Niedner, M.B., & Liller, W., 1987, The IHW Island Network, *Sky & Tele.* **73** 258–263.

Perrine, C.D., 1934, Observaciones Del cometa Halley Durante Su Aparicion En 1910, *Result. Obs. Nacional Argentino* **25** pp. 108 + 99 plates.

Sekanina, Z., & Aronsson, M. (1985) Archive of Observations of Periodic Comet Crommelin, Made During Its 1983–84 Apparition, JPL. Pub. **86–2** pp. 250 + xv.

APPENDIX 1a

The original Steering Group members of 1981 with their affiliations at that time were:

Dr M.K. Vainu Bappu, Indian Inst. Astrophys., Bangalore, India
Dr Michael J.S. Belton, Kitt Peak Nat'l Obs., Tucson, AZ, USA
Dr Jacques Blamont, Lab. d'Aeronomie CNRS, Verrieres-le-Buisson, France
Dr Geoffrey Briggs, Nat'l Aeronautics & Space Admin., Washington, DC, USA
Dr Armand Delsemme, University of Toledo, Toledo, Ohio, USA
Dr Bertram Donn, Goddard Space Flight Center, Greenbelt, MD, USA
Dr Hugo Fechtig, Max Planck Institut für Kernphysik, Heidelberg, FRG
Dr Ian Halliday, Herzberg Institute of Astrophysics, Ottawa, Canada
Dr George H. Herbig, University of California, Santa Cruz, CA, USA
Dr Y. Kozai, Tokyo Astronomical Observatory, Tokyo, Japan
Dr Reimar Lüst, Max Planck Gesellschaft, Munchen, FRG
Dr A. Massevitch, Academy of Sciences, Moscow, USSR
Dr Charles Robert O'Dell, Marshall Space Flight Center, AL, USA
Dr Rudiger Reinhard, European Space Research and Technology Center, Noordwijk, The Netherlands
Dr Hans Emil Schuster, European Southern Observatory, Santiago, Chile
Dr Vladimir Vanysek, Charles University, Praha, Czechoslavakia
Dr Joseph F. Veverka, Cornell University, Ithaca, New York, USA
Dr Kurt W. Weiler, National Science Foundation, Washington, DC, USA
Dr George Wetherill, Carnegie Institution of Washington, Washington, DC, USA
Dr Fred L. Whipple, Smithsonian Astrophysical Observatory, Cambridge, MA, USA
Dr Laurel L. Wilkening, University of Arizona, Tucson, AZ, USA
Dr Ya. S. Yatskiv, Main Astronomical Obs., Kiev, USSR

APPENDIX 1b

Over the years there were many changes in the Steering Group. The death of our valued friend and colleague Vainu Bappu in 1982 was a special loss. Jack Meadows of the University of Leicester in the UK was a great help from 1983–1986. The membership in mid-1987 was:

Dr W.I. Axford, M.P.I. für Aeronomie, Katlenburg-Lindau, FRG
Dr Michael J.S. Belton, Kitt Peak Nat'l Obs., Tucson, AZ, USA
Dr Jacques Blamont, Laboratoire d'Aeronomie CNRS, Verrieres-le-Buisson, France
Dr Geoffrey Briggs, NASA Headquarters, Washington, DC, USA
Dr William Brunk, U.S.R.A., Washington, DC, USA
Dr Armand Delsemme, University of Toledo, Toledo, OH, USA
Dr Bertram Donn, Goddard Space Flight Center, Greenbelt, USA
Dr Hugo Fechtig, Max Planck Institut für Kernphysik, Heidelberg, FRG
Dr Louis Friedman, The Planetary Society, Pasadena, CA, USA
Prof. Shu-Mo Gong, Academia Sinica, Nanjing, People's Republic of China

Dr Ian Halliday, National Research Council of Canada, Ottawa, Canada

Dr George H. Herbig, University of California, Santa Cruz, CA, USA

Prof. Kunio Hirao, Inst. Space and Astronautical Science, Tokyo, Japan

Dr Yoshihide Kozai, Tokyo Astronomical Observatory, Tokyo, Japan

Dr Lubor Kresak, Slovak Academy of Sciences, Bratislava, Czechoslavakia

Dr Reimar Lüst, European Space Agency, Paris, France

Prof. Alla Massevitch, Academy of Sciences of the USSR, Moscow, USSR

Dr Charles Robert O'Dell, Rice University, Houston, TX, USA

Dr Vernon Pankonin, National Science Foundation, Washington, DC, USA

Dr Rudeger Reinhard, European Space Research and Technology Center, Noordwijk, The Netherlands

Dr R.Z. Sagdeev, Academy of Science, USSR, Moscow, USSR

Dr Hans Emil Schuster, European Southern Observatory, Santiago, Chile

Prof K.R. Sivaraman, Indian Institute of Astrophysics, Bangalore, India

Dr Vladimir Vanysek, Charles University, Praha, Czechoslovakia

Dr Joseph F. Veverka, Cornell University, Ithaca, NY, USA

Dr George Wetherill, Carnegie Institution of Washington, Washington, DC, USA

Dr I.P. Williams, Queen Mary College, London, UK

Dr Fred L. Whipple, Smithsonian Astrophysical Observatory, Cambridge, MA, USA

Prof. Ya.S. Yatskiv, Academy of Sciences of the Ukrainian Republic, Kiev, USSR

APPENDIX 2

Discipline Specialist Teams

Astrometry

D.K. Yeomans	Jet Propulsion Laboratory
R.M. West	European Southern Observatory
R.S. Harrington	U.S. Naval Observatory
B.G. Marsden	Smithsonian Astrophysical Observatory

Infrared Spectroscopy and Radiometry

R.F. Knacke	SUNY (Stony Brook)
T. Encrenaz	Observatoire de Paris

Large-scale phenomena

J.C. Brandt	University of Colorado
M.B. Niedner	Goddard Space Flight Center
J. Rahe	Astronomical Inst., U. Erlangen-Nürnberg

Meteor studies

P.B. Babadzhanov	Academy of Sciences of the Tajik SSR
A. Hajduk	Astronomical Inst. SAV
B.A. Lindblad	Lund Observatory
B.A. McIntosh	Herzberg Institute of Astrophysics

Near-nucleus studies

S. Larson	University of Arizona
Z. Sekanina	Jet Propulsion Laboratory
J. Rahe	Astronomical Inst., U. Erlangen-Nürnberg

Photometry and Polarimetry

M. A'Hearn	University of Maryland
V. Vanysek	Charles University

Radio Studies

W. Irvine	University of Massachusetts
F.P. Schloerb	University of Massachusetts
E. Gerard	Observatoire de Paris
R.D. Brown	Monash University
P. Godfrey	Monash University

Spectroscopy and spectrophotometry

S. Wyckoff	Arizona State University
P. Wehinger	Arizona State University
M.C. Festou	Observatoire de Besançon

APPENDIX 3

R.M. West, General Secretary of the IAU sent the following text to the IHW after the Patras General Assembly.

The 18th General Assembly of the International Astronomical Union, in session at Patras, Greece, on 26 August 1982, adopted the following Resolutions:

The International Astronomical Union

recognizing that it is particularly desirable that preselected comet Halley Days for coordinated observation over a limited time be supported

recommends that observatory directors and observing programme committees give high priority to comet Halley observation during the interval 1985–1987.

The International Astronomical Union

noting that in order to organize and marshall groundbased observations of comet Halley throughout its 1986 perihelion passage and to coordinate them with space missions, an international programme, the International Halley Watch, has been established

and wishing to avoid duplication of effort at the international level and to encourage participation in this programme

endorses the International Halley Watch as the international co-ordinating agency for comet Halley observations.

APPENDIX 4

Lead Center Personnel are:

Western Hemisphere, Pasadena, CA, USA
Leader,	Ray L. Newburn, Jr.
Deputy Leader,	Murray Geller
Amateur Coordinator	Stephen Edberg
Archive Editor	Zdenek Sekanina
Software Specialist	Mikael Aronsson

Eastern Hemisphere, Bamberg, FRG
Leader,	Jürgen Rahe
Deputy Leader,	Horst Drechsel
Amateur Coordinator	Rudeger Knigge

Part II
Plasma

3

The large-scale plasma structure of Halley's comet, 1985–1986

John C. Brandt

1 INTRODUCTION

Two comets, Giacobini–Zinner and Halley, have been probed by spacecraft, and comet Halley has been extensively observed from the Earth's surface. The data obtained relevant to the large-scale structure are extensive, and researchers are beginning to go beyond establishing the basic model of the solar wind/comet interaction and straightforward measurements of distances, velocities, and accelerations. The next few years should be very profitable, with a strong exchange between theory and the observations. We should also see a parallel effort in the areas of plasma physics accessible in the useful cosmic laboratories that comets provide.

The plasma tails of comets were studied extensively by Barnard around the turn of the century. He established the outlines of the cyclic morphology and clearly understood the need for organized monitoring of comets because of their rapid and often dramatic changes (Barnard 1905). Biermann (1951) used the plasma tails of comets to establish the existence of the solar wind. An important ingredient of this work was Hoffmeister's (1943) observational study of their orientations, which are determined by the plasma tails acting as natural 'wind socks' in the solar wind. These orientations can be analyzed to determine global models of the solar wind velocity field (Brandt et al. 1972, Brandt & Mendis 1979). Alfvén (1957) included the magnetic field of the solar wind in the comet/solar wind interaction, and this interaction was studied quantitatively by Biermann et al. (1967).

The model developed historically gave us our picture of the overall plasma structure of comets as produced by the solar wind. Molecular ions produced around the comet are captured onto solar wind field lines. The field lines near the comet are decelerated, causing the field lines with their cometary molecular ions (H_2O^+, CO^+, CO_2^+) to wrap around the comet to produce the plasma tail. The magnetic structure of the tail consists of two lobes of opposite magnetic polarity separated by a current sheet. Because the comet is an obstacle in a supersonic and super-alfvénic flow, a bow shock was expected in order to allow a transition from supersonic to subsonic flow.

Many details of the *in situ* measurement are given elsewhere in this volume. In broad outline, our pre-encounter view of the basic physical processes and the formation of the plasma structures was confirmed.

Our expectations and goals were roughly as follows, based primarily on the author's position as a Discipline Specialist for the Large-Scale Phenomena Network of the International Halley Watch (IHW); see Niedner et al. (1982) and Brandt et al. (1984). Photographs of the comet were desired with wide-field imaging systems (field of view $\gtrsim 5°$), such as Schmidt cameras, at time intervals preferably not exceeding a few hours. Experience has clearly shown that the study of transient events in comets is severely hampered if the time interval between photographs is ~1 day. Approximately 100 observers or institutions joined the Large-Scale Phenomena Network. The distribution of longitudes in the northern hemisphere indicated a relatively high probability of good coverage when the comet was a northern hemisphere object (i.e., before perihelion) given some good luck and favourable weather. The situation after perihelion was another matter. The relative lack of land masses and observatories in the southern hemisphere meant that the goal of essentially continuous coverage could not be achieved. An 'Island Network' was created to fill the gaps. After considerable evolution, the 'Island Network' consisted of six Celestron Schmidt cameras operating from Tahiti, Easter Island, Antarctica, South Africa (two units), and Réunion Island. On the whole, this effort was very successful, and we are grateful for the cooperation of many individuals who made this possible; see Niedner & Liller (1987).

The total yield of the Network is expected to be well over 3000 images, possibly as many as 4000. In addition, we continue to discover individuals, not formally associated with the IHW, who have substantial numbers of good images. Observers with wide-field images of Halley's comet who are not in contact with the IHW are encouraged to write to the author. We would like to preserve as many of your observations for posterity as possible.

The ultimate expectation in terms of the physical understanding of comets is to combine *in situ* measurements of the solar wind with the wide-field images. This approach has proven to be profitable through the years. For approximately one week in March 1986, scientists interested in cometary plasmas and large-scale structure had the ultimate luxury of *in situ* measurements of the comet and extensive imaging coverage. This period will be extensively discussed below.

Halley's comet is now solidly established as the most thoroughly studied comet in history, and it is likely to retain that title for quite some time. As the various aspects of the large-scale structure are discussed, it is well to remember that comets are individuals, and that our ideas may have limited applicability to other comets.

2 PANORAMIC REVIEW

This section is a descriptive review of large-scale

COMET HALLEY

Fig. 1 — Images of Halley's comet on 15 November 1985 showing the turn-on of plasma tail activity (Bulgarian National Observatory photographs).

Fig. 2 — Halley's comet on 7 December 1985 showing a developed plasma tail and a typical helical structure (photograph from Joint Observatory for Cometary Research, operated by NASA-Goddard Space Flight Center and the New Mexico Institute of Mining and Technology).

plasma activity in Halley's comet during 1985 and 1986. For earlier descriptions, see Brandt & Niedner (1986a, b, 1987). Later, the physics will be discussed.

Plasma activity was clearly in evidence by mid-November 1985 (Fig. 1). The plasma 'plume' seems to have come and gone during this period because it was visible on some days and not on others. We tentatively adopt mid-November 1985 as the time of 'turn-on,' corresponding to a heliocentric distance, r, of 1.8 AU. Note, however, that there are some reports of possible earlier tail activity.

By early December, the plasma tail was well developed (Fig. 2). On 4 and 5 December, one of the first clearly visible disconnection events (DEs) was recorded.

January 1986 produced the first dramatic DE (see Fig. 3). The comet then moved to perihelion on 9 February, 1986. Most observations were not possible near the time of perihelion because the sky location was near the Sun. The gap may be partly filled by observations from the Solar Maximum Mission.

The comet in late February was spectacular (Fig. 4). The photograph shows a plasma tail and a DE, multiple dust tails, and an anti-tail.

The comet was in full splendour in March for the arrival of the Halley armada. The colour image on 8 March (Fig. 5*) clearly shows the dust tail as pale yellow (reflected sunlight) and the plasma tail with a prominent ray as blue (sunlight scattered by CO^+ molecules). The next day, 9 March, showed a DE (Fig. 6), and by 10 March (Fig. 7) the DE was well developed. The community of comet scientists was fortunate that the comet showed so much activity at the time of the spacecraft encounters. The activity

* to be found in the colour illustration section in the centre of the book.

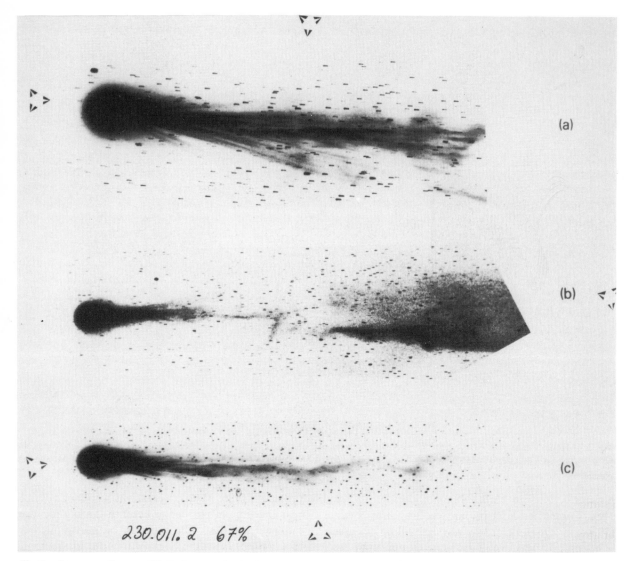

Fig. 3 — Sequence of images of Halley's comet clearly showing a Disconnection Event (DE). Images are from the Calar Alto Observatory of the Max-Planck Institut für Astronomie on 9 and 10 January 1986, (a) and (b); and from the Haute-Provence Observatory on 11 January, (c).

continued after the encounters, as illustrated by the sequences in Figs. 8 and 9.

Late March of 1986 and early April were the best times for casual viewing of the comet. The colour view (Fig. 10*) shows the comet near the globular cluster ω Centauri on 14 April; compare with the more detailed black and white image obtained on the same date (Fig. 11).

The comet continued to show a long, slender tail

well into May (Figs 12, 13) and at least until 14 June (Fig. 14). On the latter date, the tail length was approximately 60 million km or 0.4 AU. (The main, visible tail in June is now known to have been composed of dust.) Several lines of evidence including photometry and spectroscopy, indicate a plasma tail 'turn-off' shortly thereafter, or roughly in mid-June. We tentatively adopt this time, which corresponds to $r = 2.3$ AU. At worst, this is a lower limit to the 'turn-off' distance.

---

* to be found in the colour illustration section in the centre of the book.

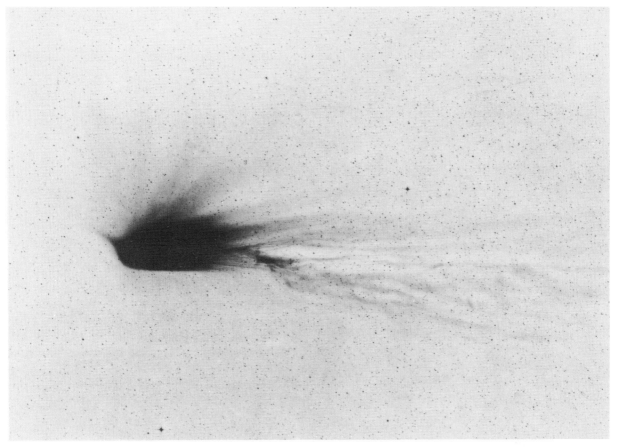

Fig. 4 — Halley's comet as photographed with the UK Schmidt Telescope on 1986 February 22.78 UT, showing an anti-tail (left), dust tail Telescope in Australia. (Photography by B.W. Hadley, © Royal Observatory, Edinburgh).

distance.

Thus, Halley's comet displayed large-scale structure for approximately seven months. The display included the general forms and colours expected, considerable fine structure, and many DEs. Apparently, DEs are a commonly observed feature of large-scale plasma structures if the coverage is sufficient. One lack seems to be the absence of good ray-turning sequences as previously observed on other comets such as comet Kobayshi–Berger–Milon (Brandt & Mendis 1979). A possible explanation is the relatively large dust emission from comet Halley which might obscure the near-nuclear region where ray-turning sequences are usually observed.

3 PHYSICAL DISCUSSION

Here we focus on two major aspects of large-scale plasma structure: (1) the 'turn-on'/'turn-off' of plasma phenomena, and (2) the disconnection events.

3.1 Turn-on/Turn-off

Mendis & Flammer (1984) extended the earlier work of Biermann *et al.* (1967) to treat the different regimes of comet/solar wind interaction as a comet approaches (and recedes) from the Sun. The parameter of interest in this discussion is the mean molecular weight of the mass-loaded solar wind as it approaches the comet. The flow is not conservative; i.e., ionized molecules are continually added to the flow. If the mean molecular weight is normalized to the value far from the comet, it cannot exceed the critical value of 4/3 without formation of an upstream bow shock.

Far from the Sun, a comet's low total gas production rate produces only a tenuous atmosphere which allows

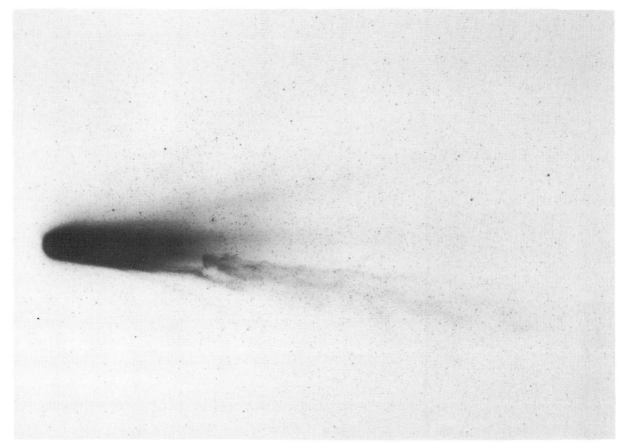

Fig. 6 — The comet on 9 March 1986 showing dust tail structures and a DE; compare with Fig. 7. Image obtained with the UK Schmidt telescope in Australia. (Photography by B.W. Hadley, © Royal Observatory, Edinburgh).

the solar wind flow to reach the nuclear surface without exceeding the critical value. Mendis & Flammer (1984) calculated that this would occur at approximately 3 AU for Halley. At this distance, the critical value would be reached one solar wind proton Larmor radius above the surface. This criterion results from the expectation that the collisionless shock would have a thickness of one ion (proton) Larmor radius, and the protons would hit the nucleus if the shock were closer than the Larmor radius.

Inside the 3 AU distance, the upstream bow shock diverts the flow around the comet without the critical value being exceeded. To form an ionopause, which separates the mass-loaded solar wind flow from the pure cometary ions by a tangential discontinuity, the outflow from the comet must supply sufficient momentum which is generally supplied by the neutrals. Houpis & Mendis (1981) showed that the solar wind flows through the cometary atmosphere until the proton–cometary neutral momentum transfer mean free path is equal to the radial distance from the nucleus. When this radial distance is greater than the Larmor radius, the ionopause forms. When the ionopause is stable, the visible tail can form. The predicted distance for this event was around 2.5 AU or less.

The observed turn-on and turn-off distances are approximately 1.8 AU and 2.3 AU, respectively. We can compare these distances with determinations of total gas production rate to check for consistency. The Larmor radius for the heliocentric distances of interest is given by $L_i = \text{const} \times r^{2/3}$ (Mendis & Flammer 1984), where r is the heliocentric distance. The ionopause distance is thought to be equal to the

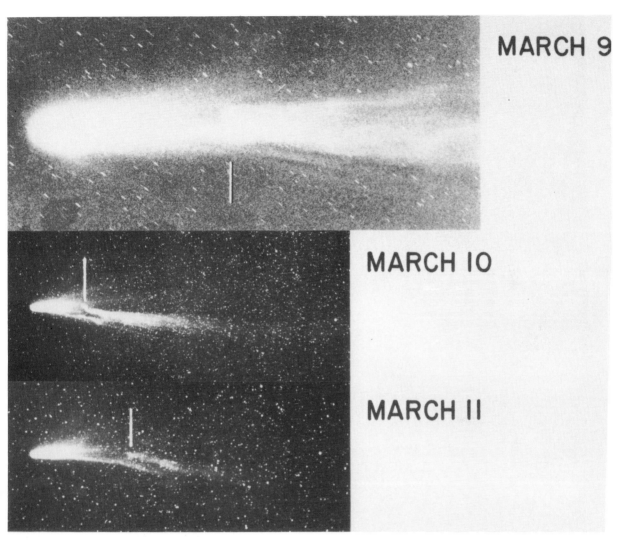

Fig. 7 — Extended sequence showing the large-scale evolution of a DE (tick marks) shown in Fig. 6. The mid-point of the exposures was within ± 15 min of 0812 UT on the days indicated. The very large distances involved are shown by the 20×10^6 km scale. On the 14 March exposure, the DE is approximately 50×10^6 km or $\frac{1}{3}$ AU from the nucleus. On this and other figures, the scale indicates distances along the prolonged radius vector. (Photographs from W.E. Celnik, Ruhr-Universität Bochum).

distance at which the ions are collisionally decoupled from the neutrals, i.e., $r_c = \alpha\, Q_n/4\pi v_s$, where α is the solar wind proton–cometary neutral momentum transfer collision cross-section ($\approx 2 \times 10^{-15}$ cm^2), Q_n is the total production rate of neutrals, and v_s is the expansion speed from the surface. If we neglect the very slow variation of v_s we can set $L_i = r_c$ and obtain the functional dependence of the turn-on/turn-off distance as

$$r_{t-o} \propto (Q_n)^{3/2} \qquad (1)$$

The ratio of the off/on distances is approximately 1.28, a value which implies a 20% greater gas production rate in mid-June of 1986 than in mid-November of 1985. Such a ratio is entirely consistent with the H_2O production rates reported by Feldman *et al.* (1986) on the basis of IUE observations.

Note that the assumption of an ionopause distance

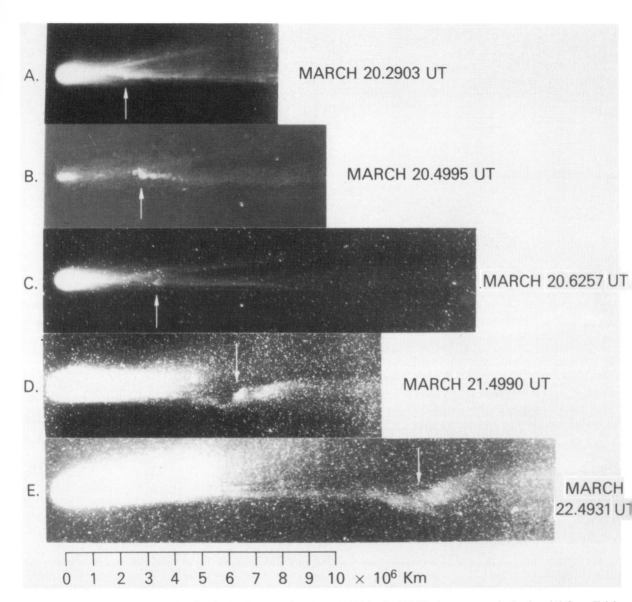

Fig. 8 — Sequence of photographs showing the development of a DE on 20–22 March 1986. The images were obtained at: (A) Cerro Tololo Interamerican Observatory (University of Michigan Curtis Schmidt); (B) Table Mountain Observatory (University College London detector); (C) Mauna Kea Observatory; (D) Easter Island as part of the IHW Large-Scale Phenomena Network's (L-SPN's) 'Island Network'; and (E) the Joint Observatory for Cometary Research. The arrows mark the near end of the disconnected tail. (From Brosius *et al*. 1987).

determined by the momentum supplied by the neutrals versus the momentum being supplied by the ions can, in principle, be checked by the shape of the ionosphere. It should be roughly spherical if neutrals dominate, and finger-shaped if ions dominate (Houpis & Mendis 1981). Finally, we note that we have not

addressed the question of the stability of the ionopause; see Ershkovich & Mendis (1983).

Fairly extensive observations of the jets in mid-November 1985 could be useful for understanding the turn-on process, but their interpretation is equivocal. Note that the comet during this period was at

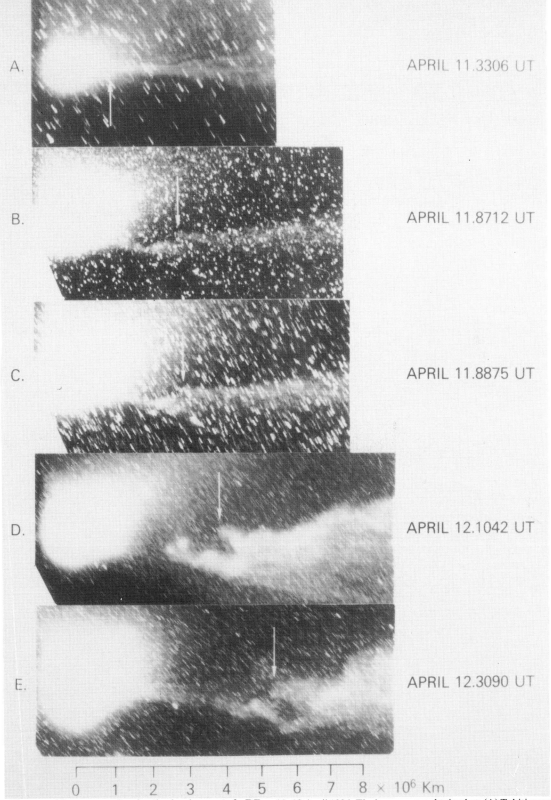

Fig. 9 — Photographic sequence showing the development of a DE on 11–12 April 1986. The images were obtained at: (A) Tahiti as part of the IHW/L-SPN's 'Island Network'; (B) and (C) Réunion Island as part of the IHW/L-SPN's 'Island Network'; (D) and (E) CTIO/University of Michigan Curtis Schmidt. The arrows mark the near end of the disconnected tail. (From Brosius *et al.* 1987).

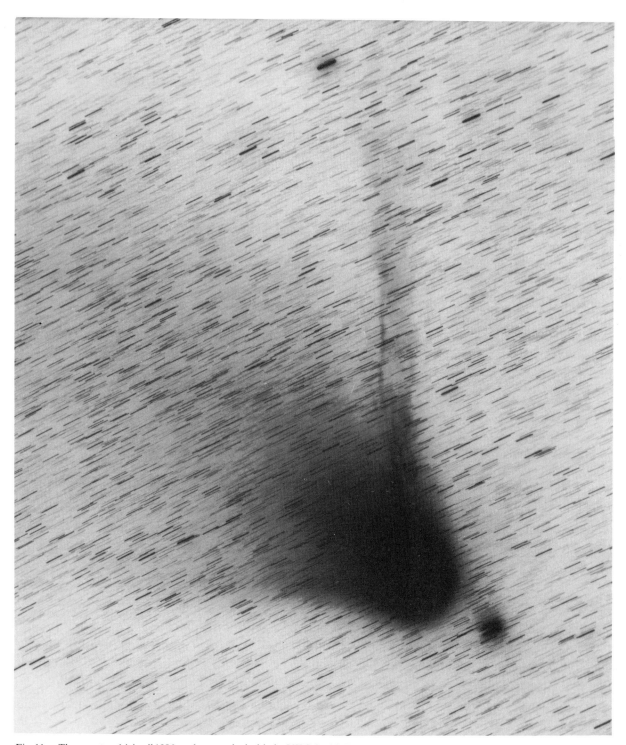

Fig. 11 — The comet on 14 April 1986 as photographed with the UK Schmidt Telescope in Australia. Note the detail. (Photography by B.W. Hadley, © Royal Observatory, Edinburgh).

Fig. 12 — Halley's comet on 12 May 1986 as photographed from the Sutherland Observatory of South Africa. (Photograph supplied by P. Maffei, University of Perugia; see Maffei (1986)).

opposition. Thus, features along the Sun–Earth–comet line (toward or away from the Sun) could show rapid changes in position angle, and such changes were observed. Shkodrov *et al.* (1986) note that the jet could be a plasma tail, but favour an interpretation that it is a dust jet connected with the rotation of the nucleus. Dobrovolsky *et al.* (1986) note the wavy structure recorded on some of the photographs, and they favour a plasma interpretation. They suggest a classical 'fountain model' situation in which the jet starts off toward the Sun and is turned back into the tail, in order to explain the rapid changes in position angle. Watanabe *et al.* (1986) argue that the feature must be a plasma tail because synchrones cannot match the turning rate. The ion tail appears to originate from the southern region of the coma; also see Grün *et al.* (1986), and Liu (1986). The situation could be improved by an analysis that uses all the data sets. This should occur in the near future.

3.2 Disconnection events (DEs)

3.2.1 History of ideas.

The general morphology of DEs was known to Barnard (1920); also see Brandt (1982) many years ago, but the knowledge of solar wind conditions necessary to work toward a physical explanation was not available. DEs were not the subject of research for approximately half a century. Their modern revival occurred in the 1978 paper by Niedner & Brandt, which was stimulated by an effort to determine the solar wind cause of certain events in comet Kohoutek during January 1974. One of these events was an

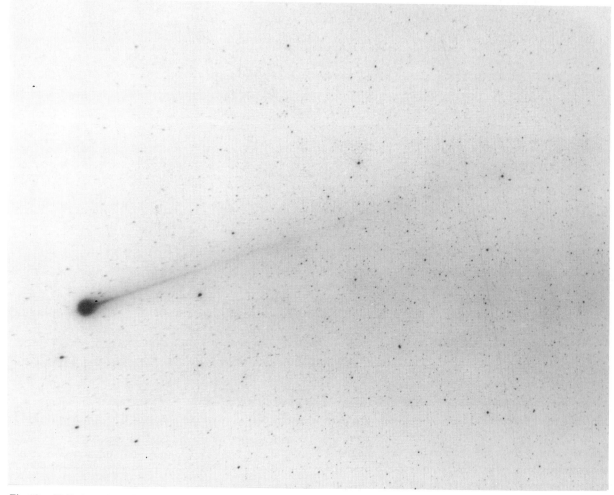

Fig. 13 — Halley's comet on 27 May 1986 as photographed with the UK Schmidt Telescope in Australia. (Photography by B.W. Hadley, © Royal Observatory, Edinburgh).

Fig. 14 — Halley's comet still with a long tail, on 14 June 1986. (Photograph taken by E.P. Moore, The Joint Observatory for Cometary Research, NASA-Goddard Space Flight Center and New Mexico Institute of Mining and Technology).

apparent detachment of the entire plasma tail followed by its rapid reformation. Such occurrences had been believed to be atypical or rare; a search of the literature revealed that this was not the case. Niedner & Brandt (1978) also noted the close correspondence between the DE and the passage of an interplanetary sector boundary, and extended Alfvén's (1957) physical picture of plasma tail formation to include the effects of the sector boundary. On their picture, magnetic reconnection would occur on the sunward side of the comet when the new magnetic field of opposite polarity is pressed into the old field, gradually severing the connection of the tail to the head, and ultimately producing the DE. As the old tail drifts away, a new tail of opposite polarity is formed as

part of a regular cyclic morphology. In principle, a DE could occur whenever a comet with a well-developed plasma tail crosses a sector boundary. The words 'sector boundary' and 'current sheet' are used somewhat interchangeably. The heliospheric current sheet is a three-dimensional structure in space separating regions of opposite magnetic polarity. When a spacecraft or a comet crosses the current sheet, it passes from a sector of one magnetic polarity to a sector of opposite polarity, i.e., it has crossed a sector boundary. Ness & Donn (1966) introduced sector boundaries as producing phenomena in comets in another connection.

Because sector boundaries were believed by some groups (in the late 1970s) to have only a small warp from the solar equator, and because DEs were observed at latitudes $\sim 40°$, there was concern about the sector boundary model. Ip & Mendis (1978) proposed that high-speed/pressure solar wind streams would produce the flute instability and cause DEs. Niedner & Brandt (1979) noted that although many high-speed/pressure streams are associated with sector boundaries, many ($\sim 40\%$) are not. The low-latitude sample showed a close correlation with sector boundaries. Since the time of these papers, the view of the extent of sector boundaries has changed. It is now generally conceded that sector boundaries can extend to solar latitudes greater than $\sim 40°-50°$, particularly at solar maximum (e.g., Hoeksema *et al.* 1982, 1983). This objection to DE mechanisms associated with sector boundaries is no longer relevant.

Another alternative disconnection event mechanism was proposed by Jockers (1981) who suggested mechanisms based on solar wind flows across the plasma tails; also see his detailed discussion (Jockers 1985).

Niedner & Brandt (1980) discussed the morphology of DE events and noted the possibility of magnetic reconnection across the tail and between the old and new plasma tails. The known sample of DEs was collected by Niedner (1981) and was used (Niedner 1982) to study the latitudinal properties of sector boundaries. The sector-boundary viewpoint was subsequently summarized by Brandt (1982) and by Niedner (1984). I explicitly note my association with the origin of one physical picture of DEs and the natural advocacy or bias that comes with it.

Russell *et al.* (1982) noted some of the similarities between Venus and comets with respect to DEs. In particular, they noted that the orientation of the Venusian magnetotail is as expected for solar wind magnetic field lines draped over the ionosphere, and that the orientation of the tail field reverses after a sector boundary crossing. Later, Ip (1985) and Russell *et al.* (1986) suggested tailward reconnection as a mechanism for producing DEs by analogy with the Earth's magnetotail. Russell *et al.* (1986) also suggested that tailward reconnection could be triggered by sudden changes in the direction of the solar wind magnetic field, which would include sector boundary crossings. Thus, DEs produced by tailward reconnection might show a correlation with sector boundary

crossings and, hence, show a statistical similarity with DEs produced by sunward reconnection. Note, however, that the morphology of the DE might be different for the different location for reconnection, and that Russell *et al.* (1986) clearly favour trigger mechanisms produced by high-speed solar wind streams.

The suggestions and model by Russell *et al.* (1986) raise interesting questions. There is general agreement that cross-tail reconnection can occur. Does it sever the entire plasma tail or does it produce a substructure in the tail? What relation does the visible plasma tail bear to the entire magnetic tail structure? And finally, if the entire tail were severed, how would the near-nuclear (toroidal) field evolve?

After the appearance of Halley's comet and the spacecraft missions, a plethora of new ideas on DEs has appeared. T. Saito and K. Saito and their collaborators have written extensively on this subject (Saito *et al.* 1986a, b, c, d, and Saito & Saito 1986). They propose that the angle between the heliospheric neutral sheet and the axis of the plasma tail is crucial to producing DEs on the Niedner–Brandt model. Thus, they view the sector-boundary model as most efficient around solar maximum when the current sheet makes a large angle with the plane of the solar equator, and as least efficient around solar minimum when the current sheet makes a small angle with the solar equator. Because the 1985/86 apparition of Halley's comet was near solar minimum, they feel that mechanisms based on sector boundary crossings would not be effective. For this reason, and because of their experience during the time of the Suisei and Sakigake encounters with Halley's comet, they have searched for alternative means to explain the many observed DEs. These alternatives, following Saito *et al.* (1986a), include (1) a process involving dynamic pressure (e.g. comet Halley, 31 Dec. 1985); (2) a slipping process (e.g. comet Halley, 13 Dec. 1985); (3) a rotating IMF process (e.g. comet Halley, 8 March 1986); (4) the Hones process, based on an analogy with auroral substorms in the Earth's magnetosphere (e.g. comet Morehouse, 30 Sept.–1 Oct. 1908); and (5) the recognition that some reported DEs are other events seen with a poor perspective. Some of these suggestions, of course, cover ideas already in the literature, and, in addition, Saito & Saito (1986) have applied some of them to Halley's comet as observed in 1910.

A new class of suggestion has been made by Caron *et al.* (1986). They propose that a simpler explanation would be if the DEs were produced by breaking off small pieces of the comet mantle which become sources of bursts of gas and dust.

Table 1 lists ideas, theories, or mechanisms for DEs, with an attempt at classification.

Table 1 — Investigations and ideas concerning DEs

Investigation	Magnetic reconnection	Ion production effects	Pressure effects	Other	Comments
Bernard (1909)				X	Disturbing medium, electrical currents
Wurm & Mammano (1972)		X			
Burlaga *et al.* (1973)				X	Not shock or high-speed stream
Jockers & Lüst (1973)				X	Interplanatary shock
Niedner & Brandt (1978, 1979, 1980)	X				Frontside reconnection
Ip & Mendis (1978)		X			Flute instability
Jockers (1981)		X	X	X	Differential acceleration
Brandt (1982)	X				Frontside reconnection
Niedner (1984)	X				Frontside reconnection
Reiff (1984)*	X			X	Tailside reconnection, slippage
Baker (1984)*	X				Tailside reconnection
Ip (1985)	X	X	X		Tailside reconnection
Brand & Niedner (1986a,b)	X				Frontside reconnection
Caron *et al.* (1986)		X			Small piece of mantle
Lundstedt & Magnusson (1986)	X		X		Tailside reconnection
Niedner & Brandt (1986)	X				Frontside reconnection
Niedner & Schwingenshuh (1986)	X				Frontside reconnection
Russell *et al.* (1986)	X				Tailside reconnection
Saito & Saito (1986)				X	1910
Saito *et al.* (1986a)		X		X	Various
Saito *et al.* (1986b,c)				X	Quasi-parallel model
Saito *et al.* (1986d)			X		
Wu & Qiu (1986)				X	Interplanetary shock
Brandt & Niedner (1987)	X				Frontside reconnection
Brosius *et al.* (1987)	X		X		Frontside reconnection

* The suggestions by Reiff and Baker appear in the discussion following Neidner's (1984) paper.

3.2.2 Philosophy of approach

Testing any of these hypotheses is very difficult. The problems lie in two general areas. The first concerns the nature of the DE process. If it is basically a single mechanism operating under the appropriate circumstances, testing hypotheses will be much more straightforward than if DEs have multiple physical causes.

The second problem area concerns the need to know solar wind conditions at the comet. Generally, spacecraft are not located in the vicinity of comets, and, hence, assumptions are required to infer conditions at the comet. The usual assumption is that conditions at the spacecraft are part of a stream, constant in time and rotating with the Sun. Latitude differences are assumed to be negligible. With these assumptions, the delay from the spacecraft to the comet is readily calculated to be,

$$\Delta T = \frac{r_s - r_c}{w} + \frac{l_s - l_c}{\Omega}, \tag{2}$$

where the subscripts s and c refer to the spacecraft and the comet, respectively, r is the heliocentric distance, w is the solar wind speed, l is the heliocentric longitude, and Ω is the synodic rate of solar rotation. The first term shows the delay due to differences in radial distance, and the second term shows the delay due to differences in solar longitude. Equation (2) gives the time differential for the same solar wind conditions to be at the spacecraft and the comet under the stated assumptions. Clearly, (2) has better applicability when the differences in heliocentric distance, longitude, and latitude are small. The same principles apply in translating geomagnetic indices or measurements from satellite spacecraft to a comet.

Normally, the translation from Sun to comet is the most difficult because a distance of roughly 1 AU is involved. This is usually a calculation of the position of the heliospheric current sheet. The magnetic polarities are determined (from solar measurements) on a source surface (2.5 $R_\odot$) and projected to the comet with constant solar wind speed, typically $w \sim 400$ km s^{-1}. Even though this calculation is uncertain, it is often used because it is the only one available. The

uncertainty can be decreased if spacecraft are available to check sector-boundary crossings.

With this background, we make several observations and assumptions about testing DE hypotheses. First, we apply 'Occam's Razor' and assume that all DEs have a common physical cause. Second, we note the difficulty in determining solar wind conditions at the comet. The uncertainties, particularly in extrapolations from the Sun, are sufficiently large that no hypothesis should fail on the basis of a single event. For an example of possible uncertainties in current sheet extrapolations, see Suess et al. (1986). Third, the period of 6 to 14 March 1986 is very special because of the spacecraft encounters and extensive ground-based coverage. A hypothesis inconsistent with this period should be viewed with suspicion. Fourth, the uncertainties dictate that each DE receive a separate, detailed study. Only then will reliable statistics suitable for definitive hypothesis testing be generated.

3.2.3 DEs in Halley's Comet

Here we attempt to discuss the physical origin of specific DEs observed in Halley's comet. Fortunately, many DEs were observed during the 1985/86 apparition, and a database for discussion exists. Table 2 lists the observed DEs. Bear in mind that the assumption of a common origin simplifies our task, and that the correlation of DEs with sector boundary crossings will be a discriminant.

Selected DEs will now be discussed in some detail.

31 December 1985

This event is of special interest because it may be the 'best counterexample' in the sense of a DE possibly not caused by a sector boundary crossing. Saito et al. (1986d) present evidence that

> '... indicates clearly that comet Halley did not cross the heliomagnetic equator (or the sector boundary) on 31 December, 1985. Hence, this DE-like knot event cannot be attributed to the sector boundary as proposed by Niedner and Brandt.'

In addition, these authors feel that the tilt angle of the neutral sheet in this phase of the solar cycle is small and stable.

Table 2 — Prominent Disconnection Events (DEs) in Halley's Comet, 1985–1986 (Events with disconnection times expressed in decimal days have been analyzed kinematically, at least preliminarily) (Courtesy of Malcolm B. Niedner, Jr 29 June 1987)

Approximate disconnection time (UT)	Sector boundary correlation[1]
1985	
December 4.7	yes (IMP-8 magnetometer data)
December 14–15	?
December 31.2	no
1986	
January 8.6	yes (IMP-8 magnetometer data)
January 9.7	yes (IMP-8 magnetometer data)
January 17	?
January 20	no
February 22.0	yes
March 1	yes (comet skimming current sheet)
March 4.5	yes (comet skimming current sheet)
March 8.4	yes (via Vega-1 magnetometer data)
March 11?[2]	yes (via Sakigake magnetometer data)
March 12.6	yes (via Giotto magnetometer data)
March 16.5	yes (via Vega-1 and PVO mag. data)
March 19.3	yes (via ICE magnetometer data)
April 1.7	yes?[3]
April 6.9	yes?
April 10.9	yes? (via ICE magnetometer data)
April 15	yes (via ICE magnetometer data)

[1] Unless otherwise noted, a correlation (or lack thereof) is deduced solely by comparing the onset time of the DE with predicted crossing times of the large-scale heliospheric current sheet calculated (at the corona) by Hoeksema and colleagues at the Wilcox Solar Observatory. A 'no' simply means that the large-scale current sheet lies at Carrington longitudes far from that of the DE, but does not preclude the existence of a local polarity reversal responsible for the DE.
[2] Existence of DE highly likely, but not guaranteed.
[3] Comet passed southernmost extent of current sheep warps; given uncertainty in computed coronal fields, DEs can be considered correlated with IMF sector boundaries.

The situation is somewhat equivocal, as can be seen in Fig. 15 which shows the heliospheric current sheet data for this event and for the DE of 10 January, 1986, to be discussed next. The dotted line shows the projected heliospheric current sheet based on the solar data, and the diamonds (◆) the crossings measured by spacecraft. Note that the current sheet extends to sufficiently high latitudes to have caused the DE. It is simply displaced in time. If we shift the projected solar data to more closely match the observed crossings, i.e. to the left, the discrepancy becomes ∼2 days. Such a discrepancy is well within the realm of possibility; see also Fig. 16 and related discussion.

The DE of 31 December is not a good fit with a sector-boundary model, but it is certainly not fatal to sector-boundary models. This event clearly illustrates the difficulty of analysis. Saito *et al.* (1986d) explain this event on the basis of a dynamic pressure model.

10 January 1986

This was the first spectacular DE and so received considerable public notice. The DE has been discussed by Caron *et al.* (1986) and by Niedner & Brandt (1986). The heliospheric current sheet data are shown in Fig. 15, and those data indicate that this DE has an excellent association with a sector boundary crossing. A small shift of the projected sector boundary to the left to match the sector boundary crossings as measured by IMP-8 produces nearly perfect agreement. Caron *et al.* (1986) have discussed this event but did not assign a specific cause.

8 March 1986

This DE is extraordinary because it occurred during the time of close spacecraft passages by the comet. This event has been extensively discussed by Niedner & Schwingenshuh (1986). The heliospheric current sheet data from their work are shown in Fig. 16. The ▲ showing the onset of the DE at the comet and the (◆) showing the sector boundary crossing measured by Vega-1 near Carrington longitude 150° are coincident, and are displaced on the diagram only for clarity.

The association of this event with a sector boundary crossing is unequivocal. The utilization of 'space truth' at the comet also illustrates the magnitude of the errors occasionally possible in using projected solar current sheet locations. At Carrington longitudes of 150° and larger, the current sheet must lie well above the projections to pass through the sector boundary crossings measured by Vega-1 and Pioneer Venus Orbiter. Between latitudes 0° to +5°, the discrepancy is substantial.

The fortunate circumstance of the 8 March 1986 DE occurring between the passages of Vega-1 and Vega-2 allows another use of 'space truth.' The polarity of the sunward magnetic barrier reversed between Vega-1 and Vega-2 as would be expected on the Niedner–Brandt model. The morphology also fits this model, and the solar wind densities were low

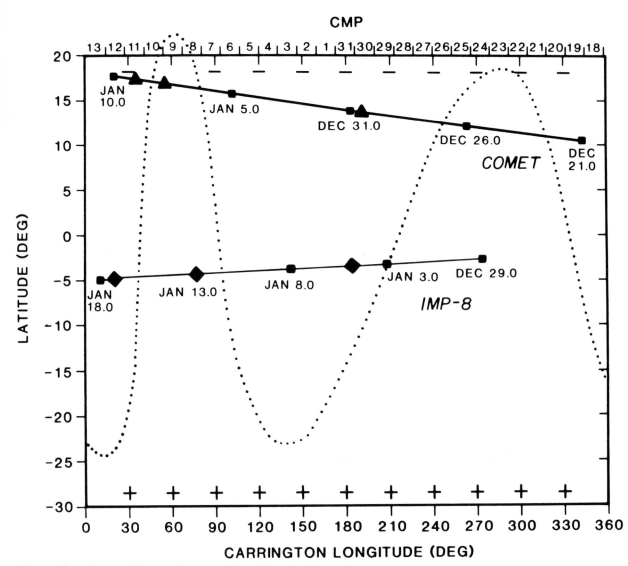

Fig. 15 — Synoptic map of the coronal 'source surface' of the solar wind and interplanetary magnetic field (after M.B. Niedner). The dotted line is the coronal neutral line across which magnetic polarity reverses and which is convected out into interplanetary space by the solar wind, giving rise to magnetic sectors. The track of the comet across the corona was obtained by computing the footprint longitude and latitude of the Archimedean spiral connecting the Sun and comet for a constant solar wind speed of 400 km s^{-1}. The track of IMP-8 was similarly computed. The triangles (▲) mark the times of prominent DEs, and the diamonds (◆) mark the times of measured sector-boundary crossings. The coronal neutral line coordinates were kindly provided by J.T. Hoeksema, Stanford University.

during this period.

Thus, with exceptional certainty for this type of investigation, we know that Halley's comet encountered a polarity reversal in the solar wind magnetic field circa 8 March 1986. Subsequently, a major DE and a polarity reversal of Halley's magnetic structure were observed. These events are in superb agreement with the Niedner–Brandt model. In addition, the minor DE of 6 March 1986 was also associated with a polarity reversal and Vega-1 measured energetic particles consistent with sunward magnetic reconnection (Verigin et al. 1987).

Of course, the models by Ip (1985) and by Russell et al. (1986) might also be associated with the sector

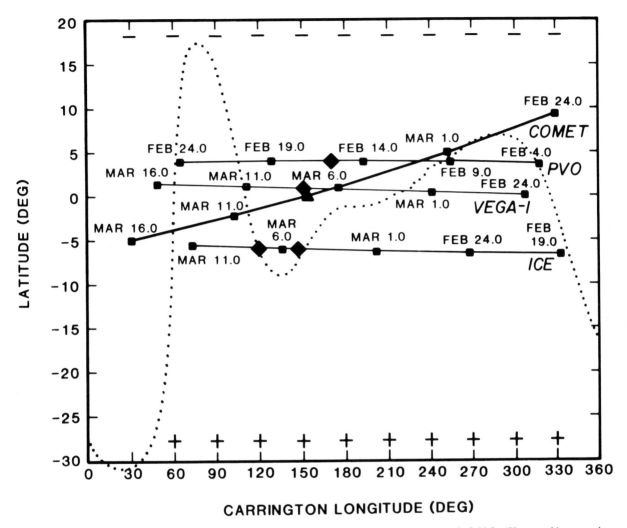

Fig. 16 — Synoptic map of the coronal 'source surface' of the solar wind and interplanetary magnetic field for 'Vega week' prepared as described for Fig. 15 (after M.B. Niedner and K. Schwingenschuh). The dotted line is the coronal neutral line, the thick solid line is the track of the comet, and the thin solid lines are the tracks of the spacecraft Pioneer Venus Orbiter (PVO), Vega-1, and the International Cometary Explorer (ICE). Measured sector-boundary crossings are marked by diamonds (♦) and the DE (8 March 1986) under discussion by the triangle (▲). The triangle and the diamond near Carrington longitude 150° are essentially coincident; they are displaced for clarity only. See text for discussion.

boundary crossing. However, the polarity reversal of the magnetic structure on Halley is not a straightforward prediction of their model. Hence, at least for this event, the evidence appears to favour sunward magnetic reconnection.

In addition to the discussion above, Saito *et al.* (1986a) suggest that the DE of 8 March was caused by a rotating IMF process. Wu & Qiu (1986) suggest that this DE was caused by an interplanetary shock wave produced by solar flares recorded at that time. No evidence for this view from the spacecraft near the comet was cited by Wu & Qiu.

20 March 1986

Brosius *et al.* (1987) have carried out a detailed investigation of this DE, using magnetometer and electron plasma data from the International Cometary Explorer (ICE). They conclude that the most likely explanation is the crossing of an interplanetary sector boundary. Note that the solar wind proton densities were uncharacteristically low at this time.

11 April 1986

Brosius *et al.* (1987) also studied this event on the basis of ICE data. Analysis was somewhat equivocal because of the heliographic latitude separation between ICE and Halley's comet. Nevertheless, this DE seems to be correlated with a polarity reversal and/or an essentially simultaneous compression region. Lundstedt & Magnusson (1986) have discussed this DE, and they attribute it to reconnection in the tail triggered by the compression region.

Jockers *et al.* (1986) have obtained narrowband filter images of the near-nuclear region during the 11 April 1986 disconnection event. Examples of their observations in CO_2^+ emission are shown in Fig. 17. The images show considerable motion (or oscillation) of the inner plasma tail, and Jockers *et al.* (1986) note that the tail disconnection first becomes visible at 0823 UT. From the published images and scales, the disconnection was at roughly 130 000 km from the nucleus at 0848 UT. Hence, we conclude that the disconnection took place at distances of $\lesssim 100\,000$ km.

Accumulation of this type of information for many events could be a discriminant for DE origins based on magnetic reconnection, because the Sunward site (Niedner & Brandt 1978) would predict a distance $\geqslant 100\,000$ km, while the tailward site (Russell *et al.* 1986) would predict a distance $> 100\,000$ km. This approach would be less of a discriminant if the site of the tailward reconnection were closer to the nucleus.

This concludes the discussion of specific DEs. Ultimately, many more DEs must be subjected to a detailed analysis before definitive statements can be made.

3.2.4 Discussion

The discussion will continue on the basis of the apparent association of DEs with sector boundaries. Table 2 clearly shows the association. The dividing line between yes and no is at approximately 2 days. Such events are marked by a ?. Events well outside 2 days are marked by no, and those well inside by yes. Events inside but close to 2 days are marked by yes?. For the 19 prominent DEs, the statistics are yes–12, yes?–3, ?–2, and no–2. Theories based on sector-boundary crossings are favoured unless all high-speed streams are closely correlated with sector boundaries during this period. Fortunately, this can be checked.

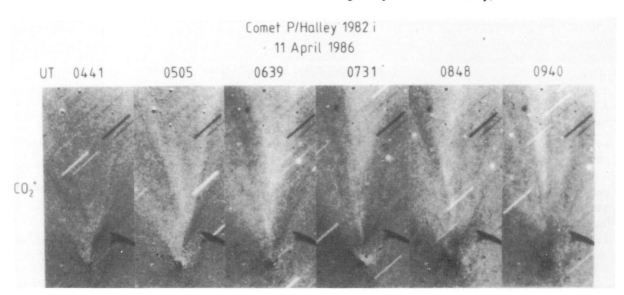

Fig. 17 — Detail of changes in the inner plasma tail as recorded through narrowband filters isolating CO_2^+ emission. An image obtained at 0417 UT was subtracted from the other images to prepare this figure; bright is positive, dark negative. See text for discussion. (Photograph supplied by K. Jockers, Max-Planck-Institut für Aeronomie).

The fact that high-speed streams are often closely associated with sector boundaries is a major complication and a source of confusion. Most sector boundaries have an associated high-speed stream, but many high-speed streams are not associated with sector boundaries. Thus, if high-speed streams are the critical element in producing a DE, approximately 40% of them should *not* be associated with sector boundaries (Niedner & Brandt 1979).

The theories consistent with the sector-boundary association are the original sunward magnetic reconnection model of Niedner & Brandt (1978) and the tailward magnetic reconnection models of Ip (1985) and Russell *et al.* (1986) *if* triggered by sector boundaries.

Note, although Russell *et al.* (1986) mention 'sudden changes in the direction of the interplanetary magnetic field' as a possible trigger of DEs (by analogy with the Earth's magnetotail), they clearly favour triggering by an interplanetary shock or a decrease in the solar wind Alfvén Mach number. They note that these two causes 'are consistent with comet tail disconnection events being well correlated with high-speed solar wind streams.' With either cause, a reversal in the polarity of the comet's magnetosphere would not result without invoking a magnetic reversal.

If tailward reconnection *is* triggered by sector boundaries and the reconnection produces a DE, at least for the 8 March 1986 DE, we would need to understand how the initially toroidal magnetic field around the comet would evolve into a magnetospheric polarity reversal, and whether this evolution was consistent with the observed morphology.

Thus, the sunward magnetic reconnection model, although not definitely proven, has survived the major test of the 1985/86 apparition of Halley's comet, and particularly the events of Vega week. The question of the physical model for DEs is, nevertheless, likely to remain an active topic of research.

The proposition by Saito *et al.* (1986b, c) that the tilt angle of the current sheet is crucial for the production of DEs does not seem to be supported by the evidence available. On their approach, the good correlation of DEs with sector boundaries would not be expected. On the sunward reconnection model, immersion in the magnetic field of opposite polarity for a minimum time of ~ 0.5 day should be sufficient to produce a DE almost independent of angle. While some DEs are associated (Table 2) with skimming the current sheet boundary, many are not; see Figs 15. and 16.

3.3 Fine structure

This review has concentrated on the large-scale plasma structures in Halley's comet, and, relatively speaking, has neglected the traditional fine structure found in plasma tails, e.g., knots, kinks, helices, etc. Halley's comet was rich in fine structure as noted by the helix shown in Fig. 2 and other fine structure shown in many figures in this paper. In particular, Celnik (1986) has obtained an impressive series of observations on which measurements of fine structure features were made to determine velocities and accelerations in the tail.

4 CONCLUSIONS

Although many advances in our knowledge of large-scale structure of comets have been made recently, the best is yet to come. The imagery and solar wind data currently available are fragmentary. Ultimately, the International Halley Watch 'Archive' and the Large-Scale Phenomena Network 'Atlas' will make the imagery available, and the solar wind data will appear via the usual channels. From this data base an understanding of DEs, other events, and fine structure should emerge.

Several approaches can be attempted in the short term. (1) Comparisons of DE data with imagery from the Near-Nucleus Network of the IHW could provide morphology useful in discriminating among DE mechanisms and stand-off mechanisms. (2) Limited solar wind data could be used to test the possible association of high-speed streams with DEs. (3) Theoretical studies with the added advantage of *in situ* comet measurements can be used to probe the validity of physical mechanisms invoked to explain large-scale features such as DEs.

In the longer term, we can continue the International Halley Watch concept through NASA's Planetary Data System (PDS). Unfortunately, the situation concerning future space missions to comets looks equivocal. Currently, except for possible retargeting of spacecraft sent to Halley's comet, there are no approved comet missions. There is reason for cautious optimism for NASA's Comet Rendezvous Asteroid Flyby (CRAF) mission. When the inevitable occurs, the data will supply new impetus for our discipline.

ACKNOWLEDGEMENTS

The breadth of this review was made possible by the gracious cooperation of many individuals and institutions. My colleague, M.B. Niedner, Jr, generously contributed materials and comments. Materials and communications have been received from W.E. Celnik, S.M. Gong, S. Larson, P. Maffei, K. Jockers, S.A. Stern, T. Saito, and the Royal Observatory, Edinburgh. I thank them all.

REFERENCES

Alfvén, H. (1957), On the theory of comet tails, *Tellus* **9** 92–96.

Barnard, E.E. (1905), On the anomalous tails of comets, *Astrophys. J.* **22** 249–255.

Barnard, E.E. (1909), Photographic observations of comet c1908 (Morehouse), *Astrophys. J.* **29** 65–71.

Barnard, E.E. (1920), On comet 1919b and on the rejection of a comet's tail, *Astrophys. J.* **51** 102–106.

Biermann, L. (1951), Kometen Schweife und solar Korpuskularstrahlung, *Zs. f. Astrophys.* **29** 279–286.

Biermann, L., Brosowski, B., & Schmidt H.U. (1967), The interaction of the solar wind with a comet, *Solar Phys.* **1** 254–284.

Brandt, J.C. (1982), Observations and dynamics of plasma tails, in *Comets* L.L. Wilkening (ed.), Univ. of Arizona Press, 519–537.

Brandt, J.C., Klinglesmith, D.A., III, Niedner, M.B., Jr, and Rahe, J. (1984), Wide-field imaging of Halley's Comet during 1985–1986 using Schmidt-type telescopes, In: *Astronomy with Schmidt-type telescopes*, M. Capaccioli (ed.), Reidel, pp 233–235.

Brandt, J.C., & Mendis, D.A. (1979), The interaction of the solar wind with comets, In: *Solar system plasma physics* **II** C.F. Kennel, L.J. Lanzerotti, and E.N. Parker (eds.), North-Holland, pp 253–292.

Brandt, J.C., & Niedner, M.B. Jr. (1986a), Tail phenomena, *Adv. Space Res.* **5** 247–253.

Brandt, J.C., & Niedner, M.B. Jr. (1986b), Plasma structures in Comet Halley, in *20th Eslab Symposium on the Exploration of Halley's Comet ESA SP-250* **1** 47–52.

Brandt, J.C., & Niedner, M.B. Jr. (1987), Plasma structures in comets P/Halley and Giacobini–Zinner, *Astron. Astrophys.* **187** 281–286.

Brandt, J.C., Roosen, R.G., & Harrington, R.S. (1972), Interplanetary gas. XVIII. An astrometric determination of solar wind velocities from orientations of ionic comet tails, *Astrophys. J.* **177** 277–284.

Brosius, J.W., Holman, G.D., Niedner, M.B., Jr., Brandt, J.C., Slavin, J.A., Smith, E.J., Zwickl, R.D., & Bame, S.J. (1987), The cause of two plasma tail disconnection events in Comet Halley during the ICE-Halley radial period, *Astron. Astrophys.* **187** 267–275

Burlaga, L.F., Rahe, J., Donn, B., & Neugebauer, M. (1973), Solar wind interaction with Comet Bennett (1969i), *Solar Phys.* **30** 211–222.

Caron, R., Guérin, P., Koutchmy, S., Sarrazin, M., & Zimmerman, J.P. (1986), The disconnection events of January 10 and April 8, 1986 in the plasma tail of Comet Halley, In: *20th Eslab Symposium on the Exploration of Halley's Comet ESA SP-250* **3** 143–147.

Celnik, W.E. (1986), The acceleration within the plasma tail, the rotational period of the nucleus and the aberration of the plasma tail of comet P/Halley 1986, In: *20th Eslab Symposium on the Exploration of Halley's Comet ESA SP-250* **1** 53–58.

Dobrovsolsky, O.V., Gerasimenko, S.I., Kiselev, N.N., & Chernova, G.P. (1986), Activity of Halley's comet in mid-November 1985, In: *20th Eslab Symposium on the Exploration of Halley's Comet ESA SP-250* **3** 31–34.

Ershkovich, A.I., & Mendis, D.A. (1983), On the penetration of the solar wind into the cometary ionosphere, *Astrophys. J.* **269** 743–750.

Feldman, P.D., Festou, M.C., A'Hearn, M.F., Arpigny, C., Buterworth, P.S., Cosmovici, C.B., Danks, A.C., Gilmozzi, R., Jackson, W.M., McFadden, L.A., Patriarchi, P., Schleicher, D.G., Tozzi, G.P., Wallis, M.K., Weaver, H.A., & Woods T.N. (1986), IUE observations of Comet Halley: Evolution of the UV spectrum between September 1985 and Geophysical Monograph **30** 79–89.

Halley's Comet ESA SP-250 **1** 325–328.

Grün, E. Graser, U., Kohoutek, L., Thiele, U., Massonne, L., & Schwehm, G. (1986), Structures in the coma of Comet Halley, *Nature* **321** 144–147.

Hoeksema, J.T., Wilcox, J.M., & Scherrer, P.H. (1982), Structure of the heliospheric current sheet in the early portion of Sunspot Cycle 21, *J. Geophys. Res.* **87** 10331–10338.

Hoeksema, J.T., Wilcox, J.M., & Scherrer, P.H. (1983), The structure of the heliospheric current sheet: 1978–1982, *J. Geophys. Res.* **88** 9910–9918.

Hoffmeister, C. (1943), Physikalische Untersuchungen an Kometen. I. Die Beziehungen des primaren Schweifstrahls zum Radiusvektor, *Zs. f. Astrophys.* **22** 265–285.

Houpis, H.L.F., & Mendis, D.A. (1981), On the development and global oscillations of cometary ionsopheres, *Astrophys. J.* **243** 1088–1102.

Ip, W.-H. (1985), Solar wind interaction with neutral atmospheres, In: *Proc. ESA Workshop on Future Missions in Solar, Heliospheric and Space Plasma Physics, ESA SP-235* 65–82.

Ip, W.-H., & Mendis, D.A. (1978), The flute instability as the trigger mechanism for disruption of cometary plasma tails, *Astrophys. J.* **223** 671–675.

Jockers, K. (1981), Plasma dynamics in the tail of comet Kohoutek 1973 XII, *Icarus* **47** 397–411.

Jockers, K. (1985), The ion tail of comet Kohoutek 1973 XII during 17 days of solar wind gusts, *Astron. Astrophys. Suppl. Ser.* **62** 791–838.

Jockers, K., Geyer, E.H., Rosenbauer, H., & Hänel, A. (1986), Narrow-band filter observations of comet Halley during a tail disconnection, In: *20th Eslab Symposium on the Exploration of Halley's Comet ESA SP-250* **1** 59–61.

Jockers, K., & Lüst, Rh. (1973), Tail peculiarities in Comet Bennett caused by solar wind disturbances, *Astron. Astrophys.* **26** 113–121.

Liu, Z.-L. (1986), The activities of Comet Halley before and during the tail formation in November of 1985, In: *20th Eslab Symposium on the Exploration of Halley's Comet ESA SP-250* **1** 145–151.

Lundstedt, H., & Magnusson, P. (1986), Two disconnection events in Comet Halley, April 1986, In: *20th Eslab Symposium on the Exploration of Halley's Comet ESA SP-250* **1** 141–143.

Maffei, P. (1986), Halley's Comet photographs from South Africa with the Schmidt telescope of the University of Perugia, In: *20th Eslab Symposium on the Exploration of Halley's Comet ESA SP-250* **3** 133–135.

Mendis, D.A., & Flammer, K.R. (1984), The multiple modes of interaction of the solar wind with a comet as it approaches the Sun, *Earth, Moon, and Planets* **31** 301–311.

Ness, N.F., & Donn B. (1966), Concerning a new theory of type I comet tails, In: *Nature et Origine des Comètes*, (Liège), 343–362.

Niedner, M.B., Jr. (1981), Interplanetary gas. XXVII. A catalog of disconnection events in cometary plasma tails, *Astrophys.J. Suppl.* **46** 141–157.

Niedner, M.B., Jr. (1982), Interplanetary gas. XXVIII. A study of the three-dimensional properties of interplanetary sector boundaries using disconnection events in cometary plasma tails, *Astrophys. J. Suppl.* **48** 1–50.

Niedner, M.B., Jr. (1984), Magnetic reconnection in comets, In: *Magnetic reconnection in space and laboratory plasmas*, AGU Geophysical Monograph 30, 79–89.

Niedner, M.B., Jr., & Brandt, J.C. (1978), Interplanetary gas. XXIII. Plasma tail disconnection events in comets: evidence for magnetic field line reconnection at interplanetary sector boundaries?, *Astrophys. J.* **223** 655–670.

Niedner, M.B., Jr., & Brandt J.C. (1979), Interplanetary gas. XXIV. Are cometary plasma tail disconnections caused by sector boundary crossings or by encounters with high-speed streams?, *Astrophys. J.* **234** 723–732.

Niedner, M.B., Jr., & Brandt, J.C. (1980), Structures far from the head of Comet Kohoutek. II. A discussion of the Swan cloud of January 11 and of the general morphology of cometary plasma tails, *Icarus* **42** 257–270.

Niedner, M.B., Jr., & Brandt, J.C. (1986), Dynamics of the 1986 January 10 disconnection event in the plasma tail of Halley's comet, *B.A.A.S.* **18** 819.

Niedner, M.B., Jr., & Liller, W (1987), The IHW Island Network, *Sky and Telescope* **73** 258–263.

Niedner, M.B., Jr., Rahe, J., & Brandt, J.C. (1982), A worldwide photographic network for wide-field observations of Halley's comet in 1985–1986, In: *Proceedings of the ESO Workshop on the Need for Coordinated Ground-based Observations of Halley's Comet*, Paris, pp 227–242.

Niedner, M.B., Jr., & Schwingenschuh, K. (1986), Plasma-tail activity at the time of the Vega encounters, In: *20th Eslab Symposium on the Exploration of Halley's Comet ESA SP-250* **3** 419–424.

Russell, C.T., Luhmann, J.G., Elphic, R.C., & Neugebauer, M. (1982), Solar wind interactions with comets: Lessons from Venus, In: *Comets*, L.L. Wilkening (ed.), Univ. of Arizona Press, 561–587.

Russell, C.T., Saunders, M.A., & Phillips, J.L. (1986), Near-tail reconnection as the cause of cometary tail disconnections, *J. Geophys. Res.* **91** 1417–1423.

Saito, T., & Saito, K. (1986), Effect of the heliospheric neutral sheet to the kinked ion tail of Comet Halley on 13 May, 1910, In: *20th Eslab Symposium on the Exploration of Halley's Comet ESA SP-250* **1** 135–140.

Saito, K., Saito, T., Aoki, T., & Yumoto, K. (1986a), Possible models on disturbances of the ion tail of comet Halley during the 1985–1986 apparition, In: *20th Eslab Symposium on the Exploration of Halley's Comet ESA SP-250* **3** 155–160.

Saito, T., Yumoto, K., Hirao, K., Nakagawa, T., & Saito, K. (1986b), Interaction between Comet Halley and the interplanetary magnetic field observed by Sakigake, *Nature* **321** 303–307.

Saito, T., Yumoto, K., Hirao, K., Nakagawa, T., & Saito K. (1986c), Quasi-parallel model for Comet Halley near the encounter of Sakigake, In: *20th Eslab Symposium on the Exploration of Halley's Comet ESA SP-250* **1** 129–133.

Saito, T. Yumoto, K., Hirao, K., Saito, K., Nakagawa, T., & Smith, E. (1986d), Dynamic Pressure model derived from an observation by Sakigake for Comet Halley on 31 December, 1985, In: *20th Eslab Symposium on the Exploration of Halley's Comet ESA SP-250* **3** 149–153.

Shkodrov, V., Ivanova, V., & Boner, T. (1986), Observations of comet P/Halley at the National Astronomical Observatory — Bulgaria, In: *20th Eslab Symposium on the Exploration of Halley's Comet ESA SP-250* **3** 195–198.

Suess, S.T., Scherrer, P.H., & Hoeksema, J.T. (1986), Solar wind speed azimuthal variation along the heliospheric current sheet, In: *The Sun and the heliosphere in three dimensions*, R.G. Marsden (ed.), (Reidel), 275–280.

Verigin, M.I., Axford, W.I., Gringauz, K.I., & Richter, A.K. (1987), Acceleration of cometary plasma in the vicinity of comet Halley associated with an interplanetary magnetic field polarity change, *Geophysical Research Letters* **14** 987–990.

Watanabe, J., Takatou, N., Kawakami, H., Tomita, T., Kinoshita, H., Nakamura, T., & Kozai, Y. (1986), On the ion tail formation of Comet Halley during November 12–13 In: *20th Eslab Symposium on the Exploration of Halley's Comet ESA SP-250* **3** 119–122.

Wu, M., & Qiu P. (1986), Activity of the plasma tail of Halley's Comet in March of 1986, In: *20th Eslab Symposium on the Exploration of Halley's Comet ESA SP-250* **3** 123–125.

Wurm, K., & Mammano, A. (1972), Contributions to the kinematics of Type I tails of comets, *Astrophys. and Space Sci.* **18** 273–286.

4

The solar wind interaction with comet Halley as seen by the Giotto spacecraft

A.D. Johnstone

1 INTRODUCTION

The physical process controlling the interaction between the solar wind and a comet is simple. Neutral gas molecules, sublimated from the nucleus, drift outwards into space. When they first leave the surface of the nucleus the density is so high that their motion is collision-dominated, but once they have travelled a few thousand kilometres the chance of a further collision is so small that they flow through the solar wind without interacting. The interaction becomes a strong one only when the neutral particles become ionized, because they must then respond to the electric and magnetic fields which permeate interplanetary space. The electric field accelerates the newly implanted ion to velocities comparable with the solar wind; the maximum possible speed is in fact twice that of the solar wind. Since these cometary ions are typical of the composition of the nucleus, e.g. C^+, O^+, OH^+, H_2O^+, H_3O^+ with masses up to, and even beyond, the 44 amu of CO_2^+, the energy and momentum they acquire in the acceleration process is therefore many times that of a typical proton in the solar wind flow. For example, if the energy of a solar wind proton is 1 keV (equivalent to a speed of 438 km s^{-1}) then a CO_2^+ ion may be accelerated to 176 keV, and it will then carry the same momentum as 88 protons. This energy and momentum must be obtained at the expense of the solar wind particles which are decelerated by the electric and magnetic fields created by the relative motion of the newly-created ion–electron pair. A small proportion of cometary

ions by number added to the flow has a large effect on the solar wind speed. The scale length of the interaction is governed by the rate at which the neutrals become ionized as they flow outward from the nucleus, i.e. by the scale length $L = V_e/v$ where V_e is the outflow velocity of the cometary neutral particles and v is their ionization rate. Since the outflow velocity is of the order of 1 km s^{-1}, and the ionisation rate of the order of 10^{-6} s^{-1}, L is of the order of 10^6 km. As is shown below, the effects of comet Halley were indeed first detected several million kilometres from the nucleus.

Even though the interaction is such a 'soft' one, involving a deceleration of the solar wind flow over millions of kilometres, a bow shock forms several hundred thousand kilometres upstream from the nucleus (Wallis 1973). Unlike the bow shock created in the solar wind flow by the planets (e.g. Earth, Venus, Jupiter, Saturn) the cometary shock is not formed to allow the flow to be deflected around an obstacle, but rather to satisfy the conservation equations of magnetohydrodynamics which, for the case of ion flow around a comet, contain source terms associated with the mass-loading.

Closer to the nucleus it was expected that the comet would create an obstacle requiring the flow to be deflected. Some form of contact surface should form, balancing the internal pressure of the outflowing cometary neutral gas and plasma with the dynamic pressure of the mixture of solar wind and cometary ions. The contact surface separates plasma of purely cometary origin from contaminated solar wind. Predictions put this surface less than 10 000 km from the nucleus of comet Halley (Schmidt & Wegmann 1982).

To summarize, the spacecraft should have encountered two major plasma boundaries, separating the plasma flow around the comet into three regions. An upstream region where the flow is already becoming mass-loaded; a region of shocked solar wind plasma heavily contaminated by cometary plasma; and a cometary ionosphere. The magnetic field, frozen into the solar wind, should become draped around the head of the comet as the flow speed is reduced, with the effect being seen most strongly on the nucleus–Sun line.

In this review, these predictions are compared with observations made by the plasma instruments on the Giotto spacecraft, and in particular, the positive ion analyzers which formed the JPA instrument. The comparison with other measurements is based on the use of published material available at the time of writing.

2 THE JPA INSTRUMENT

The instrument consisted of two sensors (Johnstone *et al.* 1987a). The Fast Ion Sensor (FIS) was designed to obtain the three-dimensional energy/charge distribution of positive ions once per spin of the spacecraft (i.e. in 4 seconds) but without mass discrimination. It covered the energy range from 10 eV to 20 keV and all angles relative to the spin axis except for the cone of $20°$ half angle around the forward velocity vector of the spacecraft. It was capable of measuring all distributions likely to be encountered from the solar wind through the bow shock to the inner coma. It was not designed to cope with the extremely high fluxes of cold plasma inside the contact surface which all appear to be coming toward the spacecraft along the relative velocity vector. These populations were covered by specialized instruments such as IMS (Balsiger *et al.* 1987). Thus, at the minor cost of not covering the cone around the velocity vector the sensor could be placed behind the dust shield and protected from dust and neutral particle impact, which could have caused damage and would have created an undesirable background flux. The second

sensor, called the Implanted Ion Sensor (IIS) was designed to measure the flux of newly-created cometary ions. It therefore covered the energy range from 90 eV/q to 90 keV/q, uniquely amongst the cometary instrumentation, and had mass discrimination. The mass-sensitive technique was time-of-flight analysis whose mass resolution was not as good as the mass-spectrometers flown on Giotto to obtain the cometary composition but was adequate for mass-loading studies. It was, however, the only sensor on any spacecraft in the Halley armada capable of determining the composition of ions in the important energy range from 10 keV to 90 keV.

3 OVERVIEW OF JPA PLASMA OBSERVATIONS

Fig. 1 shows data from the Implanted Ion Sensor (IIS). The time scale of this energy spectrogram covers the period from 1710 SCET (Spacecraft Event Time) on 13 March to 0400 SCET on 14 March during which the spacecraft travelled from 1.7 million km before, to 1.0 million km after closest approach. Each panel is an energy spectrogram, covering the energy range 90 eV/q to 90 keV/q, for one of the individual sensors within IIS placed at different angles to the spin axis. There are two main lines in the energy spectrum; the upper one is the spectrum of cometary ions and the lower one is the spectrum of the solar wind protons. Although the instrument separates ions of different masses from each other, the intensity of solar wind protons is so high that it leads to pulse pile-up in the time channel and hence to a breakthrough of protons into the channel for masses from 12 to 22 amu (Wilken *et al.* 1987a). Rather than remove the effect in the processing it has been left to show the relation of the cometary ions to the solar wind in a simple way. One should be aware, however, that it does not represent the true intensity of the solar wind which is very much higher than it appears here.

The cometary ions, whose average mass is 18 amu, have energies of between 30 and 40 keV, much greater than the solar wind protons ($E_p \sim 700$ eV). As the intensity of the cometary ions increases closer to the comet it can clearly be seen in the lower line that the energy, and therefore also the speed, of the solar wind decreases. This is the direct result of mass-loading by

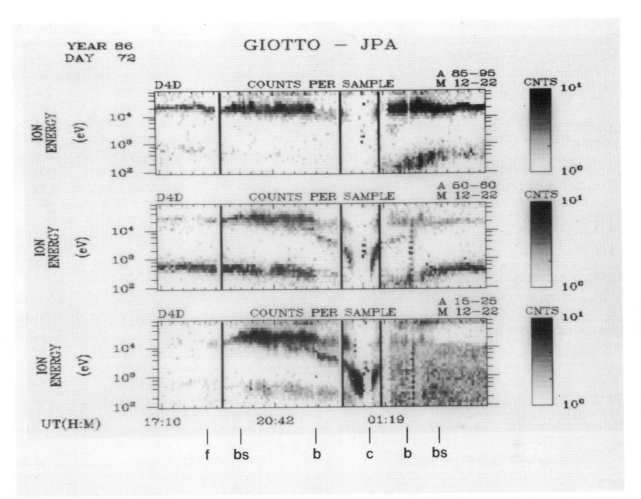

Fig. 1 — An energy–time spectrogram giving an overview of positive ion distributions during Giotto's encounter with comet Halley. Each panel shows data from one of the individual sensors in the array which make up the IIS. They are placed at different angles with respect to the spin axis of the spacecraft. The bottom panel covers the angular range closest to the relative velocity vector ($A = 15$–$25°$). The middle panel covers the range in which the solar wind is found, $A = 50$–$60°$ (lower energy line in the spectrogram). The top panel, viewing perpendicular to the spin axis, $A = 85$–$95°$, also views perpendicular to the garden-hose magnetic field direction and therefore detects the implanted cometary ions in the region upstream from the bow shock. In each case the data are averaged over one spin of the spacecraft. The plot covers the period the spacecraft was travelling from 1.7 million km before to 1.0 after closest approach (marked 'c'). At closest approach the instrument was turned off by electrical noise associated with the dust bombardment and could not be restored for 30 min. After that, one sensor was permanently noisy (bottom panel). Other points marked are the foreshock 'f', the bow shock 'bs', and the transition at 2152 SCET 'b'.

the cometary ions of the upper line. At 1922 SCET, at a distance of 1.15 million km from the nucleus on the inbound leg, and at a time of 0300 SCET and a distance of 0.74 million km outbound, the angular and energy distribution of both the solar wind and the cometary ions changes suddenly. Going inbound, the cometary ion distribution becomes more intense and broadens in angle and energy. This is the point which we

currently believe to be the cometary bow shock.

Closer to the comet, at 0.54 million km inbound and 0.47 million km outbound, another sharp boundary was encountered. At this point it appears that the cometary ion distribution suddenly splits into two lines; the upper branch continues at the same energy, while the lower branch decreases rapidly in energy as the nucleus is approached. The transition is a real one although the appearance is partly an artefact of the grey-scale display. The details of this transition are discussed in section 7, but unlike the bow shock it had not been expected and does not yet have an explanation.

4   OVERVIEW OF MAGNETIC FIELD OBSERVATIONS

An overview of the magnetic field measurements (Neubauer 1987) is shown in Fig. 2, with the position of a number of the boundaries marked as well. The diagram shows the magnitude of the field, and the direction in Halley-centred, solar ecliptic coordinates. The direction of the field both inside and outside the shock is very variable although it is close to the spiral angle of the undisturbed interplanetary field (PHI = 130°) for some of the time, and it is difficult to see clear evidence of the draping of the field lines around the comet. One feature of the interplanetary conditions at the time of the encounter which contributed to the variability was that the spacecraft, and therefore also the comet, was close to the heliospheric current sheet. Giotto crossed it several times, with the result that the magnetic field changed direction by as much as 180° on a number of occasions. On the outbound leg the draping is more obvious, perhaps because the angle between the draping and the normal spiral angle is much greater. Soon after the outbound bow shock crossing the field clearly rotates back to the spiral angle. The magnitude of the field increases steadily after the bow shock, but the most dramatic increase starts at a distance of 135 000 km (PB on the diagram). Since this is associated with sharp changes observed in the other plasma instruments it appears to be yet another structural feature in the plasma flow. The same feature can be observed in the plasma flow on the outbound leg at 188 000 km (also marked PB),

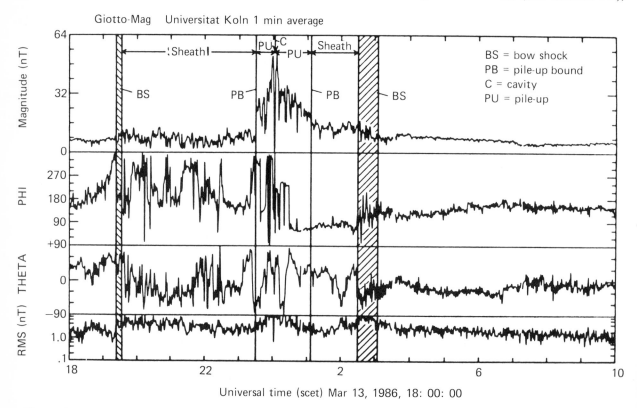

Fig. 2 — One minute average vector magnetic field observations in Halley-centred solar ecliptic coordinates. THETA is the elevation angle with respect to the X, Y plane and PHI the azimuth of the projection with PHI = 0 towards the Sun. RMS is the Pythagorean mean of the component rms values (from Neubauer 1987).

although the corresponding change in magnetic field intensity is not as noticeable. Inside this new boundary the magnitude of the field continues to increase, reaching a maximum of 60 nT at a distance of 16 500 km. Then the field drops to zero (within the accuracy of the magnetometer) as the spacecraft enters a magnetic cavity around the nucleus, at 4 600 km. This corresponds to the predicted contact surface, since the nucleus has no intrinsic magnetic field, and the field carried by the solar wind is deflected around the surface. The spacecraft emerged from the magnetic cavity at 3 900 km on the outbound leg.

These observations show that while the solar wind interaction with the comet was close to what had been predicted, there have been some surprises in the gross character. In addition to the two expected boundaries, the bow shock and contact surface, Giotto found at least two more, which have yet to be explained. These boundaries and the best estimate of their positions are summarized in Table 1. In the following sections the structure of these new boundaries is described in more detail, and we examine what we have learned so far about plasma processes in the five regions of the solar wind–cometary interaction.

5 THE UPSTREAM REGION

The most convenient location for studying the process by which cometary ions are accelerated to solar wind velocities is the upstream region outside the bow shock. The pick-up process has two stages: first the ion is accelerated by the interplanetary electric and magnetic fields perpendicular to the magnetic field; then, as a result of plasma instabilities generated by the highly anisotropic distribution of the cometary ions the ions are diffused in pitch angle and acquire a component of velocity parallel to the magnetic field. Now let us examine the two stages of the ion pick-up process in more detail. If the angle between the

interplanetary magnetic field B_y and the solar wind velocity vector U_s is ϕ (in the yz plane), then the trajectory of a newly-implanted cometary ion is a cycloid of the form,

$$
\begin{aligned}
V_x &= U_s \sin \phi \sin \omega_c t \\
V_y &= 0 \\
V_z &= U_s \sin \phi (1 - \cos \omega_c t)
\end{aligned}
\tag{1}
$$

where,

$$
\omega_c = \frac{eB}{m_c}
$$

The motion consists of a steady drift in the z direction perpendicular to B_y at a velocity $U_s \sin \phi$, combined with a gyration at the same velocity about the magnetic field. Fig. 3 illustrates ion pick-up in a velocity–space reference frame moving with the spacecraft. The circle is the section of a sphere cut by the plane of the diagram which contains the spin axis of the spacecraft and the solar wind velocity vector. The sphere has a radius U_s and is centred on the solar wind velocity vector. Implanted ions are created with a velocity of $- V_{sc}$ in this frame and picked up into an orbit which follows a small circle, or circle of latitude, on the sphere whose polar axis is parallel to the magnetic field. The distribution of cometary ions forms a ring in velocity space on this small circle after pick-up. Superimposed on the diagram are the fields-of-view of the three IIS sensors whose data are plotted in the three panels of Fig. 1. The diagram shows that upstream from any cometary influence the solar wind should be detected by sensor 4 (the centre panel of Fig. 1) as in fact it is. The cometary ions, injected into the ring, should appear perpendicular to the magnetic field in the field-of-view of sensor 3 (top panel). Fig. 1 shows that, at large distances from the comet, both inbound and outbound, the peak of the distribution is detected in sensor 3 with a maximum energy of 32 keV. From equations (1) it can be shown that the

Table 1 — Boundaries in the plasma flow

| Boundary | Cometocentric distance in km | | Corresponding times SCET | |
	inbound	outbound	inbound	outbound
Bow shock	$1.15 - 0.99 \times 10^6$	0.74×10^6	1922	0300
Unnamed	0.54×10^6	0.47×10^6	2152	0157
Cometopause (1)	135 000	188 000	2330	0049
Cometopause (2)	61 000	65 000	2348	0021
Magnetic cavity	4 700	3 800	0001:54	0003:59

maximum energy should be,

$$E_r = 4m_c E_p \sin^2\phi \qquad (2)$$

Assuming that m_c, the mass of the cometary ions, is 18, and taking the measured value of the energy of the solar wind protons $E_p = 640$ eV one derives a value for $\phi = 56°$. This is close to the angle expected from the Parker spiral of the interplanetary magnetic field and consistent with the peak of the ion distribution being observed by sensor 3 in Fig. 1. So far, it appears that the observations are in good agreement with the basic theory. The truth, as usual, is more complicated, because the observed direction of the magnetic field is more variable than the steady observations of the energy and direction of the implanted ions would

imply, as can be seen in Fig. 2. It turns out that the intensity of the implanted ions drops as the field becomes more nearly parallel to the flow. Significant fluxes are observed only when the magnetic field is close to the spiral angle. It is not yet clear to what extent this is a result of the variation of the instrumental sensitivity, which is proportional to E^2 as for all electrostatic energy analyzers, or a result of the nature of the pick-up process itself.

Once the ions have been injected into the ring, a process which takes a time of the order of a gyroperiod, $t_c = 2\pi/\omega_c$, or approximately 100 s, they become subject to pitch-angle diffusion, the second stage of the pick-up process. In the spacecraft frame, in Fig. 3, as they diffuse they will become distributed over the pick-up sphere and will then be detectable up to an energy of E_s,

$$E_s = 4m_c E_p \qquad (3)$$

and will have a wider angular distribution. For the observed solar wind conditions $E_s = 46$ keV. This is obviously happening after the spacecraft crosses the bow shock at 1922 SCET because the cometary ion distribution spreads to other sensors and is seen at higher energies. The peak flux is found in sensor 5 (bottom panel Fig. 1), which is closest to the direction of the relative velocity vector. Pitch-angle scattering is also detectable outside the bow shock, but is less obvious in the spectrogram of Fig. 1.

The pitch-angle diffusion is driven by the unstable characteristics of the ring distribution of the cometary ions. The nature of the interaction is illustrated by Fig. 4 which shows the velocity distribution in a frame moving with the solar wind. The condition for a cometary ion to be resonant with a wave is,

$$\omega - \boldsymbol{k.v} = \pm \omega_c$$

where $\boldsymbol{v}$ is the particle velocity in the solar wind frame. From equations (1) the ion has no component of velocity parallel to the magnetic field in an inertial frame. In the solar wind frame its parallel velocity is equal to the component of the solar wind velocity parallel to the magnetic field, but in the upstream direction, so the resonance condition is,

$$\omega = \boldsymbol{k.U_s} \pm \omega_c \qquad (4)$$

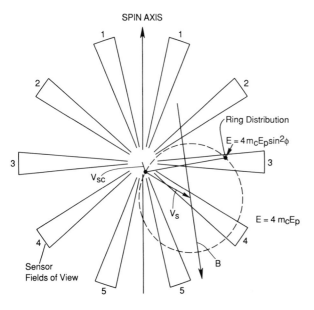

SPIN AXIS

Fig. 3 — This diagram shows the process of ion pick-up in the space craft frame of reference. The plane chosen is the one containing the spin axis and the solar direction. The triangular segments show the fields of view of the five sensors in the IIS array on the two occasions per spin when they rotate through this plane. The dotted circular line shows the intersection of this plane with the spherical ion pick-up shell. In the initial pick-up process ions are created at $-V_{sc}$ (the spacecraft velocity) and injected into a cycloidal trajectory which lies on a small circle of the shell (marked Ring Distribution) whose plane is perpendicular to the magnetic field direction (B). The ions are then seen in sensor 3 (top panel of Fig. 1). The solar wind V_s is seen in sensor 4 (middle panel). When the bow shock is crossed and the ions become diffused in pitch angle around the shell they are then observed by sensor 5 (bottom panel) as well.

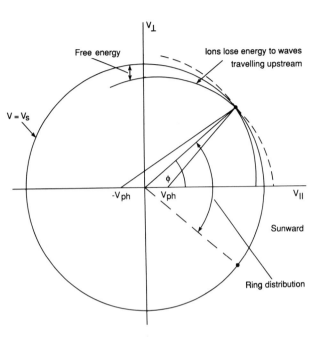

Fig. 4 — This diagram illustrates the ion pick-up process in a frame of reference moving with the solar wind. Once picked-up into a cycloidal orbit (ring distribution) the ions, which appear in this frame to be moving upstream, can interact with MHD waves propagating through the solar wind. As they interact the ions move through velocity space on a circle centered either on V_{ph} (waves moving parallel to the ions) or $-V_{ph}$ (antiparallel waves). Of the four interactions possible, the solid curves show those in which the waves gain energy from the ions, and the dashed curves those in which the ions gain energy from the waves. The ion distribution is unstable to the former which lead to the generation of turbulence and pitch-angle diffusion of the ions.

In the inertial frame, which is the same as the spacecraft frame except for its own relatively small velocity, the frequency is seen Doppler shifted by $-\mathbf{k}.U_s$ so that the observed frequency is,

$$\omega \sim \omega_c$$

Since the cometary ions are much more massive than the solar wind protons, these frequencies are well below the ion gyrofrequency in the solar wind. The resonant waves are therefore propagating in the Alfvén or magnetosonic modes. When an ion resonates with the wave it exchanges energy and momentum in such a way that its energy remains constant in the frame travelling with the wave, i.e.

$$V_\perp^2 + (V_\parallel - V_{ph})^2 = \text{constant} \qquad (5)$$

where V_{ph} is the phase velocity of the wave, $V_\perp$ is the

component of the ion velocity perpendicular to the magnetic field and $V_\parallel$ is the parallel component. Referring to Fig. 4, these two surfaces are circles centered on $V_\parallel = \pm V_{ph}$. Four types of interaction are possible, depending on whether the waves are travelling parallel ($+V_{ph}$) or antiparallel ($-V_{ph}$) to the particles, and on whether the particle gains or loses energy in the interaction. In the situation in which we are interested the energy of the waves is derived from the ions themselves. The interaction which extracts energy from the ion and which reduces the relative velocity between ion and solar wind is most important. The energy loss of an ion in the solar wind frame, while it is scattered in pitch angle around the diffusion path indicated, is the order of (Sagdeev *et al.* 1986),

$$\text{energy loss} = \tfrac{1}{2}m_c \, V_{ph} \, U_s \qquad (6)$$

The waves with which it interacts are propagating upstream in the solar wind frame, i.e. sunwards, though in an inertial frame they are in fact being convected downstream because the solar wind velocity is much greater than V_{ph}.

The basic features of the wave particle interaction are confirmed by the observations. Turbulence is generated in the solar wind at frequencies of the order of 3 to 10 mHz (i.e. periods of 100 to 300 s) (Johnstone *et al.* 1987b), and the relationship between the waves in the magnetic field and in the solar wind velocity show that it is basically Alfvénic in character (Johnstone *et al.* 1987c). Figs 5 and 6 show the waves in the velocity and magnetic field over a period of one hour at an average distance of 1.85 million km from the nucleus. For these plots the mean velocity, or the mean magnetic field, has been subtracted from the measured value by a low-pass filter, and the resulting wave vectors has then been transformed into a frame which has the y-axis in the mean $\mathbf{B}$ field direction and the z-axis in the mean $\mathbf{E} = -\mathbf{V} \times \mathbf{B}$ direction. The amplitude of the y (or field-aligned) component in both wave fields, and particularly in the magnetic field, is much smaller than the other two components. This shows that the waves are predominantly transverse to the magnetic field, as for Alfvén waves. A close inspection of the transverse components shows a high degree of correlation between the two sets of waves also indicating the presence of Alfvén waves.

There is a fairly high level of turbulence in the solar

Giotto JPA FIS solar wind proton fluctuations
Year 1986 day 72 FLD coordinates

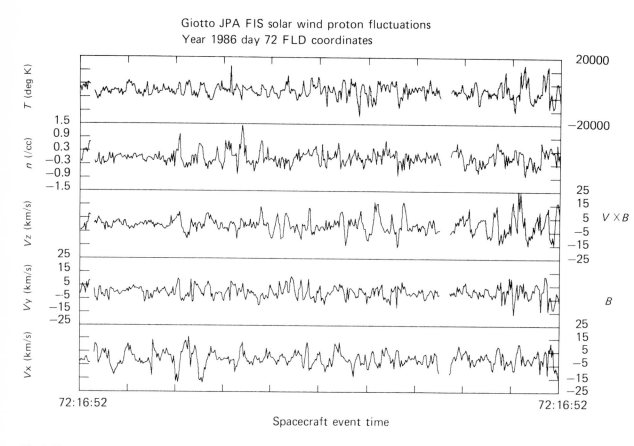

Fig. 5. Fluctuations in the solar wind proton parameters, after being passed through a high-pass filter with a cut-off at 3 MHz, plotted in FLD coordinates. Y is parallel to the magnetic field; Z is parallel to the $V \times B$ electric field. (from Johnstone *et al.* 1987c).

wind all the time, but Fig. 7 shows that the rms amplitude of the waves increases significantly as the comet is approached. At the same time the solar wind speed is decreasing because of mass-loading. There appears to be a relationship between the presence of cometary ions and the generation of solar wind turbulence. This relationship can be tested in a quantitative way by comparing the local wave energy density with the free energy given up by the local population of cometary ions since their creation. The cometary ion density can be calculated accurately from its effect on the speed of the solar wind. Applying conservation of momentum to the mass-loaded solar wind gives the cometary ion density ρ_c. If we take as a measure of the free energy $\mathscr{F}$ available for wave generation from the implanted cometary ions their kinetic energy in the solar wind frame then

$$\mathscr{F} = \frac{1}{2} \rho_c U_s^2 \cos^2 \phi = \rho_\infty U_\infty (U_\infty - U_s) f(\phi) \quad (7)$$

where $\rho_{\infty 0}$ is the solar wind density far upstream any interaction with the comet (Johnstone *et al.* 1987b).

The wave energy density $\mathscr{W}$ is given by

$$\mathscr{W} = \frac{1}{2} m_c \langle v^2 \rangle + \frac{1}{2} \frac{\langle b^2 \rangle}{\mu_0} \quad (8)$$

where b and v are the magnetic and velocity wave amplitudes and μ_0 is the permeability of free space.

If we assume equipartition of energy between the magnetic and kinetic energy components and then equate $\mathscr{W}$ to a fraction K of the free energy density the relationship,

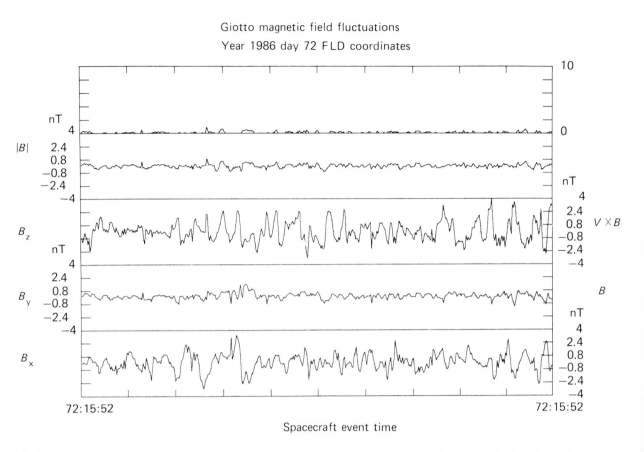

Fig. 6 — Fluctuations in the magnetic field components processed in the same way as the solar wind parameters in Fig. 5.(from Johnstone *et al.* 1987c).

$$\frac{\langle v^2 \rangle}{U_s^2} = \frac{K}{3 \tan^2 \phi} \left\{ \frac{U_x}{U_s} - 1 \right\} \qquad (9)$$

is obtained. Fig. 8 shows the result of the comparison of the two sides of equation (9). The LHS is basically a measure of wave energy; the RHS gives the free energy. The wave energy occurs at levels up to the line drawn on the plot, which represents the maximum proportion of the ion kinetic energy going into the waves. This is not actually the free energy which is the energy between the circle $V = U_s$ and the diffusion path in Fig. 4. The amount of energy released by the ion depends on the angle through which it is diffused, but the maximum proportion available is of the order of $f = V_{ph}/U_s$ (equation (6)). Setting $V_{ph} =$ Alfvén velocity and evaluating it from the measured quantities in the solar wind gives $f = 48/350 = 0.14$, while

the value of K obtained from the straight line is $K = 0.16$. The excellence of the agreement between these values, and the linear relationship shown by Fig. 8 is confirmation that our understanding of the mechanism for wave generation is correct.

6 THE BOW SHOCK

The existence of a shock in the solar wind flow around a comet was predicted by magnetohydrodynamic studies (Wallis 1973, Biermann *et al.* 1967). There are two significant differences between the cometary shock and the better-known cases of planetary bow shocks. First, it is not caused by the deflection of the flow by an obstacle but by mass-loading; second, the mass-loading slows down the flow before the shock is reached. Wallis (1973) showed that a shock must form

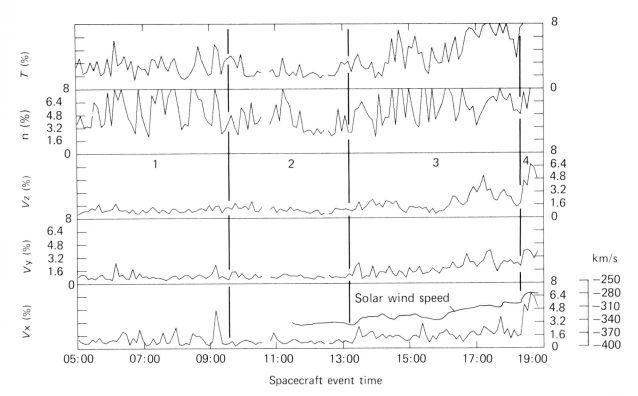

RMS deviation normalized
Giotto JPA FIS solar wind proton parameters
Year 1986 day 72 HSE coordinates

Fig. 7 — The normalized rms deviation of the solar wind proton parameters calculated for 45 consecutive 8-second samples for the period while the spacecraft travelled from 5 million to 1.25 million km from the nucleus. The solar wind speed is shown for comparison to demonstrate that the turbulence increases as the speed decreases (from Johnstone *et al.* 1987b).

before a critical point is reached in the flow where the total mass flux is,

$$\rho U = \frac{\gamma^2}{\gamma^2 - 1} \rho_x U_x \qquad (10)$$

Putting $\gamma = 2$ shows that the shock should form before the mass-flux has increased by a factor 1.33. Most of the theoretical analysis has been performed for the special case of flow along the comet–Sun line. All the spacecraft approached the comet along a line nearly perpendicular to the comet–Sun line, and therefore crossed the shock at the flanks. Since the shock is not produced primarily by a deflection it can be analyzed in terms of one-dimensional flow (Galeev *et al.* 1985). This analysis ought not to be restricted to the comet–

Sun line but should be valid over a wider region. Thus the shock should form along any streamline before the mass flux reaches the critical level given in equation (10).

Fig. 9 shows normalized flow parameters for the solar wind component across the bow shock (Coates *et al.* 1987b). The measured values have been normalized to their upstream values, corresponding to solar wind conditions before the comet had any influence. These values could not be measured directly, so we have used the values measured by Giotto before any cometary effect was detectable and assumed that after that time the conditions remained constant. If this were true then the normalized mass flux (for the solar wind component alone) shown in the second panel from the top in Fig. 9 would be equal to unity

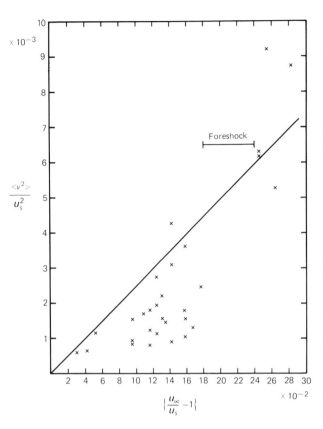

Fig. 8 — A plot of the wave energy density measured by the normalized velocity variance against the free energy density which is proportional to the normalized velocity decrease. The gap in the distribution of points is due to the decrease in mean velocity on passing through the outer limit of the foreshock (see Fig. 9) at 1825 SCET (from Johnstone *et al.* 1987b).

throughout the pass. Up to the bow shock it is, in fact, reasonably close to this value. Inside the shock it becomes a more variable quantity, partly because the type of analysis used to obtain the parameters from the raw measurements of counts is no longer as accurate once the temperature rises and the flow direction becomes widely variable (Johnstone *et al.* 1987a). The normalized velocity begins to decline well before this plot starts. After 1822 SCET, the speed decreases sharply and varies over a wider range than before. This point, marked F in Fig. 9, has been considered to be a foreshock, similar to that found on the dusk side of the Earth's bow shock, but no backstreaming ions have been found to complete the identification. The shock itself is now thought to be the region between the lines marked S_1 at 1922 and S_2

at 1930 SCET. Just after S_1 there is a very large decrease in the solar wind speed, accompanied by a deflection of the flow by almost 20°. This appears to be a precursor wave, and the flow recovers to its previous values after two minutes. At S_2 the flow speed decreases again and attains the normalized value of 0.6 for the downstream speed. In order to calculate the total mass flux so that equation (10) can be tested it is first necessary to obtain the cometary ion flux. It can be reliably estimated from the decrease in the solar wind speed by using the conservation of momentum as has already been done in section 5. Fig. 10 shows the solar wind speed, the inferred cometary ion density, and the derived total normalized mass flux. It can be seen that although the mass flux crosses the 1.33 level just about coincident with the precursor wave, it has passed this level briefly before and it is not possible to determine that 1.33 is the critical level from these data. As the region S_1 to S_2 is crossed, the distributions of all the major populations become broadened, i.e. it can be said that their temperature increases, although the distributions are not Maxwellian in character. There are four main populations: solar wind ions, electrons, heavy cometary ions, and protons of cometary origin. The heavy cometary ions, i.e. water group, are the first to broaden at 1922 SCET, just *before* the precursor wave (Wilken *et al.* 1987b). The cometary protons are the next to thermalize, four minutes later and just *after* the precursor wave. The solar wind protons do not thermalize until after S_2 and even then show a partial recovery between 1946 and 1953 SCET (Formisano *et al.* 1986). The electrons begin to increase in density and temperature during the precursor wave (Fig. 11) (Coates *et al.* 1987a). The shock has a very complex structure, as indeed it was bound to have, with four very different particle populations taking part in the dynamics.

One obvious question is whether the precursor wave is a real spatial structure or whether the appearance is caused by an encounter with a shock front which is moving backwards and forwards over the spacecraft. We note, however, that although the solar wind protons behave reversibly as the precursor is crossed, there are irreversible changes taking place as well. First the cometary ions become thermalized; and second, there is an increase in electron and ion density. When the flow deflection, which takes place

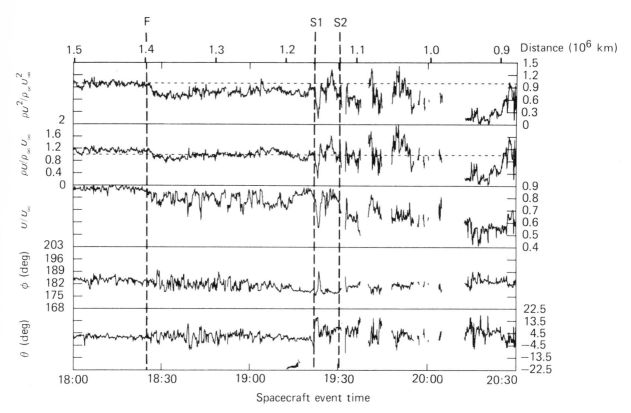

Fig. 9 — Normalized solar wind bulk parameters in the bow shock region. The values are normalized by using the upstream values at 1200 SCET on 13 March, i.e. $u_\infty = 345\ \mathrm{km\ s^{-1}}$ and $n_\infty = 6.4\ \mathrm{cm^{-3}}$. The foreshock is marked F, the beginning of the precursor wave is marked S_1, and the permanent speed change is marked S_2. The shock structure extends from S_1 to S_2 (from Coates *et al.* 1987b).

during the precursor, is compared with that observed during the later velocity decrease (at S_2), another difference is found (Fig. 12). During the precursor, the flow vector swings toward the comet, but, in the later, permanent, change, it swings away from the comet as one would expect for deflection around an obstacle. These two pieces of evidence, i.e. irreversible changes and a different deflection, indicate that the precursor is a real structure and an essential part of the shock. The magnitude of the permanent deflection is approximately 15° which shows that, to this extent at least, the one-dimensional analysis of the flow is inadequate.

7 THE UNNAMED TRANSITION NEAR 500 000 KM

At 2152 SCET, there is a sudden change in the characteristics of the ion distribution (Fig. 1). It

appears that the energy spectrum of the cometary ions splits into two components. In fact, as Fig. 13 shows, the two components exist from the time that the bow shock is crossed; after 2152 SCET the energy gap between them begins to widen much more rapidly. Several other plasma instruments on Giotto also registered changes at the same time; the electron detector (Fig. 14) showed that the electron density and temperature decreased across the transition (Reme *et al.* 1987) and the ram ion detector PICCA (Korth *et al.* 1987) found that the cold ions began a sustained increase (Fig. 14). The most noticeable effect occurs in the energetic electrons ($0.8 < E < 3.6$ keV) which drop to the level they previously had in the upstream solar wind inside the transition after having been active and up to 10 times as intense since the bow shock. There is no obvious effect in the magnetic field (Fig. 2).

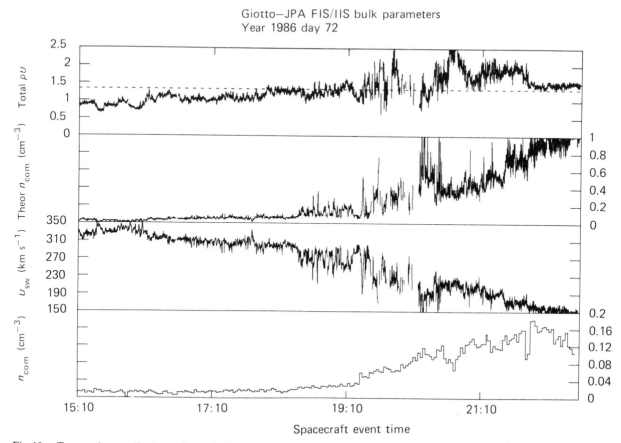

Fig. 10 — Top panel, normalized mass flux, calculated as the addition of the measured solar wind mass flux $\rho_s U_s$ to the calculated cometary ion mass flux $\rho_i U_i$. The horizontal broken line shows the critical value 1.33. Second panel, inferred cometary ion density. Third panel, solar wind speed (U_{sw}). Fourth panel, IIS measurements of the cometary water group ion density (from Coates *et al.* 1987b).

The transition in the cometary ion distribution has been studied in some detail by Thomsen *et al.* (1987). Fig. 13 shows stacked spectra in which the ion intensity is given simply in terms of the detector count rate. The transition is obvious as a decrease in the count rate and through the development of two peaks of almost equal count rate in the spectrum. The lower energy peak can also be seen before the transition; although it is not clear in this diagram it can be confirmed by other presentations. Furthermore, the main decrease in count rate seems to occur in the peak at the higher energy.

The explanation of these observations is as follows. The cometary ion spectrum is made up of ions produced in all positions from the immediate vicinity of the spacecraft to the far upstream region tracking back along the trajectory of the ions. The energy of the ions depends on the flow velocity and the magnetic field direction at their point of creation and on the flow velocity at the point of observation. If the flow field were smoothly varying, then the spectrum would be a broad feature with a single peak. The double-peaked structure arises because there is a jump in velocity variation at the bow shock. Very few ions are created at velocities within the range of the jump, and this leads to a gap in the spectrum and hence a double peak. Thomsen *et al.* (1987) show how the spectrum should vary from the shock through to the inner coma, taking into account the field-of-view of the sensor as well as the variation of the flow, and they find that it agrees well with the observations. However, this does not explain why there is sudden transition at 2152

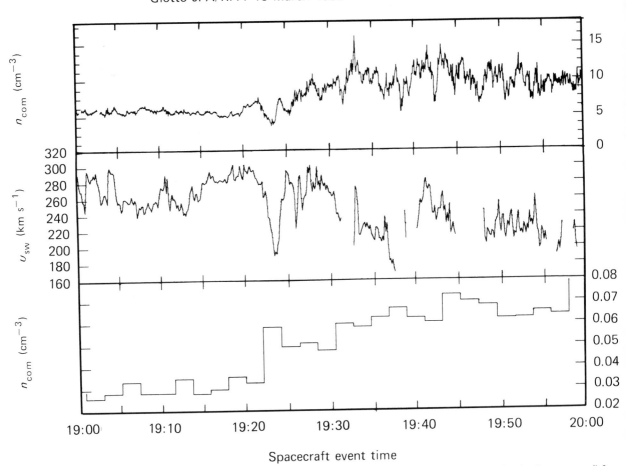

Giotto JPA/RPA 13 March 1986

Spacecraft event time

Fig. 11 — Electron density (top panel), solar wind proton speed (middle panel), and cometary water-group ion density (bottom panel) for the hour on 13 March 1986 containing the comet Halley bow shock crossing (from Coates *et al.* 1987a).

SCET. First, it is necessary to determine exactly what happened. It can be shown that there is a sudden decrease in velocity at the transition by approximately 90 km s^{-1}. Since the phase space density of the particles is obtained from the count rate by dividing by a factor proportional to V^4 the decrease in count rate can be accounted for entirely by the decrease in velocity of a population of ions which maintains a constant density along its trajectory. The nature of this velocity change is not yet understood.

After the transition the energy of the lower peak begins to decrease more quickly than the energy of either the upper peak or the solar wind. Its intensity increases while that of the upper one decreases,

suggesting that the lower curve consists of locally produced ions while the upper one consists of ions produced some distance upstream. In Fig. 15 the ratio of the energy of the lower peak to that of the solar wind is shown. An obvious explanation for the decrease in the value of this ratio is that the angle ϕ between the field and the flow is decreasing, and so the ion pick-up energy also decreases (equation (2)). The lower curve (Wilken *et al.* 1986) gives the derived value of ϕ (or θ_{VB} as marked on the diagram) which decreases from a value near 50° as in the upstream region to a value of approximately 25° at the edge of the magnetic pile-up region. This decrease in ϕ should be reflecting the draping of field lines around the

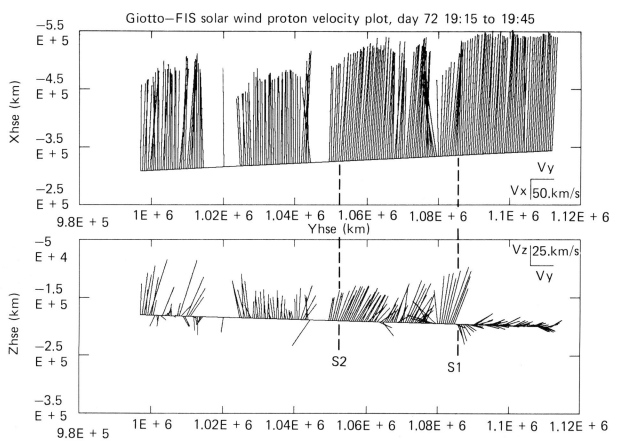

Fig. 12 — Projection of the solar wind proton velocity vector along the spacecraft trajectory. The upper panel shows the x–y plane, i.e. the view from above the ecliptic plane. The lower panel shows the z–y plane, i.e. the view from the Sun. The vectors are plotted once every 8 s according to the velocity scale in each plot (from Coates *et al.* 1987b).

comet which in turn should be apparent in the magnetic field measurements (Fig. 2). The measured magnetic field direction is much more variable than the rather smooth variation of the angle ϕ inferred from the ion energy spectrum, and it is difficult to see the draping until the spacecraft enters the pile-up region itself. Kimmel *et al.* (1987) attempted to model the spectral shape by calculating the trajectories of large numbers of test ions in a model of the magnetic field and plasma flow obtained from a magnetohydrodynamic simulation, and achieved a fair degree of success. Their result also depends on there being draped field lines around the comet in this region. To verify the explanation, or possibly to eliminate it, awaits a detailed comparison of the spectrum and the field direction.

8 THE INNER COMA

Fig. 16 shows an energy spectrogram from the FIS sensor for a distance of approximately 0.25 million km either side of the closest approach. At 2330 SCET there is a sharp decrease in the intensity of the solar wind protons coincident with the boundary of the pile-up region in the magnetic field. There is also a simultaneous decrease in the electron density (Fig. 14). It has been suggested that this boundary (Reme *et al.* 1987) is the same as the cometopause described by Gringauz *et al.* (1986) from the data of the Soviet spacecraft Vega 1. There is a similar change in the FIS spectrogram at a time of 0049 SCET on the outbound leg. There is another sharp change at 2348 SCET on the inbound leg matched by a similar one at

JPA/IIS/D4D
12–22 amu/e
POLAR DETECTOR 5
All azimuths

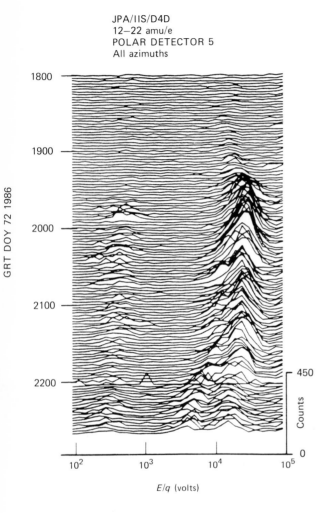

Fig. 13 — Count rate in the water group mass channel of polar detector 5 of the IIS, which looks closest to the ram direction, during the inbound portion of the Halley encounter. The count rate is shown as a function of E/q, with each 128-second spectrum plotted slightly below the previous one so that time increases in the downward direction. The peak near 400 eV is due to pile-up of solar wind protons in the time-of-flight unit. The character of the water group population changes at 2200 GRT (Ground Received Time = SCET + 8 min), with the emergence there of a second, lower energy peak in the count rate and a concurrent decline in the count rate of the higher energy peak. (from Thomsen et al.1987)

determined what the true nature of this feature is; for example, it may be associated with a rapid decrease in velocity which moves the distribution out of the field of view of the sensor. It does not seem to be associated with a coincident change in another instrument, but it may be related to a general increase in the cold ions shown in Fig. 17 (Balsiger et al. 1986). The change at 2330 SCET occurs as the count rate of the ion group O^+, OH^+, H_2O^+, increases past 20 counts s^{-1} while the change at 2348 SCET occurs as the group S^+, H_3O^+, starts to become significant at the same level. Both these transitions may be related to the cometopause as observed by the Vega spacecraft. Further study is required to resolve these issues.

Fig. 18 shows a similar spectrogram to Fig. 16 with the time scale expanded to show the measurements made within the magnetic cavity. The first interesting feature is a short burst of ions coincident with the edge of the cavity itself. The second is an extremely intense burst centred on the point of closest approach. This burst included the highest count rates encountered in the mission. In an early publication (Johnstone et al. 1986) it was speculated that these ions were produced by the impact of dust particles on the spacecraft itself. These ions were also detected by other instruments, and it has been found that their composition is just the same as the other cometary ions observed, and does not include any material that would be expected to come from a dust impact such as aluminium from the spacecraft itself or silicon from the dust particles (D.T. Young, private communication). They remain unexplained.

9 CONCLUDING REMARKS

The plasma measurements made at Comet Halley have been extremely successful in providing a picture of the important processes taking place in the neighbourhood of a comet. They have confirmed the fundamental ideas concerning mass-loading by cometary ions and the formation of a contact surface and a bow shock. Much still remains to be done in the analysis and interpretation of the data obtained from the encounter of March 1986, particularly concerning the unexpected features of the ion flow around the comet, such as the transition at 0.5 million km, the cometopause and related structure at 100 000 km, and

0021 SCET on the outbound leg. At these two positions, the intensity of the cometary ions suddenly decreases. The *upper* energy band in this spectrogram (Fig. 16) is the remains of the *lower* energy band in the cometary ion spectrum shown in Fig. 1. It appears to terminate at 2348 SCET. It has not yet been

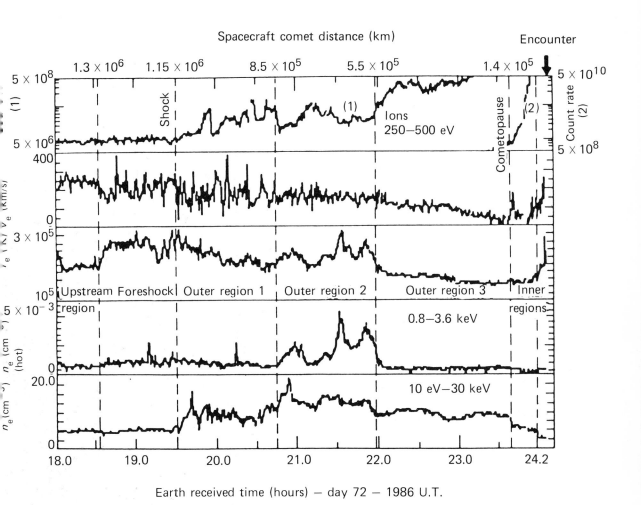

Fig. 14 — Plasma data during the last 6.2 h before the closest approach. From bottom to top panels: n_e, the density of 10 eV to 30 keV electrons, n_e(hot), the density of 0.8 to 3.6 keV electrons, T_e, the electron temperature estimated from the slope of the energy spectra near 70 eV, V_e, the electron bulk velocity calculated from the electron distribution and the count rate of the RPA2-PICCA in the ram direction for 250 to 500 eV/q ions (note a change of scale from count rate (1) to count rate (2) at time 2330 UT). Times given are SCET + 8 min. (Reme *et al.* 1987)

the hot ions seen inside the contact surface. There are two further measurements which are required for a complete study of the plasma environment of a comet; first, it is important to measure the composition of the energetic cometary ions; second, measurements of electrons in the energy range 0–10 eV are needed. With regard to the former, although the IIS obtained some interesting information on the mass distribution of these ions its resolution was not good enough to obtain the chemical composition. Some of its data indicate that the composition is significant. There are

several regions where electrons in the very low energy range may have played an important rôle in ionization or in the plasma dynamic processes.

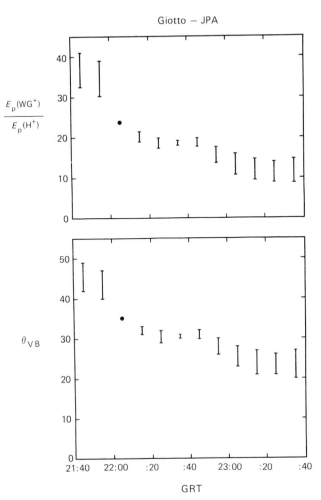

Fig. 15 — The ratio between the energies of water group ions (E_p (WG)[+]) and solar wind ions E_p (H[+]) as the comet is approached interpreted as a change in the angle θ_{VB} between the magnetic field and the flow direction. (from Wilken *et al.* 1986)

ACKNOWLEDGEMENTS

Many people contributed, not only their effort, but a great deal of enthusiasm to the acquisition of this unique data set. I would like to acknowledge here the contributions of the JPA team members J.A. Bowles, A.J. Coker, A.J. Coates, R.A. Gowen, F. Little, S.J. Kellock, J. Raymont, at the Mullard Space Science Laboratory; H. Borg and S. Olsen at the Kiruna Geophysical Institute; B. Wilken and W. Weiss at the Max Planck Institut, Lindau; J.D. Winningham and C. Gurgiolo at the Southwest Research Institute; R. Cerulli at the Istituto di Fisica dello Spazio Interplanetario; M. Thomsen at the Los Alamos National Laboratory; and T. Edwards at the Rutherford Appleton Laboratory.

Thanks are also due to the project team at the European Space Agency, particularly D. Dale, J. Credland, R. Reinhard, C. Berner, J. Noyes, A. Parkes, and H. Nye for their support to the experiment team and the skill with which they carried out the challenging task of placing a spacecraft close to the nucleus of a comet.

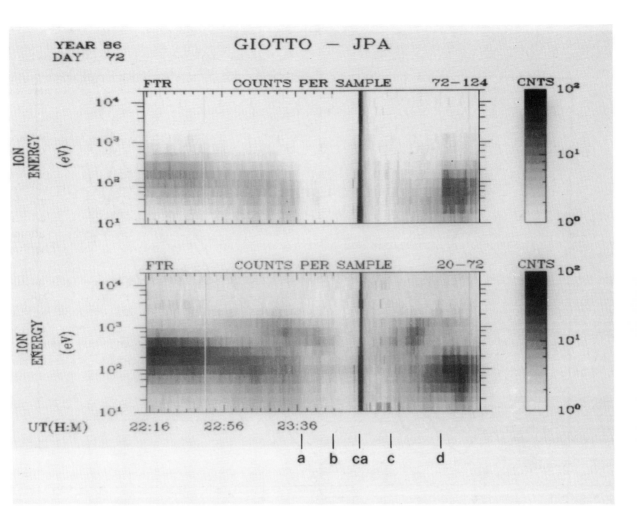

Fig. 16 — An energy–time spectrogram of data from the Fast Ion Sensor (FIS) covering the range from 0.5 million km before to 0.35 million km after closest approach ('ca'). The two panels cover different angular ranges; the upper is 72–124° from the velocity vector, the lower is 20–72°. The upper band in the energy spectrum of the lower panel is the lower energy branch of the cometary ion spectrum seen in Fig. 1. The more prominent band, seen in both panels, is the shocked, decelerated solar wind plasma. Two sharp changes are seen in the distributions on both the inbound and outbound legs. At the first ('a'), 2330 SCET inbound (note that the time scale on the plot is GRT = SCET + 8 min) and 0049 outbound, the solar wind plasma seems to decrease suddenly. At the second ('b'), 2348 SCET inbound, and 0021 outbound, the cometary ions decrease.

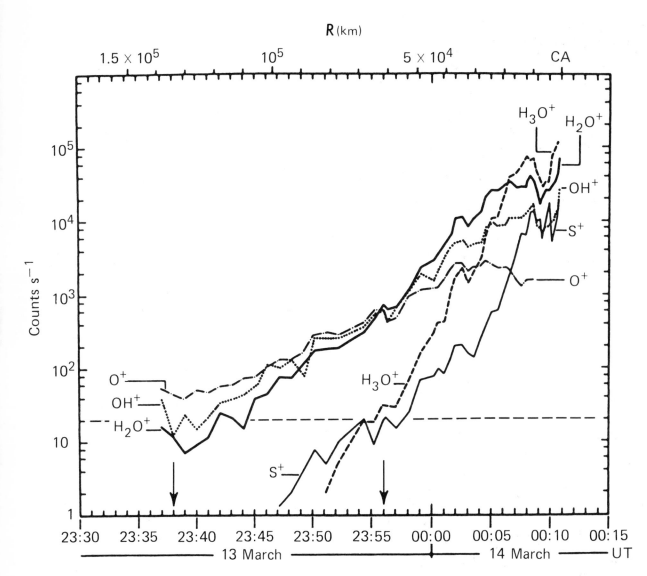

Fig. 17 — Time profiles of the HIS mass spectrometer count rates for masses 16, 17, 18, and 19, as well as the mass 32–34 group. Time is GRT; R is the distance of Giotto from the nucleus. (from Balsiger *et al.* 1986)

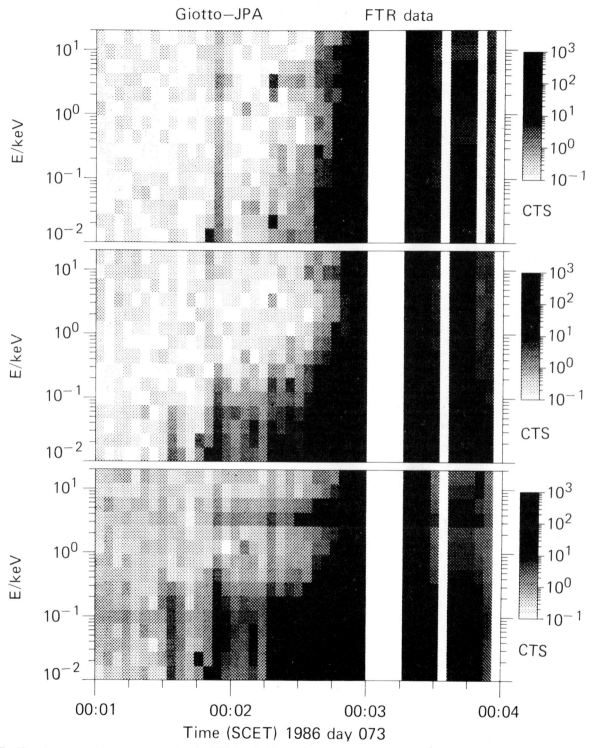

Fig. 18 — An energy–time spectrogram showing FIS data for three minutes around closest approach. The panels cover different angular ranges relative to the spin axis; 20–72° -bottom, 72–124° -middle, 124–180° -top. A short burst of particles is observed just before 0002 SCET coincident with the edge of the magnetic cavity. An extremely intense burst of particles is seen centred on the point of closest approach. A few seconds before the actual point, dust impacts caused nutation of the spin axis so that the telemetry signal was lost for approximately 20 s, and was intermittent for some time after.

REFERENCES

Balsiger, H., Altwegg, K., Buhler, F., Geiss, J., Ghielmetti, A.G., Goldstein, B.E., Goldstein, R., Huntress, W.T., Ip, W.H., Lazarus, A.J., Meier, A., Neugebauer, M., Rettenmund, U., Rosenbauer, H., Schwenn, R., Sharp, R.D., Shelley, E.G., Ungstrup, E., & Young, D.T. (1986) Ion composition and dynamics at Comet Halley. *Nature* **321** 330.

Balsiger, H., Altwegg, K., Benson, J., Buhler, F., Fischer, J., Geiss, J., Goldstein, B.E., Goldstein, R., Hemmerich, P., Kulzer, G., Lazarus, A.J., Meier, A., Neugebauer, M., Rettenmund, U., Rosenbauer, H., Sager, K., Sanders, T., Schwenn, R., Shelley, E.G., Simpson, D., & Young, D.T. (1987) The ion mass spectrometer on Giotto. *J. Phys. E* **20** 759.

Biermann, L., Brosowski B., & Schmidt, H.U.: 1967. *Solar Physics* **1** 254.

Coates, A.J., Lin, R.P., Wilken, B., Amata, E., Anderson, K.A., Borg, H., Bryant, D.A., Carlson, C.W., Curtis, D.W., Formisano, V., Jockers, K., Johnstone, A.D., Korth, A., Mendis, D.A., Reme, H., Richter, A.K., Rosenbauer, H., Sauvaud, J.A., Studemann, W., Thomsen, M.F., d'Uston, C., & Winningham, J.D. (1987a). Giotto measurements of cometary and solar wind plasma at the Comet Halley bow shock, *Nature* **327** 489.

Coates, A.J., Johnstone, A.D., Thomsen, M.F., Formisano, V., Amata, E., Wilken, B., Jockers, K., Winningham, J.D., Borg, H., & Bryant, D.A. (1987b) Solar wind flow through the Comet P/Halley bow shock. *Astron. Astrophysics* **6** 55–60.

Formisano, V., Amata, E., Wilken, B., Jockers, K., Johnstone, A.D., Coates, A., Heath, J., Thomsen, M., Winningham, J.D., Borg, H., & Bryant, D.A. (1986) Giotto observations of the bow shock at Comet Halley. *Adv. Space Res.* **6** 229–233.

Galeev, A.A., Cravens, T.E., & Gombosi, T.I. (1985) Solar wind stagnation near comets. *Astrophys. J* **289** 807–819.

Gringauz, K.I., Gombosi, T.I., Tatrallyay, M., Verigin, M.I., Remizov, A.P., Richter, A.K., Apathy, I., Szeremey, I., Dyachkov, A.V., Balakina, O.V., & Nagy, A.F., Detection of a new 'chemical' boundary at Comet Halley, *Geophys. Res. Lett.* **13** 613–616.

Johnstone, A., Coates, A., Kellock, S., Wilken, B., Jockers, K., Rosenbauer, H., Studemann, W., Weiss, W., Formisano, V., Amata, E., Cerulli-Irelli, R., Dobrowolny, M., Terenzi, R., Egidi, A., Borg, H., Hultquist, B., Winningham, J., Gurgiolo, C., Bryant, D., Edwards, T., Feldman, W., Thomsen, M., Wallis, M.K., Biermann, L., Schmidt, H., Lust, R., Haerendel, G., & Paschmann, G. (1986) Ion flow at Comet Halley. *Nature* **321** 344.

Johnstone, A.D., Coates, A.J., Wilken, B., Studemann, W., Weiss, W., Cerulli-Irelli, R., Formisano, V., Borg, H., Olsen, S., Winningham, J.D., Bryant, D.A., & Kellock, S.J. (1987a). The Giotto three-dimensional positive ion analyser. *J. Phys. E* **20** 795.

Johnstone, A.D., Coates, A.J., Heath, J., Thomsen, M.F., Wilken, B., Jockers, K., Formisano, V., Amata, E., Winningham, J.D., Borg, H., & Bryant, D.A. (1987b) Alfvénic turbulence in the solar wind flow during the approach to Comet P/Halley. *Astron. Astrophys* **187** 25–32.

Johnstone, A., Glassmeier, K., Acuña, M., Borg, H., Bryant, D., Coates, A., Formisano, V., Heath, J., Mariani, F., Musmann, G., Neubauer, F., Thomsen, M., Wilken, B., & Winningham, J. (1987c) Waves in the magnetic field and solar wind flow outside the bow shock at Comet P/Halley. *Astron. Astrophys.* **187** 47–54.

Kimmel, C.D., Luhmann, J.G., & Phillips, J.L. (1987) Characteristics of cometary picked-up ions in a global model of Giacobini–Zinner. *Geophys. Res.* **92** 8536–8544.

Korth, A., Richter, A.K., Loidl, A., Guttler, W., Anderson, K.A., Carlson, C.W., Curtis, D.W., Lin, R.P., Reme, H., Cotin, F., Cros, A., Medale, J.L., Sauvaud, J.A., d'Uston, C., & Mendis, D.A. (1987) The heavy ion analyser PICCA for the Comet Halley fly-by with Giotto. *J. Phys. E* **20** 787.

Neubauer, F.M. (1987) Giotto magnetic-field results on the boundaries of the pile-up region and the magnetic cavity. *Astron. Astrophys.* **187** 73–79.

Reme, H., Sauvaud, J.A., d'Uston, C., Cros, A., Anderson, K.A., Carlson, C.W., Curtis, D.W., Lin, R.P., Korth, A., Richter, A.K., & Mendis, D.A. (1987) General features of Comet P/Halley: solar wind interaction from plasma measurements. *Astron. Astrophys.* **187** 33–38.

Sagdeev, R.Z., Shapiro, V.D., Shevchenko, V.I., & Szego, K. (1986.) MHD turbulence in the solar wind comet interaction region. *Geophys. Res. Lett.* **13** 85–88.

Schmidt, H.U., & Wegmann, R. (1982.) In: *Comets,* ed. L.L. Wilkening (Tucson: University of Arizona Press), 538.

Thomsen, M.F., Feldman, W.C., Wilken, B., Jockers, K., Studemann, W., Johnstone, A.D., Coates, A., Formisano, V., Amata, E., Winningham, J.D., Borg, H., Bryant, D., & Wallis, M.K. (1987) *In situ* observations of a bi-modal ion distribution in the outer coma of Comet P/Halley. *Astron. Astrophys.* **187** 141–148.

Wallis, M.K. (1973) Weakly-shocked flows of the solar wind plasma through atmospheres of comets and planets. *Planet. Space Sci.* **21** 1647–1660.

Wilken, B., Jockers, K., Johnstone, A.D., Coates, A., Heath, J., Formisano, V., Amata, E., Winningham, J.D., Thomsen, M., Bryant, D.A., & Borg, H. (1986) Comet plasma tail formation — Giotto observations. *Adv. Space Res.* **6** 337–341.

Wilken, B., Weiss, W., Studemann, W., & Hasebe, N. (1987a) The Giotto implanted ion spectrometer (IIS): physics and technique of detection. *J. Phys. E* **20** 778.

Wilken, B., Johnstone, A.D., Coates, A., Borg, H., Amata, E., Formisano, V., Jockers, K., Rosenbauer, H., Studemann, W., Thomsen, M.F., & Winningham, J.D. (1987b) Pick-up ions at Comet P/Halley's bow shock: observations with the IIS spectrometer on Giotto. *Astron. Astrophys.* **187** 1–7.

5

Magnetic field regions formed by the interaction of the solar wind plasma with comet Halley

Fritz M. Neubauer

1 INTRODUCTION

The interaction between the solar wind and the atmospheres of comets is one of the few visible manifestations of space plasma physics. In fact the solar wind has been detected by visual observation of plasma tails (Biermann 1951). This detection was made, even though there was only a poor physical understanding of the complex plasma physics of the interaction. In the last decades strong effects have been made by theorists and Earth-based observers to physically understand how the flow of plasma around a comet works. With the encounters of spacecraft with comets P/Giacobini–Zinner in September 1985 and P/Halley in March 1986, a test of these physical ideas and a quantum jump in our knowledge on the plasma physics of comets became possible.

For an ideal observational characterization of the flow of a magnetoplasma it is necessary, in principle, to know the plasma and neutral gas properties, i.e. ideally the distribution functions of all ion and neutral gas species as well as electrons and the electromagnetic field at every location in space and time. For a single flyby at a comet these quantities are known only along a trajectory. However, a combination of some basic physical ideas and concepts with these one-dimensional observations makes it possible to deduce the most important characteristics of the flow.

One of the fundamental concepts of the physics of cosmic plasmas is the 'frozen-in-field' theorem, which describes the close relationship between the kinematics of the plasma flow and the geometry of magnetic field lines. This theorem is an important tool in using magnetic field observations to describe the flow of a magnetoplasma.

There is general agreement on the basic processes determining the interaction between the solar wind and a cometary atmosphere. Hence these basic physical processes can be used as the framework for the interpretation of observations by spacecraft and from the Earth.

As the solar wind, which initially consists mainly of protons and alpha particles as well as electrons, flows into the cometary atmosphere, it is continuously modified and slowed down by the addition of cometary ions due to the ionization (and charge exchange) of cometary neutral particles. This 'ion pick-up' process implies a mass loading of the plasma which is continuously enriched in cometary ions. Because the portion of a field line close to the comet is slowed down by these processes in contrast to the free moving part of the field line at large distances, the field lines are folded over the comet in a process generally referred to as draping (Alfvén 1957). The addition of cometary ions to the plasma also implies velocity space distribution functions, which are far from equilibrium and therefore strongly conducive to the development of plasma instabilities.

As one approaches a comet on the sunward side or on the flanks from a distance of several million kilometres, the first indication of its vicinity is then the development of electromagnetic wave and turbulence fields due to instabilities. Also, a weak draping effect sets in as a consequence of the slowing down of the plasma flow. After the total mass flux reaches a critical value a bow shock must form according to the flow theory, leading to a transition from super-fast flow in the MHD-sense to sub-fast flow. In the 'magnetosheath' inside the cometary bow shock the plasma is slowed down further and additional free energy becomes available for the development of plasma turbulence. As the cometary nucleus is approached even closer, the magnetic field starts to pile-up in the vicinity of the comet. Theoretical models before the spacecraft observations have generally predicted a smooth transition from the magnetosheath to the pile-up region. (e.g. Schmidt & Wegmann 1982). After the maximum in the magnetic field magnitude the field tends to zero, thereby providing a magnetic pressure gradient force which is necessary to balance the increasing frictional force due to the neutral gas flow (Ip & Axford 1982). Inside the pile-up region a region of space with a magnetic field $B = 0$, i.e. a magnetic field cavity has been

deduced by theoretical reasoning. The boundary can be called a 'contact surface' or ionopause. It is the only region not connected magnetically to the Sun by interplanetary magnetic field lines.

The idealized magnetoplasma system just described, was based on the tacit assumption that the solar wind is not varying as a function of time. This assumption has also been made in most theoretical modelling efforts. In reality, however, the solar wind is a highly dynamical flow system. Magnetically, it is divided into regions of varying polarity separated by complex sector boundaries. A sector contains one or more high-speed stream systems with characteristic variations in magnetic field, density, temperature, velocity vectors, etc. This is apart from more complex sporadic disturbances and broadband fields of plasma turbulence.

In this paper we mainly review observations by the Giotto magnetometer experiment and discuss these observations in the framework of the physical ideas developed above. We shall use some results obtained by the   Vega -spacecraft at Comet Halley and by the

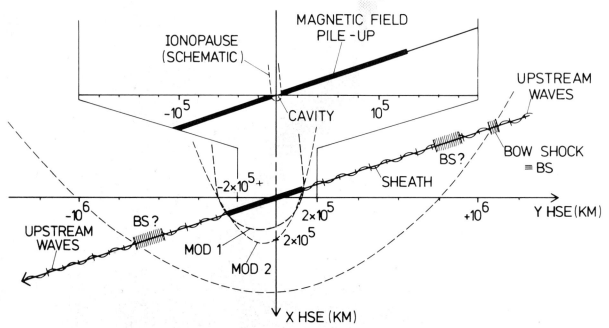

Fig. 1 — The encounter geometry of Giotto at Comet Halley with the flyby trajectory projected on the Halley centred solar ecliptic X-Y plane (see text). The most important interaction regions are marked on the trajectory; the tick marks are 1 h apart. The important boundaries deduced from the data are shown as dashed lines. They are all axisymmetric with respect to the X'-axis, which is obtained from the X-axis by a 6° anticlockwise rotation around the Z-axis due to aberration. The null field cavity is shown on an expanded scale. The pile-up boundary has been fitted by two model shapes MOD 1 and MOD 2 assuming different flaring (Modified after Neubauer *et al.* (1986).

ICE spacecraft at Comet Giacobini–Zinner as necessary.

The interplanetary magnetoplasma was not in a simple state during the Giotto encounter interval on 13 and 14 March 1986. It occurred when the spacecraft was moving close to and mostly south of the heliospheric current sheet for several days with frequent crossings of it and after Comet Halley had passed through the current sheet on its orbit from northern heliographic latitudes to southern heliographic latitudes.

2 UPSTREAM WAVES AND THE BOW SHOCK

Fig. 1 shows the flyby trajectory of Giotto in Halley centred solar ecliptic coordinates with X_{HSE} pointing toward the Sun from the nucleus, Y_{HSE} in the ecliptic plane, and Z_{HSE} pointing to the north pole of the ecliptic. In this coordinate system Giotto was moving about 2.5×10^5 km h^{-1}.

Starting on 12 March, magnetic field fluctuations were observed, the appearance of which suggested an origin due to pick-up ions from Comet Halley. An unambiguous association of these early waves with the comet will, however, have to await studies that also involve plasma data. The activity of these wave fields was increasing and decreasing on a time scale of hours with a superposed trend to generally increase as the comet was approached, with a similar behaviour in reverse order during the outbound pass. After about 1600 SCET (Spacecraft Event Time) on 13 March the waves were clearly of cometary origin even with the diagnostics based on magnetic field data only. As an example Fig. 2 shows one hour of magnetic field fluctuations due to cometary electromagnetic waves presumably produced by resonant interactions with water group ions.

The deviation from the expected steady increase in wave activity as the distance to the comet decreased

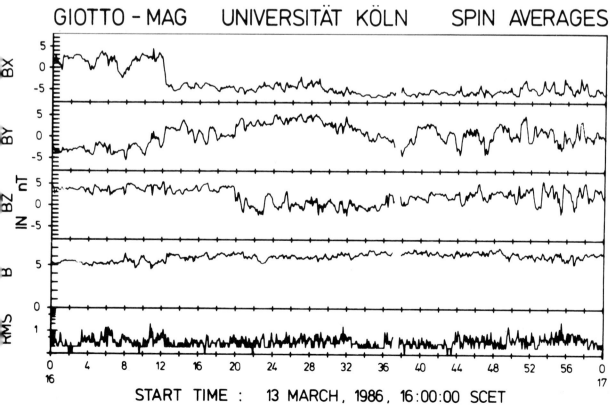

Fig. 2 — A one-hour interval of wave observations before the inbound bow shock crossing. The components are given in the Halley-centred solar ecliptic coordinate system.

can be attributed to the varying solar wind conditions.

Solar wind turbulence associated with sector boundaries was sometimes mixing with the waves of cometary origin. Also, the variation of solar wind parameters changed the conditions for wave growth.

Close to the bow shock crossing, the increase in wave activity was interrupted in the foreshock region, which was characterized magnetically by a rotating magnetic field of somewhat increased magnitude and a somewhat reduced level of fluctuations in magnitude and direction. The foreshock feature is also clearly seen after the crossing of the outbound bow shock.

Inbound, the bow shock transition started at 1923 SCET at a distance of 1.15×10^6 km from the comet. The first shock transition with magnetic fields increasing from about 8 nT to 18 nT was followed by variations possibly due to an oscillatory motion of the shock structure back and forth across the spacecraft. The shock was quite turbulent, and appeared much thicker than shocks observed at the planets visited in the past. The outbound bow shock is much less well defined than the inbound shock. It has been identified in the time interval from 02 30 to 0305 SCET on 14 March 1986 at a distance range from 6×10^5 km to 7.5×10^5 km. A parabola with the focus at the nucleus can be fitted to the bow shock crossing locations.

In three dimensions the paraboloid describing the bow shock as the boundary between the upstream wave and draping region and the magnetosheath is given by

$$X' = 4.5 \times 10^5 \text{ km}\left(1 - \frac{Y'^2 + Z'^2}{(9 \times 10^5 \text{ km})^2}\right) \quad (1)$$

where the primed coordinates X', Y', Z' take into account an aberration angle of $6°$, i.e. the X'-axis and Y'-axis are obtained from the X-axis and Y-axis by an anticlockwise rotation of $6°$ around the Z-axis. This aberrated paraboloid constitutes a physical improvement of the bow shock shape derived by the Giotto investigators (Neubauer *et al.* 1986, Rème *et al.* 1986). When the shock weakens at tailward values of X, i.e. increasingly negative X, it could better be described by a hyperboloid, the flaring of which is determined by the fast mode wave characteristics of the solar wind.

3 THE MAGNETOSHEATH REGION AND THE PILE-UP BOUNDARY

Fig. 3 gives an overview of the magnetic field observations based on one minute averages from 1800 SCET (UT) on 13 March until 1000 SCET on 14 March 1986. Inbound, strong variations in magnetic field magnitude prevail throughout the magnetosheath (or more briefly 'sheath') without a generally increasing trend in magnitude until 2330 SCET when the magnitude jumps up to about 30 nT and stays at higher levels with superimposed fluctuations. We call this sharp boundary the pile-up boundary. It constitutes the inner boundary of the magnetosheath region.

The magnetosheath is further characterized by appreciable directional variations most of the time.

Closer inspection of the variations indicates that the magnetosheath can be further subdivided according to the characteristics of the magnetic turbulence (Glassmeier *et al.* 1987). For example, the interval from 2230 to 2252 SCET is characterized by a strong compressional wave feature with only little directional variations. The discussion of these nonlinear wave regions is outside the scope of this paper.

On the outbound pass it is much more difficult to identify the pile-up boundary and the bow shock transition region. The magnetosheath is also characterized by strong magnitude variations but by much reduced variations in direction.

Comparison of the inbound and outbound magnetosheath immediately leads to the following question. Are the substantial differences due to a strong asymmetry of the global cometary plasma flow between positive and negative Y-coordinates, or is it at least partly due to a solar wind structure convected or propagating into the cometary environment? The large rotation of the field direction before the inbound bow shock, the strong directional variation in the inbound magnetosheath, and the regular arrangement of magnetic slabs of opposite polarity throughout the inner pile-up region suggest that a major part of the asymmetry is due to disturbed interplanetary magnetic field conditions associated with the heliospheric current sheet affecting the inbound observations. Complex variations in magnetic field structure including strong discontinuities in magnitude can sometimes occur near interplanetary sector boundaries (Behannon *et al.* 1981). The dramatic inbound pile-up

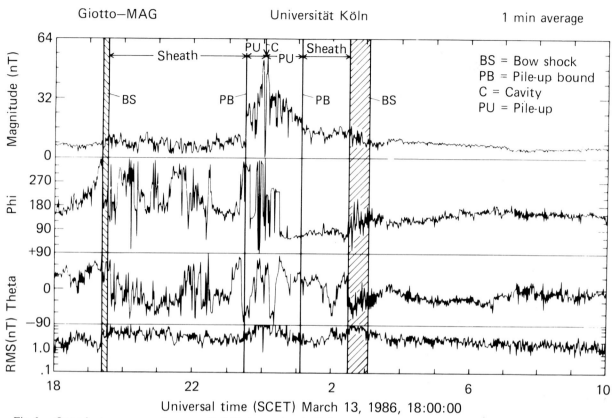

Fig. 3 — One-minute average vector magnetic field observations in Halley-centred solar ecliptic coordinates. The definition of PHI and THETA is standard, i.e. THETA is the elevation angle with respect to the X, Y-plane and PHI the azimuth of the projection on the X, Y-plane with PHI = 0 toward the Sun. RMS is the Pythagorean mean of the component rms values (From Neubauer 1987).

boundary may then be explained as a tangential discontinuity (Landau & Lifschitz 1967) moving slowly toward the comet. The plausibility of this explanation is enhanced by the absence of similar structure in the   Vega magnetic field observations (Riedler *et al.* 1986), which apparently occurred under different interplanetary conditions.

Whether the boundary of the pile-up region is discontinuous or gradual, the comparison of the observations by Giotto and the   Vega 's on one side and the tail observations by ICE at Giacobini–Zinner on the other side suggest that the pile-up boundary becomes the tail boundary as one goes tailward, i.e. to negative values of X_{HSE}. The three-dimensional shape of the pile-up boundary has also been discussed by Neubauer (1987). The extreme observational models, assuming no flaring and strong flaring, are also shown in Fig. 1.

As an alternative to the tangential discontinuity interpretation the observations of the fine structure of the pile-up boundary could also be explained by the somewhat exotic version of a rotational or Alfvénic discontinuity. A rotational discontinuity with plasma isotropy on both sides, however, requires magnetic field magnitude conservation.

A way out of this problem can be that the plasma is highly anisotropic on at least one side of the discontinuity (Neubauer 1987). The magnitude must generally jump for a rotational discontinuity in this case (Hudson 1970, Neubauer 1970).

Although the outbound pile-up boundary is much less dramatic and much less well defined, the same considerations could apply in this case.

4 THE MAGNETIC PILE-UP REGION AND THE MAGNETIC CAVITY

The magnetic pile-up region is characterized by a magnetic field magnitude much greater than the undisturbed solar wind values. It is generally also quieter than the sheath region, with regular variations in direction and in magnitude. Strong dips in magnitude sometimes occur. On the inbound trajectory a maximum in magnitude occurred at 2359 SCET on 13 March, i.e. at a distance of 16 500 km from the comet with a maximum magnitude of 57 nT. After the maximum the magnetic field decreased rapidly until it reached $B = 0$ at 0001:54 SCET on 14 March.

The magnetic field cavity region (Neubauer *et al.* 1986, Neubauer 1987) is followed by an outbound increase in magnetic field magnitude starting at 0003:59 SCET and reaching the absolute maximum of 65 nT at 0005 SCET.

The inbound and outbound magnetic field profiles are symmetrical except that the outbound profile appears somewhat compressed. The inner boundary of the pile-up region or outer boundary of the cavity is sometimes referred to as the 'contact surface'. We call it the ionopause, since it bounds the ionospheric region proper which is not connected by the magnetic field to the interplanetary medium. The inbound and outbound ionopauses occur at distances from the nucleus of 4660 km and 3930 km, respectively.

The ionopause transition layers turned out to be extremely thin. Both for the inbound and outbound ionopause transitions the thicknesses were determined to be 25 km, i.e. approximately a thermal ion gyro radius for the ion temperature determined outside the inbound ionopause by the Giotto ion mass spectrometer (Balsiger *et al.* 1986). The drop in magnetic field magnitude was 20 nT and 18.3 nT for the inbound and outbound ionopause, respectively (Neubauer 1988).

The magnetic field vectors during the transition can be used to determine the normal vectors of the ionopause surface, using the minimum variance analysis (MVA) introduced by Sonnerup & Cahill (1967). Together with the crossing points on the flyby trajectory these normals give important constraints on the geometry of the ionopause. Because, however, of the many possibilities for a distortion of the ionopause

the interpretation is not unambiguous. In any case, it follows that the ionopause is not a simple smooth surface centred at the nucleus. The observations can be explained only by a combination of a displacement of the whole ionopause in a northern direction by possibly more than 1000 km and ripples on the surface due to an instability (Neubauer 1987).

A further question that cannot be answered by the observations from one spacecraft alone is that of the tailward extension of the cavity. The ICE observation at Giacobini–Zinner (e.g. Slavin *et al.* 1987) yielded a field of 5–10 nT in the plasma sheet between the lobes with high magnetic field. The presence of this field may be due to the penetration of magnetic fields into the inner ionosphere, because instabilities prevent the formation of a stable ionopause. We suggest, as an alternative, that the ionopause surface extended to a distance less than the distance of closest approach, with the approximate shape of a tadpole. The shape is sketched in Fig. 4* together with other important geometrical features of the comet solar wind interaction.

Returning now to the inner part of the pile-up region, theory and observations suggest that at least the inner few thousand kilometres can be described as a stagnant magnetoplasma held in place by the magnetic $j \times B$-force against the frictional force of the neutral gas (Ip & Axford 1986, Cravens 1986) and a substantial contribution by an outward pressure force (Neubauer 1988).

In spite of the approximate stagnation the magnetic field lines must slowly move through this region until they slip sideways over the ionopause obstacle. Physically, the only property of the initial interplanetary magnetic field lines dragged through the cometary atmosphere, that is 'remembered' by the field lines in the inner pile-up region, is the direction of the projection on a plane perpendicular to the direction of initial undisturbed flow, i.e. the Y', Z'-plane. This implies that the original interplanetary polarity is remembered.

This physical expectation is borne out in an interesting way by Fig. 5*, which shows unit vectors giving the observed magnetic field directions along the inner part of the flyby trajectory (Raeder *et al.* 1987). There is a one-to-one relationship between

---

* to be found in the colour illustration section in the centre of the book.

slabs of opposite magnetic field polarity on the inbound and outbound trajectory. This relation is shown by the different shades in Fig. 5*. It can easy be explained by short interplanetary magnetic field polarity sectors swept up by the comet, when it passed through the heliospheric current sheet and its super-posed ripples. This is consistent with the fact that the Giotto encounter with the comet going from north to south occurred just below the heliospheric current sheet, i.e. south of it, and not very long after the comet must have passed through the current sheet (Neubauer *et al.* 1986).

5 SUMMARY

The Giotto encounter with Comet Halley has led to a picture of the spatial structure of the cometary magnetoplasma which is characterized by four major spatial regions with definite magnetic signatures. From a far distance inward these are the upstream wave region, the magnetosheath, the magnetic pile-up region, and the magnetic cavity region. These regions are separated by the bow shock, pile-up boundary, and ionopause. A further subdivision is possible if one uses the fluctuation or wave characteristics and plasma data. A unique aspect of the Giotto encounter is that it occurred very close to the heliospheric current sheet. This had two consequences: first, the inner pile-up region contained several magnetic sectors swept up by the comet, when it passed through the heliospheric current sheet; second, there is some evidence that an interplanetary disturbance region of the type characteristic for the vicinity of sector boundaries was moving through the cometary neighbourhood at the time of the Giotto pass.

In a three-dimensional picture, as sketched in Fig. 4* the pile-up boundary becomes the tail boundary at negative values of the HSE coordinate X_{HSE}. The cavity boundary, i.e. the ionopause, may form a closed tadpole type surface. In certain circumstances not encountered during the Giotto flyby at Comet Halley, instabilities may prevent a sharp ionopause.

ACKNOWLEDGEMENTS

We acknowledge important contributions by the members of the Giotto magnetometer team: M.H. Acuña, L.F. Burlaga, B. Franke, K.H. Glassmeier, B. Gramkow, F. Mariani, G. Musmann, N.F. Ness, M. Pohl, H.J. Raeder, H. Schmidt, E. Ungstrup, and M.K. Wallis, and by the members of the Giotto spacecraft team.

This work has been supported financially by the Federal Republic of Germany Bundesministerium für Forschung und Technologie.

---

* to be found in the colour illustration section in the centre of the book.

REFERENCES

Alfvén, H. (1957) On the theory of comet tails, *Tellus* **9** 92–96

Balsiger, H., Altwegg, K., Bühler, F., Geiss, J., Ghielmetti, A.G., Goldstein, B.E., Goldstein, R., Huntress, W.T., Ip, W.-H., Lazarus, A.J., Meier, A., Neugebauer, M., Rettenmund, U., Rosenbauer, H., Schwenn, R., Sharp, R.D., Shelley, E.G., Ungstrup, E., Young, D.T. (1986) Ion composition and dynamics at comet Halley, *Nature* **321** 330–334

Behannon, K.W., Neubauer, F.M., & Barnstorf, H. (1981) Fine-scale characteristics of interplanetary sector boundaries, *J. Geophys. Res.* **86** 3273–3287

Biermann, L. (1951) Kometenschweife und solare Korpuskular-strahlung, *Z. Astrophys.* **29** 274–286

Cravens, T.E. (1986) The physics of the cometary contact surface *Proc. 20th ESLAB Symposium on the Exploration of Halley's Comet, ESA SP 250* 241–246

Glassmeier, K.H., Neubauer, F.M., Acuña, M.H., & Mariani, F. (1987) Low frequency magnetic field fluctuations in comet P/Halley's magnetosheath: Giotto observations, *Astron. and Astrophysics* **187** 65–68

Hudson, P.D. (1970) Discontinuities in an anisotropic plasma and their identification in the solar wind, *Planet. Space Sci.* **18** 1611–1622

Ip. W.H., & Axford, W.I. (1982) Theories of physical processes in the cometary comae and ion tails in *Comets*, ed. L.L. Wilkening, Univ. of Arizona Press, 588–634

Ip, W.H. & Axford, W.I. (1987) The formation of a magnetic field-free cavity at comet Halley, *Nature* **325** 418–419

Landau, L.D. & Lifschitz (1967) *Elektrodynamik der Kontinua*, Akademie-Verlag Berlin

Neubauer, F.M. (1970) Jump relations for shocks in an anisotropic magnetized plasma, *Z. Phys.* **237** 205–223

Neubauer, F.M. (1987) Giotto magnetic field results on the boundaries of the pile-up region and the magnetic cavity, *Astronomy and Astrophysics*, **187** 73–79

Neubauer, F.M. (1988) The ionopause transition and boundary layers at comet Halley from Giotto magnetic field observations, *J. Geophys. Res.*, **93**, 7272–7281

Neubauer, F.M., Glassmeier, K.H., Pohl, M., Raeder, J., Acuña, M.H., Burlaga, L.F., Ness. N.F., Musmann, G., Mariani, F., Wallis, M.K., Ungstrup, E., & Schmidt, H.U. (1986) First results from the Giotto magnetometer experiment at comet Halley, *Nature* **321** 352–355

Raeder, H.J., Neubauer, F.M. Ness, N.F., & Burlaga L.F., (1987) Macroscopic perturbations of the IMF by P/Halley as seen by the Giotto magnetometer, *Astronomy and Astrophysics* **187** 61–64

Rème, H., Sauvaud, J.A., d'Uston, C., Cotin, F. Cros, A., Anderson, K.A., Carlson, C.W., Curtis, D.W., Lin, R.P., Mendis, D.A., Korth, A., & Richter, A.K., (1986) *Comet Halley* — solar wind interaction from electron measurements aboard Giotto, *Nature* **321** 349–352

Riedler, W., Schwingenschuh, K., Ye. Yeroshenko, G., Styashkin, V.A., Russell, C.T. (1986) Magnetic field observations in comet Halley's coma, *Nature* **321** 288–259

Schmidt, H.U. & Wegmann, R. (1982) Plasma flow and magnetic fields in comets. In *Comets*, Wilkening, L.L. ed., Univ. of Arizona Press, 538–560

Slavin, J.A., Smith, E.J., Daly, P.W., Flammer, K.R., Gloeckler, G., Goldberg, B.A., McComas, D.J., Scarf, F.L., & Steinberg, J.L. (1987) The P/Giacobini–Zinner magnetotail, *Proc. 20th ESLAB Symposium on the Exploration of Halley's Comet, ESA SP 250*, 81–87

Sonnerup, B.U.Ö., & Cahill, L.J., (1967) Magnetopause structure and attitude from Explorer 12 observations, *J. Geophys. Res.* **72** 171–183

6

Regions of interaction between the solar wind plasma and the plasma environment of comets

H. Rème

The passage of the International Cometary Explorer (ICE) through the tail of comet Giacobini–Zinner on 11 September 1985, and the flyby of comet Halley in March 1986 by a fleet of spacecraft, have provided for the first time an *in situ* study of the interaction of the solar wind with a cometary plasma under different conditions (Table 1).

In this paper, devoted to the regions of interaction between the solar wind plasma and the plasma environment of comets, the data base provided by the RPA-COPERNIC experiment (Rème *et al.* 1987a; Korth *et al.* 1987) aboard the European Giotto spacecraft will be used. The closest approach of this spacecraft to comet Halley occurred on 14 March 1986, giving the first opportunity to study the innermost part of the interaction region. Comparisons will be made with results obtained by various experiments on other spacecraft.

The RPA-COPERNIC experiment consists of two detectors:

- an electron electrostatic analyzer (RPA1-EESA) able to provide a full 4π electron distribution twice per spacecraft spin (2 s) from 10 eV to 30 keV (Rème *et al.* 1987a);
- a thermal positive ion composition analyzer (RPA2-PICCA) giving the distribution of 10–213 amu positive ions in the coma of the comet (Korth *et al.* 1987).

Two seconds before closest approach, the spacecraft's telemetry signal was lost. When it returned

~ 30 min later, it was apparent that the RPA-COPERNIC experiment had received some damage, presumably from dust impact. Data from the post-encounter period require special analysis, and will not be presented here.

Before the *in situ* measurements, models of interactions of the solar wind with a negligible gravity and no magnetic field body were founded upon ground-based cometary measurements and theoretical calculations. Owing to this negligible gravity of the small cometary nucleus, the sublimation gases expand supersonically and interact with the solar wind over a length scale that is typically 5–6 orders larger than the size of the nucleus, when it is at about 1 AU, through charge exchange, photoionization, and impact ionization which increases as cometocentric distance decreases. The newly created ions 'mass load' the solar wind and slow it down. Theoretical models predict a gradual slowing and heating of the solar wind until the Mach number of the solar wind flow is reduced to a value of $M \sim 2$. (Biermann *et al.* 1967, Brandt & Mendis 1979). At that point, a weak bow shock forms which further slows, heats, and deflects the flow. Inside the shock, mass loading must continue as the solar wind penetrates closer to the nucleus. A strong deceleration of the solar wind is expected to begin at a distance at which the momentum-transfer collision mean-free path of a solar wind ion is of the order of its radial distance from the nucleus (Mendis *et al.* 1986). This distance has been termed the 'collisionopause'. Inside this boundary, the solar wind would decelerate rapidly and also cool owing to exchange of energetic ions formed upstream with new, less energetic ions continuously being formed in the decelerating flow (Galeev *et al.* 1985). This would lead to the formation of a magnetic barrier. At the inner edge of this barrier is the ionopause, a tangential discontinuity between the inflowing contamined solar wind and the purely cometary plasma. The slowing down of the magne-

tized plasma near the comet also leads to the draping of the interplanetary magnetic field around the comet to form a magnetotail with two lobes separated by a cross-tail current sheet.

In fact the particle and field environment of comet Halley encountered by the spacecraft in March 1986 displays a very complex structure (Balsiger *et al.* 1986, Galeev 1986, Gringauz *et al.* 1986, Johnstone *et al.* 1986, Mukai *et al.* 1986, Neubauer *et al.* 1986, Rème *et*

Table 1 — Parameters for the different spacecraft–comet encounters

Spacecraft	Comet	Distance to comet nucleus at closest approach (km)	Encounter geometry	Date	Solar wind speed (km s^{-1})	Comet production	Sun–Comet distance (AU)
ICE	Giacobini–Zinner	7800 7800	Tailside	11 September 1985	Between 400 and 500	Between 2×10^{28} and 5×10^{28} water, molecules s^{-1} (Stewart *et al.* 1985)	1.05
Vega-1	Halley	8890 8890	Sunside	6 March 1986	510	Total gas production rate: 1.3×10^{30} molecules s^{-1} (Gringauz *et al.* 1986)	0.79
Vega-2	Halley	8030 8030	Sunside	9 March 1986	620	OH Gas production rate $\sim 9 \times 10^{29}$ molecules s^{-1} (Moreels *et al.* 1986). OH production rate: $\sim 2 \times 10^{30}$ molecules s^{-1}. Water production $5.6 \times 4 \times 10^{29}$ molecules s^{-1} (Krasnopolsky *et al.* 1986). Water production: 5.6×10^{29} molecules s^{-1} (Festou *et al.* 1986)	0.83
Suisei	Halley	151 000	Sunside	8 March 1986	~ 500		0.81
Sakigake	Halley	6.99×10^6	Sunside	11 March 1986	~ 450		0.86
Giotto	Halley	600	Sunside	14 March 1986	350 ~ 400	Total gas production: 6.9×10^{29} molecules s^{-1}. Water production: 5.5×10^{29} molecules s^{-1} (Krankowsky *et al.* 1986). Water production: 5.2×10^{29} with nucleus near a minimum of activity (Festou *et al.* 1986)	0.89

al. 1986). In the midst of rapid fluctuations, one also distinguishes fairly well defined regions separated by sharp transitions.

During its approach to the comet, the Giotto RPA-COPERNIC plasma experiment identified several regions and boundaries in which the solar wind interaction with the cometary plasma displayed characteristic features:

- Beginning $\gtrsim 4.6 \times 10^6$ km from the comet, there is an upstream region with sporadic magnetic connection to the comet.
- An electron foreshock is present up to 2.5×10^5 km away from the bow shock.
- A bow shock is detected at 1.15×10^6 km from the comet.
- Between the bow shock and the magnetic pile-up boundary, the outer cometosheath can be divided into three parts.
- A magnetic pile-up boundary (MPB) is found at $\sim 1.35 \times 10^5$ km, and density decreases are detected at $\sim 8.2 \times 10^4$ km and at $\sim 4.5 \times 10^4$ km from the comet.
- An ionopause is detected at $\sim 4\,700$ km from the comet nucleus.

Fig. 1, from Rème *et al.* (1987b), summarizes some plasma data measured by the RPA-COPERNIC experiment during the last 6.2 hours before the Giotto encounter with comet Halley when the spacecraft goes from the upstream region to the closest approach to the comet: n_e is the density of 10 eV–30 keV electrons; n_e (hot) the density of 0.8–3.6 keV electrons; T_e the electron temperature estimated from the slope of the energy spectra near 70 eV; V_e the electron bulk velocity calculated from the electron distribution. In the top panel the RPA2-PICCA count rate in the ram direction is shown for 250–500 eV/q ions.

THE UPSTREAM REGION

Among the parameters that can be calculated by the electron sensor is the heat flux ratio, or the ratio of minimum to maximum heat flux computed parallel and antiparallel to the magnetic field direction, the latter taken as the axis of symmetry of the electron momentum flux tensor. Before 1900 UT (4.6×10^6 km from the nucleus) before the Giotto encounter with comet Halley, effects are seen in the heat flux ratio, but it is impossible to link them unambiguously to the presence of the comet because during this

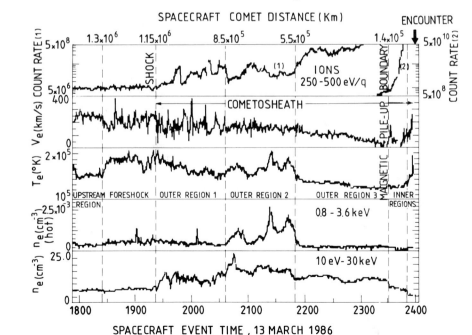

Fig. 1 — Some plasma electron data and count rate of the 250–500 eV/q ions in the ram direction during the last 6.2 hours before the Giotto encounter with comet Halley. Note a change of scale from count rate (1) to count rate (2) at time ~ 2330 SCET in the top panel.

period Giotto is very near the heliosphere current sheet (Neubauer *et al.* 1986).

Fig. 2 shows the first effect on the electron data that seems clearly linked to the presence of the comet, between 0622 and 0652 UT SCET (Spacecraft Event Time) at a comet distance of about 4.3 × 10⁶ km. Here the following quantities are shown from top to bottom: the total electron density from 10 eV to 30 keV, the heat flux ratio, and two impact parameters X and R calculated from the magnetic field direction estimated from the axis of symmetry of the electron momentum flux tensor: X is the distance along the Sun–comet line, from the comet to the point of closest approach along this direction (positive on the Sunward side), and R is the distance of closest approach between this direction and the Sun–comet line. The calculation assumes that the field lines are straight and that the transport of the electrons is simply by their kinetic motion. No effect on the 10 eV–30 keV electron density is noted at these times but around 0632 SCET a large increase appears in the heat flux ratio, correlated with X becoming negative and R diminishing strongly. At this time the heat flux ratio is close to unity (equal forward and reverse heat fluxes),

an indication that electrons may be streaming back along the interplanetary magnetic field, in the direction opposite to the heat flux from the solar corona. Around 0643 SCET the heat flux ratio decreases, associated with X becoming positive, but the heat flux ratio is still above the level found before 0632 SCET, R staying relatively small. Thus this effect on the heat flux ratio seems to be due to the presence of the comet. In fact this event is the beginning of a large turbulent region extending upstream from the bow shock for ~4.6 × 10⁶ km from the nucleus. In this region, fluctuations in the electron distribution occur, characterized by abrupt variations in the electron heat fluxes, with the heat flux ratio reaching a value of unity when the measurement region seems magnetically connected with the comet. This region, with sporadic connections to the comet, is the *upstream region* which extends from 4.6 × 10⁶ or further to 1.4 × 10⁶ km from the comet.

The upstream region is characterized by variations in magnetic field strength and plasma parameters and a high level of microturbulence, much of which is caused by the pick-up of cometary ions by the solar

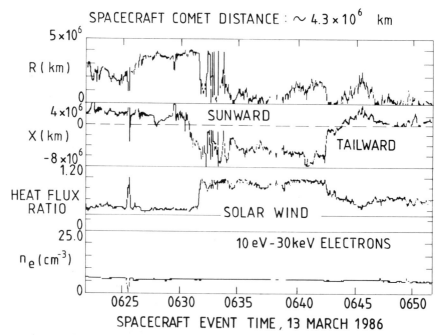

Fig. 2 — Effect on the electron heat flux ratio linked with variations of the R and X impact parameters without any effect on the 10 eV–30 keV electron density, at about 4.3 × 10⁶ km from comet Halley.

wind. Pick-up ions are observed out to 7.8×10^6 km from the nucleus (Johnstone *et al.* 1986, Neubauer *et al.* 1986).

Cometary ions picked up by the solar wind are also detected by Vega-1 (Somogyi *et al.* 1986) at a distance of 10^7 km from the comet. These ions start to drift across magnetic field lines with the solar wind plasma in a self-consistent electric field, and they have a very small initial velocity (of the order of 1 km s^{-1}) in the direction of the magnetic field lines. Thus they form a beam in the solar wind plasma exciting Alfvén waves due to the ion-cyclotron instability. Alfvén waves with frequencies of $\sim 10^{-2}$ Hz excited by picked-up ions are detected by Riedler *et al.* (1986) on Vega-1, starting at a distance of $\sim 3.5 \times 10^6$ km from the comet. From the Sakigake magnetometer, the ion cyclotron waves in the ionized water of cometary origin can be presumed to exist even at a distance of 7×10^6 km from the nucleus.

At 1000 SCET on 13 March 1986, at about 3.5×10^6 km from the comet, the solar wind speed determined on Giotto from the electron distribution is ~ 400 km s^{-1}. Afterwards the flow speed begins to decrease but it is difficult for the first few hours to separate solar-interplanetary effects (Oyama *et al.* 1986, Apáthy *et al.* 1986) from mass loading effects resulting from the ionization of cometary molecules. The speed decrease is generally systematic over an interval of several hours, but there are significant shorter term variations (Anderson *et al.* 1987).

THE FORESHOCK

An electron foreshock is present from 1.4×10^6 to 1.15×10^6 km from the comet. The entry into this foreshock at about 1826 SCET (Fig. 1) is marked by a large decrease of the plasma flow speed which after 10 minutes has decreased by about 100 km s^{-1}. The flow direction changes by about 15° over the same interval and in addition varies rapidly. Later, the flow speed becomes quite variable but recovers to around its previous value just before encounter with the bow shock. In the foreshock there is a slight increase in the total electron density n_e, but a larger increase in the energetic electron density n_e (hot). In fact, the percentage increase in the electron densities after 1826 SCET systematically increases as the electron energy in-

creases. The maximum intensity is quickly reached at the highest energy, but more slowly at the lower energies (Anderson *et al.* 1987). The reverse heat flow, in the sunward direction, increases also (not shown), with the heat flux ratio rising to nearly unity; the temperature in the foreshock reaches a value of about 2.7×10^5 K, and the electron distributions are not strictly Maxwellian at these times. Some ions are also detected in the ram direction.

Beginning about 20 minutes before the bow shock traversal, spikes of keV electrons appear. The energetic electrons in these spikes may have been accelerated in essentially the same way electrons are accelerated at the Earth's bow shock, to energies of 1 to 10 keV (Anderson *et al.* 1979). Alternatively, these electrons could have escaped from deep in the cometosheath where the electrons intensities are much higher (Rème *et al.* 1987b).

THE BOW SHOCK

The bow shock is detected around 1922 SCET at 1.15×10^6 km from the comet with large changes in the flow speed and direction (Fig. 3). The flow speed begins a sharp decline of about 100 km s^{-1}, thus defining the location of the shock quite precisely and lending support to the belief that comet Halley indeed possesses a bow shock. The electron temperature reaches values as high as 3×10^5 K, and the density, after an increase beginning at 1920 SCET, corresponding to a region of solar wind electron compression, falls to a very low value at 1924 SCET corresponding to a region of rarefaction seen also by the magnetometer (Neubauer *et al.* 1986). Then a second rise in electron density occurs followed by a drop to upstream values, after which the density rises and remains elevated above upstream densities. The density variations from 1920 to about 1950 SCET suggest large amplitude wave disturbances. No later than 90 s after 1922 SCET, the electron distribution function becomes isotropic.

In this boundary region, many plasma electron parameters are very similar to those in planetary bow shocks: large and abrupt changes in the plasma density, large fluctuations in density indicating strong plasma turbulence, enhanced temperature and nonthermal components in the plasma electrons.

The important rôle of heavy ions in the shock structure is emphasized in the work of Coates *et al.* (1987). This structure is highly complex because the gyroradius of a heavy ion of cometary origin is much larger than that of a solar wind proton: the shock transition starts with a cometary ion 'foot' seen at a distance of the order of cometary ion gyroradius upstream from the location of a sharp decrease in the solar wind proton speed. The total solar wind thermal and magnetic pressure is dominated by the relatively small population of cometary ions throughout the shock region.

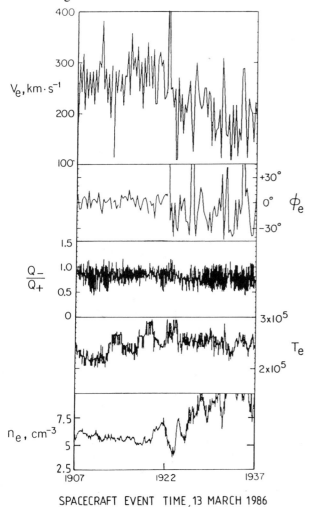

SPACECRAFT EVENT TIME, 13 MARCH 1986

Fig. 3 — Plasma electron data at the Halley bow shock. From bottom to top: n_e is the 10 eV–30 keV density, T_e the temperature, $Q-/Q+$ the heat flux ratio, ϕ_e the flow direction, and V_e the flow speed.

Fig. 4, from Coates *et al.* (1987), shows, from top to bottom, the total electron density between 10 eV and 30 keV from the RPA–COPERNIC experiment, the solar wind proton speed, and the cometary ion density from the Giotto JPA experiment (Johnstone *et al.* 1986), during a 1 hour period including the shock traversal. In the rarefaction region (1922:30–1924 SCET) there is an initial decrease in solar wind speed, and the pick-up ion density reaches a peak. The 'permanent' drop in solar wind proton speed does not occur until ~1931 SCET, ~11 minutes and ~45 000 km after the first compression in the solar wind. The pick-up ion density appears to increase with each decrease in the solar wind speed, and is generally proportional to the solar wind electron density. Thus, the shock transition is highly structured and extends over several pick-up ion gyroradii but with sharp jumps in solar wind velocity on scales of a solar wind proton gyroradius.

Coates *et al.* (1987) have shown that about half of the plasma thermal pressure both upstream and downstream from the shock resides in the pick-up water-group ions. So a minor species in terms of number density plays the dominant rôle in the pressure balance.

With the evidence provided by the Vega-1 and -2 spacecraft of a broad (~10^5 km) and heavily structured bow shock at a distance of ~1.1–1.2×10^6 km (Gringauz *et al.* 1986), and from the results of the Suisei spacecraft (Mukai *et al.* 1986), the existence of a well developed bow shock at Halley is now well established.

Taking into consideration: (1) that the bow shock can be given by an equation of paraboloidal form with a flaring ratio ~2 (Mendis *et al.* 1986), and (2) the Giotto flyby geometry (the Giotto trajectory is at an angle of ~$107°$ to the Sun–comet Halley line) the estimate of the stand-off distance of the bow shock (that is, its distance from the nucleus in the sunward direction) is ~4×10^5 km, which corresponds to a total production rate of neutrals, $\dot{Q}_n \simeq 3.8 \times 10^{29}$ s^{-1} if the shock Mach number is 2 and $\dot{Q}_n \simeq 5.6 \times 10^{29}$ s^{-1} if the shock Mach number is 1.5 according to the model of Mendis *et al.* (1986).

On ICE, Bame *et al.* (1986) did not find evidence for a conventional bow shock but rather the beginning of a 'transition region' inside which the solar wind is heated, compressed, and slowed.

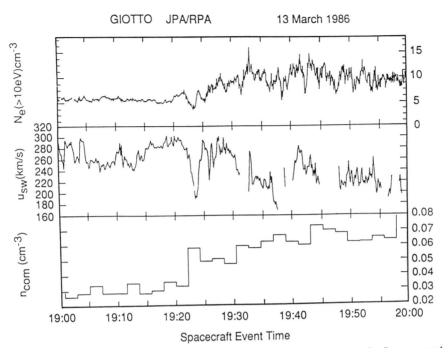

Fig. 4 — Electron density (top panel), solar wind proton speed (middle panel), and cometary ion density (bottom panel) for the hour on 13 March 1986 containing the comet Halley bow shock crossing, from Coates *et al.* (1987).

To compare the sizes of comet Halley and comet Giacobini–Zinner solar wind interaction, Fig. 5 shows a single curve which represents the locations of the bow shock in the case of Halley and the bow wave corresponding to the entry of the Giacobini–Zinner transition region with different distance scales. It is assumed that these two boundaries are given by an equation of paraboloidal form with the same flaring ratio of 2. Thus, the stand-off distance of the bow wave for Giacobini–Zinner is $\sim 6 \times 10^4$ km deduced by the ICE bow wave–nucleus distance of 1.3×10^5 km. The linear factor relating the solar wind–comet interaction regions of the two comets at the time each was observed is thus in the range 6–7, this difference being mainly due to a larger neutral production rate from the Halley nucleus.

Another difference in the comparison of the solar wind–comet interactions at Halley and Giacobini–Zinner is that these density fluctuations are much smaller at Halley. Just outside Halley's bow shock the density variations $\Delta n/\langle n\rangle$ do not exceed about 10% compared to variations of $\geqslant 100\%$ at Giacobini–Zinner. The great difference between the two comets

around the bow shock waves is shown in Fig. 6 (Anderson *et al.* 1987). In that figure the time scale has been made the same. Owing to the greatly different spacecraft–comet relative velocities the spatial scales are quite different, as indicated in the figure. Note also that the density changes at Halley are much more suggestive of a bow shock. The greater plasma turbulence at Giacobini–Zinner is due to the higher rate of injection of ions there because the bow shock is located at a cometocentric distance of $\sim 10^6$ km for comet Halley and $\sim 10^5$ km for comet Giacobini–Zinner, and therefore that the plasma turbulence associated with the solar wind–comet interaction is probably due to the strong phase space anisotropy associated with newly ionized molecules rather than the presence of the ions which formed much earlier and are now drifting by the spacecraft.

The lower level of fluctuations at comet Halley has allowed the bow shock and its structure to be identified, while at Giacobini–Zinner no clear shock signature was clear in the electron measurements (Bame *et al.* 1986). However, a more recent coordinated analysis of all the ICE data has confirmed the

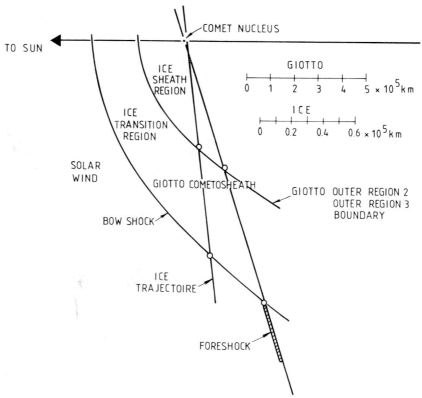

Fig. 5 — Comparison of the size of the comet Halley and comet Giacobini–Zinner solar-wind interaction regions. The bow shock, the Halley mysterious transition, and the Giacobini–Zinner transition–sheath boundary, and the distance scale appropriate to each comet are shown. The plane of projection is defined by the spacecraft trajectory and the Sun–comet direction.

existence of a weak shock at comet Giacobini–Zinner (Smith *et al.* 1986).

THE OUTER COMETOSHEATH

Between the bow shock and the magnetic pile-up boundary there is a large region, the outer cometosheath, where the 10 eV–30 keV electron density is much higher than in the foreshock or close to the comet, and which can be divided into three distinct regions called the 'outer regions' (Rème *et al.* 1987 b,c).

Outer region 1 extends from the shock to 8.5×10^5 km from the comet. In this region (Fig. 1) there is a large increase in the electron density with rapid and strong fluctuations on time scales of the order of one minute. The electron temperature and bulk velocity

also show fluctuations. The average temperature decreases toward the comet, whereas the average density tends to increase. In this region, the electron distribution function may fluctuate rapidly from quiet solar wind-like to nearly isotropic, which explains the variations of the heat flux ratio between 0.45 and 1.0. Some keV electrons are detected, but their fluxes decrease toward the comet. Inside this region, low-energy ions (protons) are detected in the ram direction, but their fluxes seem anticorrelated with the electron bulk velocity. The more the electrons are slowed, the more the solar wind protons are disturbed and seen in the ram direction.

Outer region 2 is crossed at distances between 8.5×10^5 and 5.5×10^5 km, and is characterized by larger electron densities, and by larger variations of electron density and temperature that occur on a slower time scale. The average electron bulk velocity is almost

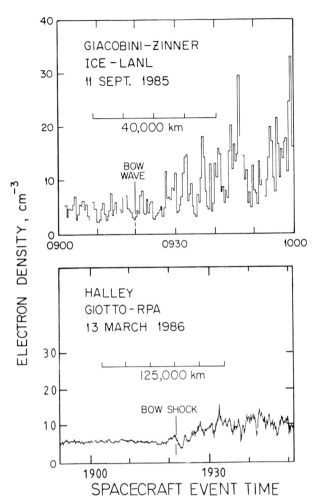

Fig. 6 — Plasma electron density at Giacobini–Zinner and Halley comet bow shocks from Anderson *et al.* (1987).

constant. The fluxes of the ions detected in the ram direction are anticorrelated with the electron density and temperature. But the main feature of this region is probably the detection of significant fluxes of keV electrons as seen in the plot of n_e (hot) in Fig. 1. These fluxes of 0.8–3.6 keV electrons detected in outer region 2 are not present after it, and are much smaller before it. The 3.9–30 keV electron density (not shown) stays at the background level. Except for the highest energy level, all the densities are maximum in this region. It is interesting to note that these 0.8–3.6 keV electron fluxes display a tendency to be magnetic field-aligned (Rème *et al.* 1987b). Owing to these

unexpected and presently not well understood high-energy aligned electron fluxes, this region is called the 'mystery region'.

At about 5.5×10^5 km from the comet, Giotto penetrates into outer region 3. The boundary between regions 2 and 3, or the 'mysterious transition', described in detail in d'Uston *et al.* (1987), occurs on a short time scale. This discontinuity separates the zone containing energetic electrons from outer region 3. Fig. 7 shows from bottom to top the electron bulk velocity, the electron temperature, the 40 eV–73 eV electron density, the 800–3600 eV electron density, and the 250–500 eV/q flux of ram ions during the traversal of this boundary around 2150 SCET. This transition is characterized by a decrease in electron bulk speed temperature and density, and an increase in the flux of ram ions. The parameters displayed here allow a determination of the width of the transition. It takes place between 2146:14 and 2156:47 SCET, thus giving a width of 45 000 km. Note that this transition corresponds also to a very important change in the behaviour of the solar wind and implanted ions detected by the JPA experiment. This is the time when the implanted ion distribution starts to divide into two components (Johnstone *et al.* 1986). This boundary is located roughly halfway between the shock position and the nucleus.

In the case of Giacobini–Zinner, Bame *et al.* (1986) identified two different broad regions behind the bow wave, and referred to them as the 'transition region' and the 'sheath'. According to their analysis the transition region is characterized by gradual increases in electron density and temperature and by decreasing flow speed. The sheath is the region in which the density and temperature decrease while the flow speed continues to decrease. The change is abrupt between the two regions.

Because of the similar behaviour observed by Giotto and ICE behind the bow shock waves of the two comets, Rème *et al.* (1986) compared the position of the abrupt change in the behaviour of the electron temperature at ~2150 UT during the Giotto Halley encounter with the position of the transition region-sheath boundary identified by Bame *et al.* (1986) at comet Giacobini–Zinner. To do this the linear scaling factor of 6.7 determined from the bow shock wave crossing locations at the two comets was used. Fig. 5

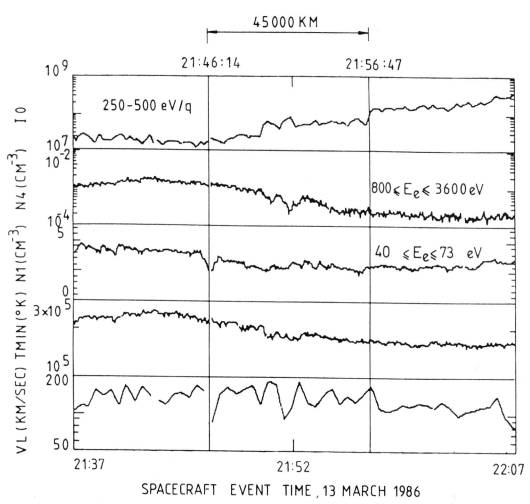

Fig. 7 — From bottom to top: electron flow speed, electron temperature, 40–73 eV electron density, 800–3600 eV electron density, and 250–500 eV/q ram ions at the comet Halley 'mysterious' boundary crossing, from d'Uston *et al.* (1987a).

shows the bow shock wave, the region 2-region 3 Halley boundary, the Giacobini–Zinner transition–sheath boundary, and the distance scale appropriate to each comet. The Halley and Giacobini–Zinner boundaries were constructed by fitting a paraboloid, with the same flaring factor as for the bow shock wave, from the two inbound boundary crossing points. The position of the abrupt change in electron temperature inside Halley cometosheath (marked 'Halley' in Fig. 5) lies remarkably close to the transition–sheath surface inferred from the ICE data at Giacobini–Zinner, implying that the two comets probably possess the same type of environmental structure, and that this structure could well be a

feature of all active comets at a distance of ~1 AU from the Sun. Such a structure has not been foreseen in comet–solar wind interaction models, and some theoretical work is clearly needed to interpret this behaviour.

Outer region 3 extends from 5.5×10^5 km to 1.35×10^5 km from the comet. In this region the average 10 eV–30 keV electron density is smaller than before and is relatively constant, and the electron temperature and bulk velocity decrease smoothly. The heat flux ratio is everywhere greater than 0.57, showing that the electron distribution functions are more isotropic than in the other outer regions. In fact the electron distributions become stable and monotonically cool. On the contrary, low-energy ions are increasingly detected in the ram direction, as seen in Fig. 1. Owing to the much smaller fluctuations of the electron properties in this region, it can be called a 'quiet region'. Here, the high-energy electron density is small (smaller than in all the regions encountered before, see Fig. 1).

THE INNER REGIONS

After traversal of these regions, the spacecraft enters the inner regions.

As observed from Giotto, in the ram direction, the first ions of cometary origin, belonging essentially to the water group, are detected at 1.5×10^5 km (Korth et al. 1986). At first the energy distribution of these ions is rather broad, indicating that their temperature is high. Slightly later, at 1.35×10^5 km from the nucleus, the Giotto magnetometer detects an abrupt increase in the magnetic field intensity, from $\sim 8\gamma$ to $\sim 30\gamma$ (Neubauer et al. 1986). The JPA experiment indicates a sudden decrease in the proton density at this time, and the RPA-COPERNIC electron detector shows a rapid change in the characteristics of the

electron plasma (d'Uston et al. 1987, 1989). At this moment the region of magnetic field pile-up begins. We note, however, that from the two Vega spacecraft, the magnetic field pile-up region appears to start around 4.5×10^5 km from the comet, well before, without a strong magnetic boundary (Galeev 1986). Owing to the short time scale of the variations at 1.35×10^5 km from the comet seen by Giotto, this boundary (the MPB: Magnetic Pile-up Boundary) is considered to be a magnetic structure, having a stand-off distance from the nucleus of $\sim 5 \times 10^4$ km. At the MPB the electron bulk velocity measured by RPA-COPERNIC tends to 0, that is, it is decreasing almost down to stagnation.

Fig. 8 shows from bottom to top the three electron temperatures given by the diagonal pressure tensor, the 10–40 eV electron density, and the 170–760 eV electron density during the MPB traversal period. Before the MPB, the three temperatures are tightly linked together meaning that the electron distribution function is isotropic. After the MPB, the parallel temperature increases while the perpendicular one decreases. This indicates that inside the MPB the electron distribution is cigar shaped with the long axis along the magnetic field direction. During the same time the ratio of minimum to maximum heat flux (not

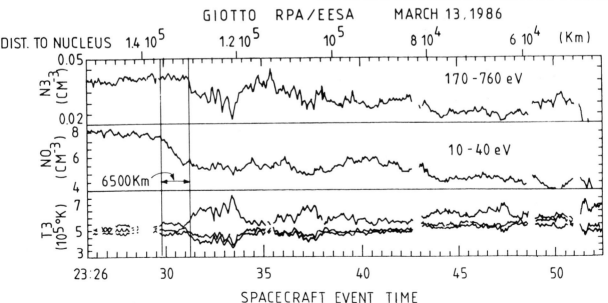

Fig. 8 — From bottom to top, three electron temperatures given by the diagonal pressure tensor, 10–40 eV electron density and 170–760 eV electron density around the comet Halley magnetic pile-up boundary.

shown) transported by the electrons parallel and antiparallel to the magnetic field, close to 1 before the MPB, drops suddenly to about 0.5 at this boundary. This indicates that the electrons leak along the field lines in one preferential direction (d'Uston *et al.* 1989). The 10–40 eV electron density begins to decrease around 2329:45 SCET, and this decrease continues until the abrupt decrease around 2331:20 SCET of the 170–760 eV electron density. The width of the boundary deduced from these electron measurements is $\sim 6\,500$ km, or more than 10 cometary iongyroradii in a 30γ field. After 2332 SCET, the suprathermal electron density, the 170–760 eV electron density, and the T3 temperatures undergo a series of low-frequency oscillations, as shown in Fig. 8. The oscillation frequency is of the order of the gyrofrequency for 18 amu ions in a 30γ magnetic field ($\sim 2.5 \times 10^{-2}$ Hz). Around 2343 SCET there is a region, at $\sim 8.2 \times 10^4$ km from the comet where the electron density decreases. The JPA experiment indicates that the solar wind flow terminates around 2348 SCET (Johnstone *et al.* 1986).

Several models predicted the formation of a magnetic barrier starting at a collisionopause where the strong deceleration of the solar wind is expected to commence, because of momentum transfer collisions with the outflowing neutrals. Mendis *et al.* (1989) has therefore identified this boundary with the so-called collisionopause, but the measurements have revealed that this transition is very sharp, of the order of 10^4 km.

From the MPB to the time of the encounter, Giotto is in the inner regions; there is a fast increase in the cometary ion fluxes detected in the ram direction. Up to 4.5×10^4 km from the nucleus the electron density decreases. At this distance from the comet another sharp density decrease is detected. After, the 10 eV–30 keV electron density has a value as low as 2 cm^{-3}. Here the ions in the ram direction are cold, and appear in the form of peaks in the E/q spectra. It is in this region that the magnetic field reaches its maximum intensity of $\sim 60\gamma$, around 2359 SCET, i.e. at about 16 000 km from the comet (Neubauer *et al.* 1986).

It is interesting to remark that all along the cometosheath the density of the ram ions as measured by RPA2-PICCA does not vary in the same way, and is almost anticorrelated with the 10 eV–30 keV electron density and increases strongly when Giotto approaches the comet. Owing to the neutrality of the plasma, it can be concluded that in the inner coma most of the electrons have an energy much lower than the threshold of RPA1-EESA, well below 10 eV. This is the same as in the case of ICE measurements at comet Giacobini-Zinner, and Meyer-Vernet *et al.* (1986) measured, by thermal noise spectroscopy, a high density of electrons, having a temperature slightly above 1 eV near the tail axis.

This ionopause traversal at about 4 700 km from the comet is marked by the entry into a zero magnetic field cavity (Neubauer *et al.* 1986), by very well mass resolved positive cold ions (Balsiger *et al.* 1986, Krankowsky *et al.* 1986) and by well mass resolved high fluxes of negative cold ions detected by RPA1-EESA in the ram direction.

All along these inner regions, keV electron fluxes are very small, close to background.

Fig. 9 summarizes RPA-COPERNIC measurements near comet Halley. The trajectory of the Giotto spacecraft makes an angle of 107° with the Sun–comet line and intersects the bow shock and the other boundaries on the flank. For example, the bow shock is intersected at a distance along the spacecraft track of 1.15×10^6 km. As in Fig. 5, the shock surface is assumed to be a paraboloid with a flaring factor of 2. In Fig. 9 all the other boundaries are assumed to be paraboloids with the same flaring factor.

CONCLUSION

The global nature of the solar wind interaction with cometary plasmas is, therefore, in general agreement with theoretical predictions, but it is more complex. A well defined bow shock appears to be formed with features different for Halley and Giacobini–Zinner. Calculations indicate that the positions of the main transitions at Halley (shock, ionopause) are consistent with the total gas production rate of the comet and the associated mass loading and neutral drag on the inflowing solar wind. There are, however, several other sharp features, whose origins, whether temporal or spatial, are unknown and which have not been identified with any obvious physical process, although the neutral interaction with inflowing contaminated solar wind plasma may play a rôle in all these

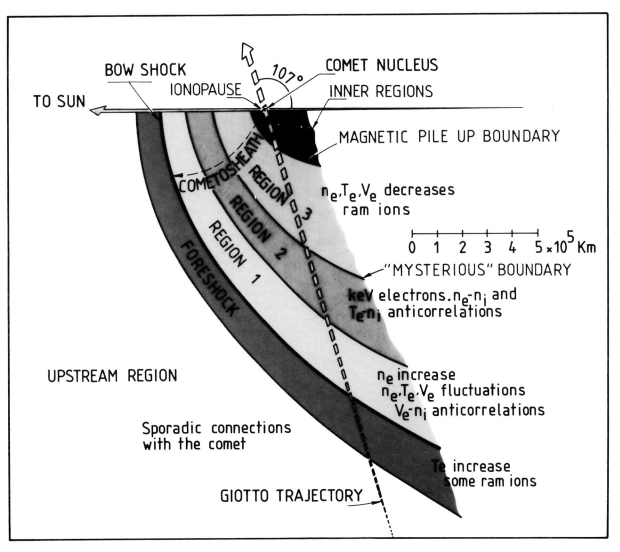

Fig. 9 — Main regions and boundaries deduced from the RPA-COPERNIC experiment during the Giotto flyby of comet Halley.

transitions. Important features are also the detection of accelerated keV electrons upstream from comet Halley, in the well-defined electron foreshock and in cometosheath regions 1 and 2 with much higher intensities in region 2 and fluctuations in the electron density ascribed to strong plasma turbulence. Near the comets, plasma is very cold, with electron temperatures well below 10 eV.

ACKNOWLEDGEMENTS

The success of the RPA COPERNIC experiment aboard Giotto is the result of a very efficient cooperation between C.E.S.R. Toulouse, MPae Lindau, S.S.L. Berkeley, and E.E.C.S. San Diego. The work was supported by C.N.E.S. under Grant No. 1212, by NASA Contract NASW-3575 and by the Max-Planck Gesellschaft für Förderung der Wissenschaften and by the Bundesminister für Forschung und Technologie under Grant No. 01 0F 052.

REFERENCES

Anderson, K.A., Lin, R.P., Martel, F., Lin, C.S., Parks, G.K., Rème, H. (1979) Thin sheets of energetic electrons upstream from the Earth's bow shock, *Geophys. Res. Lett.* **6** 401

Anderson, K.A., Carlson, C.W., Curtis, D.W., Lin, R.P., Rème, H., Sauvaud, J.A., d'Uston, C., Korth, A., Richter, A.K., & Mendis, D.A. (1987) The upstream region, foreshock and bow shock wave at Halley's Comet from plasma electron measurements, *Astron. and Astrophysics* **187** 290–292

Apáthy, I., Remizov, A.P., Gringauz, K.I., Balebanov, V.M., Szemerey, I., Szendrö, S., Gombosi, T., Klimenko, I.N., Verigin, M.I., Keppler, E., & Richter, A.K. (1986) PLAZMAG-1 experiment solar wind measurements during the closest approach to Comet Giacobini–Zinner by the ICE probe and to Comet Halley by the Giotto and Suisei spacecraft, *ESA SP 250* **I** 65

Balsiger, H., Altwegg, K., Bühler, F., Geiss, J., Ghielmetti, A.G., Goldstein, B.E., Goldstein, R., Huntress, W.T., Ip, W.H., Lazarus, A.J., Meier, A., Neugebauer, M., Rettenmund, Y., Rosenbauer, H., Schwenn, R., Sharp, R.D., Shelley, E.G., Ungstrup, E., & Young, D.T. (1986) Ion composition and dynamics at Comet Halley, *Nature* **321** 330

Bame, S.J., Anderson, R.C., Asbridge, J.R., Baker, D.N., Feldman, W.C., Fuselier, S.A., Gosling, J.T., McComas, D.J., Thomsen, M.F., Young R.D., & Zwickl, D.T. (1986) The Comet Giacobini–Zinner: Plasma description, *Science* **232** 356

Biermann, L., Brosowski, B. & Schmidt, H.U. (1967) The interaction of the solar wind with a comet, *Solar Phys.* **1** 254

Brandt, J.C. & Mendis, D.A. (1979) The interaction of the solar wind with comets, *Solar system plasma physics*, edited by North-Holland Publishing Company, Vol II 253

Coates, A.J, Lin, R.P., Wilken, B., Amata, E., Anderson, K.A., Borg, H., Bryant, D.A., Carlson, C.W., Curtis, D.W., Formisano, V., Jockers, K., Johnstone, A.D., Korth, A., Mendis, D.A., Rème, H., Richter, A.K., Rosenbauer, H., Sauvaud, J.A., Stüdemann, W., Thomsen, M.F., d'Uston, C., & Winningham, J.D. (1987) Giotto measurements of cometary and solar wind plasma at the Comet Halley bow shock, *Nature* **327** 489

Festou, M.C., Feldman, P.D., A'Hearn, M.F., Arpigny, C., Cosmovici, C.B., Danks, A.C., McFadden, L.A., Gilmozzi, R., Patriarchi, P., Tozzi, G.P., Wallis, M.K., & Weaver, H.A. (1986) IUE observations of Comet Halley during the VeGa and Giotto encounters *Nature* **321** 361

Galeev, A.A. (1986) Theory and observations of solar wind/cometary plasma interaction processes, *ESA SP 250* **I** 3

Galeev, A.A., Cravens, T.E., & Gombosi, T.I. (1985) Solar wind stagnation near comets, *Astrophysics. J.* **289** 807

Gringauz, K.I., Gombosi, T.I., Remizov, A.P., Apáthy, I., Szemerey, I., Verigin, M.I., Denchikova, L.I., Dyachkov, A.V., Keppler, E., Klimenko, I.N., Richter, A.K., Somogyi, A.J., Szegö, K., Tátrallyay, M., Varga, A. &, Vladimirova, G.A. (1986) First *in situ* plasma and neutral gas measurements at Comet Halley, *Nature* **321** 282

Johnstone, A., Coates, A., Kellock, S., Wilken, B., Jockers, K., Rosenbauer, H., Studemann, W., Weiss, W., Formisano, V., Amata, E., Cerulli-Irelli, P., Dobrowolny, N., Terenzi, R., Egidi, A., Borg, H., Hultqvist, B., Winningham, J., Gurgiolo, C., Bryant, D., Edwards, T., Feldman, W., Thomsen, M., Wallis, M.K., Biermann, L., Schmidt, H., Lust, R., Haerendel, G., & Paschmann, G., (1986) Ion flow at comet Halley, *Nature* **321** 344

Korth, A., Richter, A.K., Anderson, K.A., Carlson, C.W., A., Curtis D.W., Lin, R.P., Rème, H., Sauvaud, J.A., d'Uston, C., Cotin, F.F., Cros, A., & Mendis, D.A. (1986) Cometary ion observations at and within the cometopause region of Comet Halley, *Adv. Space Res.* **5** 221

Korth, A., Richter, A.K., Loidl, A., Güttler, W., Anderson, K.A., Carlson, C.W., Curtis, D.W., Lin, R.P., Rème, H., Cotin, F., Cros, A., Médale, J.L., Sauvaud, J.A., d'Uston, C., & Mendis, D.A. (1987) The heavy ion analyzer Picca for the Comet Halley flyby with Giotto, *J. Phys. E: Sci. Instrum.* **20** 787

Krankowsky, D., Lämmerzahl, P., Herrwerth, I., Woweries, J., Eberhardt, P., Dolder, U., Herrmann, U., Schulte, W., Berthelier, J.J., Illiano, J.M., Hodges, R.R., & Hoffman, J.H. (1986) *In situ* gas and ion measurements at Comet Halley, *Nature* **321** 326

Krasnopolsky, V.A., Gogoshev, M., Moreels, G., Moroz, V.I., Krysko, A.A., Gogosheva, Ts, Palazov, K., Sargoichev, S., Clairemidi, J., Vincent, M., Bertaux, J.L., Blamont, J.E., Troshin, V.S., & Valníček, B. (1986) Spectroscopy study of Comet Halley by the VeGa 2 three-channel spectrometer, *Nature* **321** 269

Mendis, D.A., Smith, E.J., Tsurutani, B.T., Slavin, J.A., Jones, D.E., & Siscoe, G.L. (1986) Comet–solar wind interaction: Dynamical length scales and comets, *Geophys. Res. Lett.* **13** 239

Mendis, D.A., Flammer, K.R., Rème, H., Sauvaud, J.A., d'Uston, C., Cotin, F., Cros, A., Anderson, K.A., Carlson, C.W., Curtis, D.W., Larson, D.E., Lin, R.P., Mitchell, D.L., Korth, A., & Richter, A.K. (1989) On the global nature of the solar wind interaction with Comet Halley, *Ann. Geophys.* **2** 99

Meyer-Vernet, N., Couturier, P., Hoang, S., Perche, C., & Steinberg, J.L. (1986) Physical parameters for hot and cold electron populations in Comet Giacobini–Zinner with the ICE Radio experiment, *Geophys. Res. Lett.* **13** 279

Moreels, G., Gogoshev, M., Krasnopolsky, V.A., Clairemidi, J., Vincent, M., Parisot, J.P., Bertaux, J.L., Blamont, J.E., Festou, M.C., Gogosheva, Ts., Sargoichev, S., Palasov, K., Moroz, V.I., Krysko, A.A., & Vanysek, V. (1986) Near ultraviolet and visible spectrometry of Comet Halley from VeGa 2, *Nature* **321** 271

Mukai, T., Miyake, W., Terasawa, T., Kitayama, M., & Hirao, K. (1986) Plasma observation by Suisei of solar-wind interaction with Comet Halley, *Nature* **321** 299

Neubauer, F.M., Glassmeier, K.H., Pohl, M., Roeder, J., Acuña, M.H., Burlaga, L.F., Ness, N.F., Mussmann, G., Mariani, F., Wallis, M.K., Ungstrup, E., & Schmidt, H.U. (1986) First results from the Giotto magnetometer experiment at Comet Halley, *Nature* **321** 352

Neugebauer, M., Lazarus, A.J., Altwegg, K., Balsiger, H., Goldstein, B.E., Goldstein, R., Neubauer, F.M., Rosenbauer, H., Schwenn, R., Shelley, E.G., & Ungstrup, E. (1986) The pick-up of cometary protons by the solar-wind, *ESA SP 250* **I** 19

Oyama, K-i, Hirao, K., Hirano, T., Yumoto, K., & Saito, T. (1986) Was the solar wind decelerated by Comet Halley? *Nature* **321** 310

Rème, H., Sauvaud, J.A., d'Uston, C., Cotin, F., Cros, A., Anderson, K.A., Carlson, C.W., Curtis, D.W., Lin, R.P., Mendis, D.A., Korth, A., & Richter, A.K. (1986) Comet Halley–solar wind interaction from electron measurements aboard Giotto, *Nature* **321** 349

Rème, H., Cotin, F., Cros, A., F., Médale, J.L., Sauvaud, J.A., d'Uston, C., Anderson, K.A., Carlson, C.W., Curtis, D.W., Lin, R.P., Korth, A., Richter, A.K., Loidl, A., & Mendis, D.A.

Rème, H., Cotin, F., Cros, A., F., Médale, J.L., Sauvaud, J.A., d'Uston, C., Anderson, K.A., Carlson, C.W., Curtis, D.W., Lin, R.P., Korth, A., Richter, A.K., Loidl, A., & Mendis, D.A. (1987 a) The Giotto Electron Plasma Experiment, *J. Phys. E: Sci. Instrum.* **20** 721.

Rème, H., Sauvaud, J.A., d'Uston, C., Cros, A., Anderson, K.A., Carlson, C.W., Curtis, D.W., Lin, R.P., Korth, A., Richter, A.K., & Mendis, D.A. (1987 b) General features of the Comet Halley-solar wind interaction from plasma measurements, *Astron and Astrophys.* **187** 33

Riedler, W., Schwingenschuh, K., Ye. Yeroshenko, G., Styashkin, V.A., & Russell, C.T. (1986) Magnetic field observations in Comet Halley's coma, *Nature* **321** 288

Smith, E.J., Slavin, J.A., Bame, S.J., Thomsen, M.F., Cowley, S.W.H., Richardson, I.G., Hovestadt, D., Ipavich, F.M., Ogilvie, K.W., Coplan, M.A., Sanderson, T.R., Wenzel, K.P., Scarf, F.L., Vinas, A.F., & Scudder, J.D. (1986) Analysis of the Giacobini–Zinner bow wave, *ESA SP 250* **III** 461

Somogyi, A.J., Gringauz, K.I., Szegö, K., Szabó, L., Kozma, Gy., Remizov, A.P., Erö Jr., J., Klimenko, I.N., Szücs, I.T., Verigin, M.I., Windberg, J., Cravens, T.E., Dyachkov, A., Erdõs, G., Faragó, M., Gombosi, T.I., Kecskeméty, K., Keppler, E., Kovács Jr, T., Kondor, A., Logachev, Y.I., Lohonyai, L., Marsden, R., Redl, R., Richter, A.K., Stolpovskii, V.G., Szabó, J., Szentpétery, I., Szepesváry, A., Tátrallyay, M., Varga, A., Vladimirova, G.A., Wenzel, K.P., & Zarándy, A. (1986) First observations of energetic particles near Comet Halley, *Nature* **321** 285

Stewart, A.I.F., Combi, M.R., & Smyth, W.H. (1985) *Bull. Am. Astron. Soc.* **17** 686

d'Uston, C., Rème, H., Sauvaud, J.A., Cros, A., Anderson, K.A., Carlson, C.W., Curtis, D., Lin, R.P., Korth, A., Richter, A.K., & Mendis, D.A. (1987) Description of the main boundaries seen by the electron experiment inside the Comet Halley-solar wind interaction regions, *Astron. and Astrophys.* **187** 137

d'Uston C., Rème, H., Sauvaud, J.A., Carlson, C.W., Anderson, K.A., Curtis, D.W., Lin, R.P., Korth, A. & Mendis, D.A. (1989) Properties of plasma electrons in magnetic pile-up region of Comet Halley, *Ann. Geophys.* **191**

Interactions between the plasma tail of comet Halley and the solar wind magnetic field

Takao Saito

1 INTRODUCTION

The first national observatory in Japan was built in Asuka, Nara prefecture, in AD 675. According to Nihon-Shoki, a series of books on ancient Japanese history, this ancient observatory reported various astronomical and geophysical events; solar, lunar, and stellar eclipses, meteors, comets, aurorae, etc. Among them one of the most important records was an early report of comet Halley, observed in Japan on 7 September during its apparition in AD 684. Fig. 1 shows this part of the record, with the origin or meaning of each of the characters. Since that year, the comet at every apparition has been reported in Japan — for 13 centuries except for the AD 760 apparition. The absence of any record for AD 760 had four probable reasons: (1) there were no trained observers; (2) the editor of Shoku-Nihon-Gi, the next series of historical books, did not have an interest in astronomical events; (3) the comet appeared during the worst of the rainy season; and (4) the comet was unfavourably placed for observation from Japan. Reasons (1) and (2) are from K. Saito (1982). All 15 later apparitions have been recorded in Japan. At the 1910 apparition, Japan was fortunately on the dayside when the comet made its transit of the solar disk.

Before the 1985–1986 apparition, Japanese scientists decided to launch two comet Halley spacecraft; Sakigake and Suisei. The present author joined this project as the principle investigator of the magnetic field experiment on Sakigake.

Many new results, including the unexpectedly wide distribution of the water group ions around the coma (T. Saito *et al.* 1986b, Yumoto *et al.* 1986) were revealed from our experiment (see the review by Saito *et al.* 1987g). In the present paper the interactions of the plasma tail with the solar wind will be discussed, with particular regard to the most spectacular aspect of such interactions, namely the disconnection events (DE).

To study the solar wind-plasma tail interaction, we have to understand the characteristics of the solar wind. To understand further the structure of the solar wind, we have to understand the heliomagnetosphere. Therefore, simultaneously with the start of the Sakigake/IMF observation on 19 February 1985, the variation of the heliosphere was studied (T. Saito *et al.* 1986a) while we were waiting for the apparition of comet Halley.

In section 2 a brief description will be given of the magnetometer on board Sakigake. The general state of the heliosphere during the apparition will be summarized in section 3. Based on about 500 photographs of the comet, a catalogue of the main plasma tail disturbances will be presented in section 4, together with possible mechanisms for these disturbances. Among these, individual outstanding instances of such disturbances will be separately

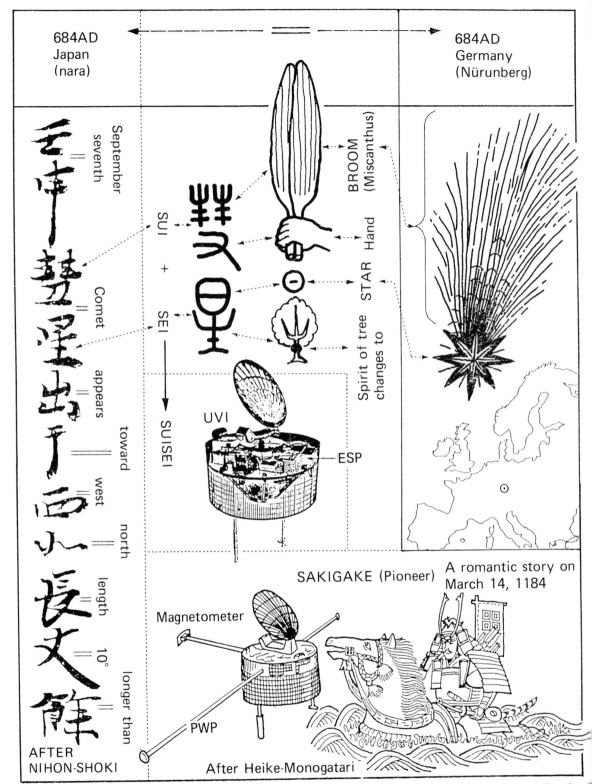

Fig. 1 — Simultaneous records of comet Halley in both Japan (left) and Germany (right) during its apparition in AD. 684. Meaning of each of characters registered in Nihon-shoki is described in the left column. Origin and meaning of Suisei are exhibited at the upper centre. A romantic story in 1184 that was registered in Heike-Monogatari is illustrated in relation to Sakigake at the bottom.

Fig. 2 — Sakigake, one of the twin Japanese spacecraft to comet Halley. The black rectangular part at the centre is the cover of the ring-core sensor of the magnetometer. Photo by ISAS.

Fig. 3 — Ricrestometer at the centre of the shielding room (cf. Fig. 4). This device was specially designed to measure sensitively the residual magnetism of spacecraft parts. Two orthogonal sets of ring-cores are fixed with 40 cm distance at the top of the right pole. The specimen is put at the top of the vertically rotating axis at the left. The three-component vectors of the residual magnetism of small specimen are obtained within 10–30 seconds. Three orthogonal sets of Helmholtz coils to simulate interplanetary space are seen on the inside wall of the 6 m diameter spherical room.

The basic experiments and the fundamental design of the ring-core magnetometer on board Sakigake were carried out by the project team from Tohoku University, Tohoku Institute of Technology, and Tokai University. The design was realized to the

discussed in relation to DE in sections 5 to 7. The findings will be discussed, and conclusions drawn, in section 8.

2 IMF EXPERIMENT OF SAKIGAKE

The Japanese-made magnetometer on board Sakigake (Fig. 2) consists of three orthogonal ring-core sensors, a pantograph-type boom antenna and an electronics section. A molybdenum-permalloy foil with 4 μm thickness and 3 mm width was wound 32 times around a ceramic bobbin 25 mm in diameter. The three sensors mounted on a Vespel block with an aluminium cover weigh 390 g. The 2 m aluminium boom was folded 10 times into a case and deployed by a d.c. motor and gearing on 19 February 1985.

Fig. 4 — Three-dimensional spherical magnetically shielding room specially designed to measure the offset field from the spacecraft body to our magnetometer. The outside diameter of the Permalloy sphere is 8 m. Sakigake is about to be measured in this room. The three-layered door at the left is shut during the measurement. Photo by ISAS.

flight model by Tohoku Metal Industries Ltd (for further details see T. Saito 1983).

To measure the interplanetary magnetic field (IMF) accurately, the offset bias field due to the magnetization of the spacecraft body had to be minimized at the magnetometer sensor at the tip of the 2 m boom antenna. Hence, residual magnetization of every part of the spacecraft body was measured by a newly-designed device, 'ricrestometer' (ring-core residual magnetometer), shown in Fig. 3 (T. Saito 1983).

A three-layered spherical shielding room with inside diameter 6 m was specially constructed for the testing of the magnetometer. The bias field due to the spacecraft was experimentally obtained in the shielding room as shown in Fig. 4 (T. Saito 1983).

Sakigake was launched on 9 January 1985, and it deployed the pantograph type boom antenna on 19 February 1985. The solar wind magnetic field has been continuously observed since that time to monitor the varying heliospheric conditions (T. Saito *et al.* 1987g).

3 THE HELIOSPHERIC CONDITION DURING THE APPARITION

It is essential for the study of solar wind–plasma tail interaction, especially of the relation between the

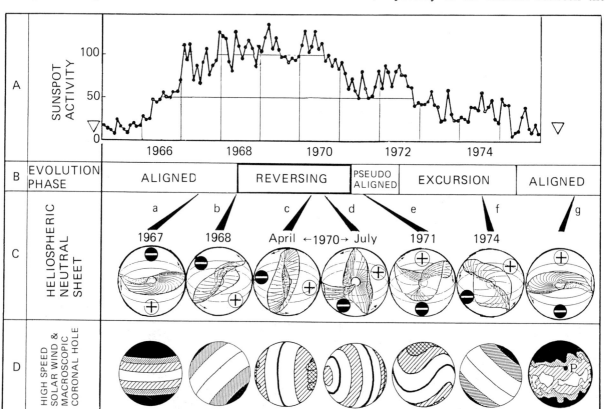

Fig. 5 — Solar cycle variation of the heliosphere.
(A) Relative sunspot number. (B) Four cyclic phases from the viewpoint of the heliomagnetospheric evolution. (C) Heliomagnetosphere as characterized by the neutral sheet. The inner and outer circles indicate the Sun and the unit sphere with one AU radius. The neutral line as calculated for the solar source surface is extended to the unit sphere by Parker spirals. Note that one neutral sheet is nearly horizontal in aligned phase, makes rotational reversal in reversing phase, through a saddle type in pseudo-aligned phase and an oblique type in excursion phase, and returns to aligned phase. (D) Distribution of high-speed solar wind and macroscopic coronal holes as expressed by the darker area. The speed is restricted by a strong looped coronal field in reversing phase. Coarse and schematic distribution is shown in a-f, while microstructure viewed from 10.6° Carrington longitude is shown, based on the IPS data in g. The point P indicates the calculated source of the solar wind stream that caused the kink in the plasma tail of comet Halley on 10–11 January 1986.

sector boundary and tail disconnection event, to understand the variation of the heliomagnetospheric structure. The general solar cycle variation of the three-dimensional heliomagnetosphere was proposed by T. Saito (1975) and T. Saito *et al.* (1978) and designated the two-hemisphere model. It is summarized in Fig. 5, which shows (A) variation in sunspot activity, (B) the heliomagnetospheric evolution phase in a typical solar cycle, (C) a representative heliosphere expressed by the magnetic neutral sheet, and (D) a schematic distribution of coronal holes and high-velocity stream sources on the solar source surface.

In Fig. 5(C), the inner and outer circles represent the solar source surface sphere with a radius of 2.5 $R_\odot$ and the unit sphere with radius of one AU, respectively. The lower ellipse represents the Earth's orbit on 8 March, while the upper one represents the orbit on 8 September. The computer display of the neutral sheet was based on the source surface neutral line obtained by the potential method. The term (heliomagnetospheric) neutral *sheet* is used in the present paper synonymously with the (interplanetary) sector boundary or the (heliomagnetospheric) current sheet (Schulz 1973).

As predicted by this model, the inclination of the neutral sheet decreased gradually from the beginning of the IMF measurement by Sakigake in the excursion phase to the present time in the aligned phase. Note that the heliosphere has only one neutral sheet basically throughout a solar cycle, especially during the aligned phase when comet Halley appeared. Sometimes an island type neutral line on the source surface and a subcone type neutral sheet in the heliosphere appear, as seen in Fig. 5(Cd). However, these are short-lived, and the appearance is limited to near sunspot maximum or the heliospheric reversing phase (T. Saito 1987). This point is important in any discussion of the DE-sector boundary relation during the aligned phase.

There are four methods (T. Saito *et al.* 1987d, T. Saito 1988b) to obtain observationally the form of the heliospheric neutral sheet; the potential method from a solar magnetogram; the scanning method from interplanetary or terrestrial data; the midline method or MBC method from K-corona data; and the minimum velocity method from interplanetary scintil-

lation (IPS) data. Generally, the heliolatitudinal range of undulation of the neutral sheet obtained by the potential method tends to be larger than those obtained by the other three methods. One of the reasons for the difference in the latitudinal range was attributed (T. Saito *et al.* 1987b) to the flattening effect due to large heliolongitudinal velocity gradients as observed by Sakigake and IPS (see Fig. 6).

In the minimum velocity method, the interplanetary neutral line is obtained by connecting the points of the velocity minimum for every heliolongitude on an IPS synoptic chart of the solar wind velocity. The magnetic neutral line on the source surface is generally compared with the heliodip equator. According to the two-hemisphere model and the IPS

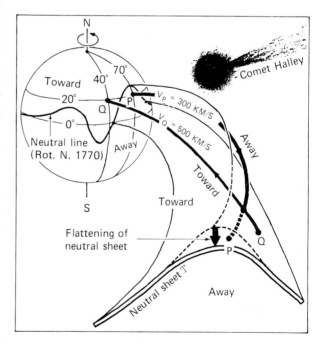

Fig. 6 — Flattening mechanism of the heliospheric neutral sheet. The neutral line on the source surface as calculated from the Stanford solar magnetograph is extended to the space by the Parker spirals forming the neutral sheet. The low-speed stream from P is surpassed by the high-speed stream from Q, causing a replacement of the 'away' IMF polarity by the 'toward' polarity at the point P′. The replacement explains the observed flattening of the neutral sheet. The speeds from P and Q are based on the IPS observation. An MHD computer simulation of the heliosphere (T. Saito *et al.* 1987h) reveals that the flattening occurs even inside the surpassing point.

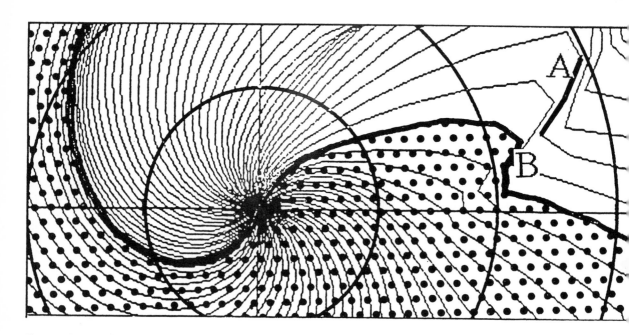

Fig. 7 — Computer simulation of the magnetic field structure of the heliosphere disturbed temporally by high-speed streams from solar flares. The dotted area indicates the field lines that are originally 'toward' on the source surface. The long thick curve shows the two-heliospheric sector boundaries. Note the apparent sector boundary near A. Stable bumps such as B are grown in the inner heliosphere owing to a co-rotating high-speed stream with a high longitudinal velocity gradient, and to co-rotation with the Sun (see section 5.5). The dots and the thick lines are added by the author on a part of the figures by Akasofu and Hakamada.

observation, the velocity gradient with respect to heliodip latitude is much larger during the aligned phase than in the reversing phase; in other words, there are many steep and local velocity gradients with respect to heliodip longitude in this phase, as schematically illustrated in Fig. 5(Dg). Taking into account the longitudinal resolution of the velocity measurements from Sakigake, the longitudinal velocity gradient must be much larger and more local than that estimated from the IPS chart (T. Saito *et al.* 1987g). This implies that low-velocity streams are surpassed by higher velocity streams in many longitudinal meridians within one AU during the aligned phase. This implication suggests further that there are many local sector polarity reversals (see A in Fig. 7) that indicate apparent solar sector boundaries, or local co-rotating bumps on the heliospheric neutral sheet (see B in Fig. 7).

These important characteristics will be used in the discussion of the sector boundary–plasma tail interaction in later sections.

4 WHOLE DISTURBANCE CATALOGUE OF THE PLASMA TAIL IN 1985–1986

4.1 Daily disturbances of the plasma tail

More than 500 photographs of comet Halley taken by Japanese professional and amateur astronomers were sent to our project team during the 1985/86 apparition. By careful examination of all the photographs, a catalogue of main disturbances of the plasma tail during the period December 1985 to May 1986 was made, as listed in Table 1, where the disturbances are classified on a daily basis into streamer, ray, condensation, helix, arcade, kink or disconnection event (DE). Typical examples of each kind of disturbance are shown in Fig. 7.9 of Saito (1989). The morphological definition of a DE is given in Brandt (1982). Although the DEs in Table 1 are based on photographs taken only by Japanese astronomers, more DEs are reported by Brandt (1989), based on photographs from worldwide. To understand the main day-to-day signatures of the plasma

Table 1 — Main disturbances in plasma tail

Date		S	R	C	H	A	K	DE1	DE2
Dec.	3		*						
	5		*						
	8		*						
	10		*	*			*		
	12		*			*	*		
	13		*			*			
	14		*						
	15		*						
	16		*						
	17		*						
	24		*						
	29								
	31	*	*	*	*	?	?		
Jan.	2		*	*	*				
	4		*		*				
	5		*			*			
	6		*	*					
	7		*	*	*	*			
	8		*	?	?		?		
	9		*		?				
	10		*		?	*			
	11		*	*	*				
	12	*	*						
	13		*	*	?				
	14	*							
	15	*							
	16	*							
	17	*							
	20	?							
Feb.	22		*		*	*			○
	25	?							
	28	*							
Mar.	1/2								○
	3	*							
	4	*	*						○
	5	*							
	6	*				?			
	7	*							
	8	*	*	*	*	*	*		○
	9		*	*	*	*			
	10		*		*	?	*		
	11	*	*		*				
	12		*		*				
	13		*		*	*			
	13/14								○
	14		*	*		*	?		
	15	*							
	16		*	*		?			
	17		*			*			
	19	*							
	20		*	*		*	?		○
	21		*	?					
	25								
Apr.	8								○
	9		*	*					
	11								○
	13		*						
	15		*	*					
	16		*			*			
	17		*	*					
	29	*							
May	6	*							
	7	*							
	8	*							
	15/16								○

S: streamer, R: ray, C: condensation, H: helix, A: arcade, K: kink, DE: disconnection event

tail throughout this apparition, 65 representative photographs out of more than 500 are displayed in Fig. 8. The date, time (UT), place where the photograph was taken, name of the person who provided it, and the reference book from which the photograph was reproduced, are given below each photograph and a key to the letter codes used is given in Table 2.

4.2 Models

4.2.1 Streamer, ray, condensation, helix, and arcade

Rays or condensations are considered to be formed when the plasma distribution near the contact surface is inhomogeneous (Schlosser et al. 1986) and/or the solar wind contains significant tangential and/or rotational discontinuities (T. Saito et al. 1987g). Where the distribution is not so inhomogeneous and the wind does not contain many irregularities, the tail forms a streamer without significant microstructures (e.g. 16 March 1986). Generally, rays tend to converge apparently not into the upstream region of the coma, but into the downstream, even though a significant DE does not occur afterwards. In a case study, 90% of the converged points distributed in the downstream region (Saito et al. 1987d). This implies that the IMF must be overdraped around the rear of the cometary ionopause as shown in Fig. 9. The overdraping model can explain not only the convergent point distribution, but also the dynamic pressure process, as will be discussed in section 5.

This model must be appropriate, since a similar tendency for overdraping was reported by Russell et al. (1982) and Marubashi et al. (1985) in the magnetosphere of Venus, which has no internal magnetic field.

For a discussion of the relationship of the helix with hydromagnetic waves and three-dimensional field-aligned current system, see T. Saito et al. (1987b). Possible mechanisms for the arcade will be presented in section 4.2.2(2), while those for a kink will be given in section 6.

4.2.2. Disconnection event (DE)

The well-known disconnection event (DE) model is that given by Niedner et al. (1978). This proposes that when a comet passes through a sector boundary (namely, the heliospheric neutral sheet), field-line

108

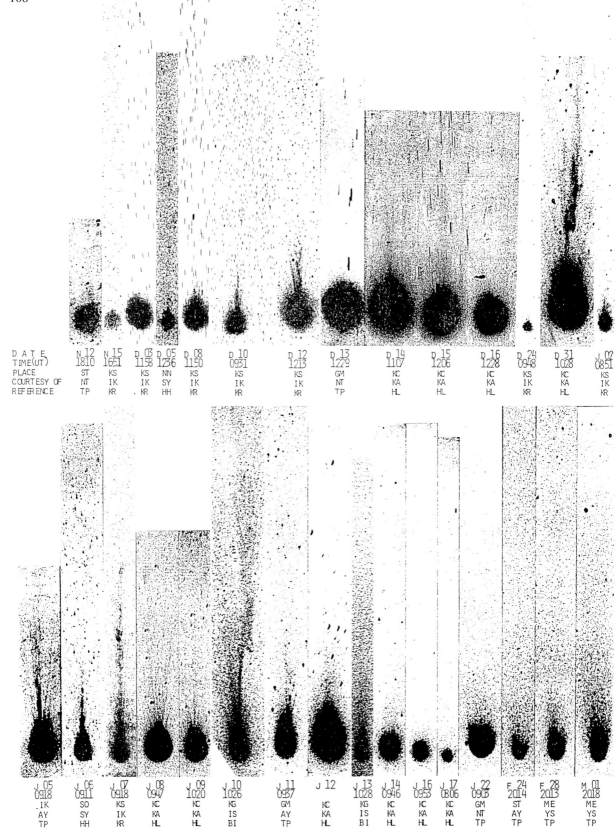

D A T E	N 12	N 15	D 03	D 05	D 08	D 10	D 12	D 13	D 14	D 15	D 16	D 24	D 31	J 02
TIME(UT)	1810	1651	1158	1236	1150	0931	1213	1229	1107	1206	1228	0948	1028	0851
PLACE	ST	KS	KS	NN	KS	KS	KS	GM	KC	KC	KC	KS	KC	KS
COURTESY OF	NT	IK	IK	SY	IK	IK	IK	NT	KA	KA	KA	IK	KA	IK
REFERENCE	TP	KR	KR	HH	KR	KR	KR	TP	HL	HL	HL	KR	HL	KR

	J 05	J 06	J 07	J 08	J 09	J 10	J 11	J 12	J 13	J 14	J 16	J 17	J 22	F 24	F 28	M 01
	0918	0911	0918	0947	1020	1026	0957		1028	0946	0953	0806	0903	2014	2013	2018
	IK	SO	KS	KC	KC	KG	GM	KC	KG	KC	KC	KC	GM	ST	ME	ME
	AY	SY	IK	KA	KA	IS	AY	KA	IS	KA	KA	KA	NT	AY	YS	YS
	TP	HH	KR	HL	HL	BI	TP	HL	BI	HL	HL	HL	TP	TP	TP	TP

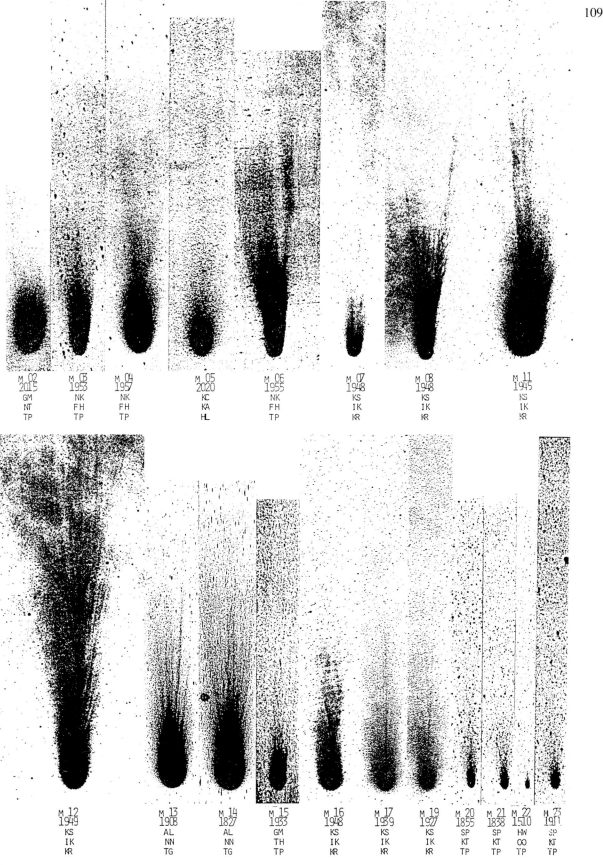

M 02
2015
GM
NT
TP

M 03
1953
NK
FH
TP

M 04
1957
NK
FH
TP

M 05
2020
KC
KA
HL

M 06
1955
NK
FH
TP

M 07
1948
KS
IK
KR

M 08
1948
KS
IK
KR

M 11
1945
KS
IK
KR

M 12
1949
KS
IK
KR

M 13
1908
AL
NN
TG

M 14
1827
AL
NN
TG

M 15
1953
GM
TH
TP

M 16
1948
KS
IK
KR

M 17
1959
KS
IK
KR

M 19
1927
KS
IK
KR

M 20
1855
SP
KT
TP

M 21
1838
SP
KT
TP

M 22
1510
HW
OO
TP

M 23
1911
SP
KT
TP

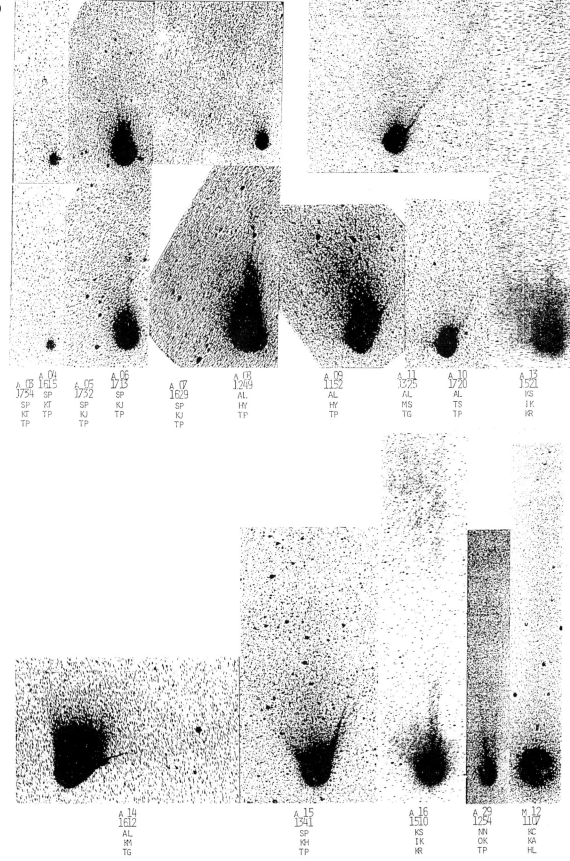

A 03
1734
SP
KT
TP

A 04
1615
SP
KT
TP

A 05
1732
SP
KJ
TP

A 06
1713
SP
KJ
TP

A 07
1629
SP
KJ
TP

A 08
1249
AL
HY
TP

A 09
1152
AL
HY
TP

A 11
1325
AL
MS
TG

A 10
1720
AL
TS
TP

A 13
1521
KS
IK
KR

A 14
1612
AL
KM
TG

A 15
1341
SP
KH
TP

A 16
1510
KS
IK
KR

A 29
1254
NN
OK
TP

M 12
1107
KC
KA
HL

Table 2 — Key to latter codes used in Fig. 8

Key

Place		Courtesy of	
KS	Kiso	NT	T. Niijima
ST	Saitama	IK	K. Ishida
KC	Kochi	KA	A. Kawazoe
IK	Ibaraki	AY	Y. Arai
GM	Gumma	YS	S. Yuasa
KG	Kagoshima	NO	Norikura
ME	Mie	KN	N. Kimura
NN	NAGANO	ST	T. Sudo
AL	Australia	TH	Tokai Univ. Comet Halley OBS Team
SP	Saipan	KT	T. Koseki
HW	Hawaii	OO	O. Okoshi
SO	Shizuoka	TJ	J. Kojima
NK	Norikura	SH	JH. Suzuki
		HY	Y. Hirose
	Reference	TS	S. Takahashi
		KM	M. Koishikawa
		KH	H. Kitamura
TP	Tokyo Univ. Press	OK	K. Ogiwara
TG	Tenmon Gaido	IS	I.S.A.S.
TB	Tokyo Astr. Obs. Bull.	SY	Y. Sakai
HL	Comet Halley '85/'86	MS	S. Murotani
HH	Hoshi to Hito	NN	N. Nishitani
BI	Bull. I.S.A.S.	FH	H. Fukushima
KR		KJ	

Fig. 8 — Daily representative photographs of comet Halley showing outline of the variation during the 1985/86 apparition. Pages 108, 109 and facing page: the photographs are restricted to those taken by Japanese astronomers. See Table 1 for the daily type of plasma tail structure and Table 2 for further details of the photographs reproduced here.

reconnection takes place near the dayside contact surface, causing disconnection of the old plasma tail from the coma, from which a new plasma tail grows out about once every week. This model will be described as the dayside-crossing model, because the most fundamental key words discriminating it from other models are 'dayside reconnection' and 'sector boundary crossing'.

The following aspects of the dayside crossing model should be noted:

(1) *Reconnection rate*

The rate (namely, the efficiency) of the reconnection must be dependent on the sheet–tail angle that is defined to be the angle between the heliomagnetospheric neutral sheet and the plasma tail axis before draping of the sheet around the cometary head. When the sheet–tail angle is sufficiently small, the efficiency (rate) of reconnection becomes small. Actually, when Sakigake made its closest approach to comet Halley, a sector boundary with nearly zero sheet–tail angle and inefficient reconnection were observed (for details see

section 7).

(2) *Symmetry of the reconnection*

Reconnection must take place symmetrically when the sheet–tail angle is 90°, but asymmetrically when the angle is sufficiently small. Since the IMF forms the Parker spiral, the statistical sheet–tail angle maximizes to $\sim 45°$ (at one AU) during the reversing phase (see Fig. 5(Cc)), but minimizes to $\sim 0°$ in the aligned phase (see Fig. 5(Ca)). Examples of the reversing and aligned phases are shown by the cases of comet Bennett in 1970 (see Fig. 7 of K. & T. Saito 1986), and of comet Halley in 1986 (Fig. 5 of K. & T. Saito 1986).

According to the computer simulation by Ogino et al. (1986) for a symmetric encounter (namely, for a sheet–tail angle of 90°), the plasma tail should separate into three plasmoids. Since no comet with three distinct plasma tails has apparently ever been observed, the difference between the simulation and observation was attributed to the asymmetric encounter of the sheet with the tail axis (Fig. 3 of T. Saito

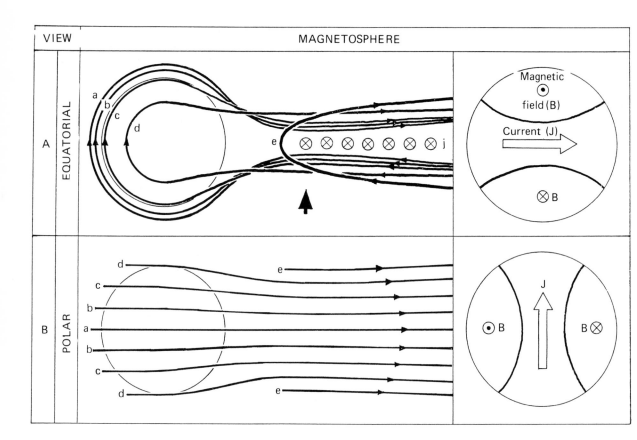

Fig. 9 — Three-dimensional overdraping model of general cometary magnetosphere. The draped field lines are considered to have slipped from a, via b, c, d, to e. The equatorial and polar views are so named conventionally in comparison with the Earth's magnetosphere. When the dynamic pressure of the solar wind increases suddenly, a reconnection of the field lines is considered to take place near the region indicated by the large arrow (see section 5). Note the arcade structure of the field line e. We are inclined to observe a comet in the polar view with the vertical current sheet when the IMF shows only the undistorted Parker spirals.

1988b or Fig. 7.34 of Saito 1989).

The draping of the sheet over the cometary head enlarges the sheet–tail angle. However, if we take the flux conservation law into consideration, the reconnection must be still asymmetric, and the reconnection rate must become smaller for sufficiently smaller sheet–tail angles. For more detailed discussion see T. Saito *et al.* (1987d).

Many scientists reported DEs during the 1985/86 apparition. When we take (1) and (2) above into consideration with the aligned phase (see section 3), the reported DEs seem to involve not only the dayside crossing process, but also the following five processes (see Fig. 7.33 of Saito 1988b for more detail):

(1) Dynamic pressure process (T. Saito *et al.* 1986c,

1987c), when the dynamic pressure of the solar wind increased suddenly (e.g. 31 December 1985, see section 5).

(2) Slipping IMF process (T. Saito *et al.* 1986c, 1987b), when high-density plasma along an arcade-shaped field line is swept away by the solar wind (e.g. 13 December 1985. See the field line e in Fig. 9).

(3) Rotating IMF process when the IMF direction rotates gradually as shown in Fig. 10. (e.g. 4 March 1986, 4 May 1910).

(4) Hone's process, when a comet crosses the sector boundary, and a plasmoid is generated by the nightside reconnection. (See section 5).

(5) Perspective effect on kink (T. Saito *et al.* 1987b).

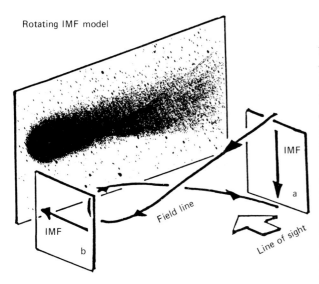

Fig. 10 — Rotating IMF model to explain a DE with a fan-shaped or fork-type plasma tail. The photograph was taken at the Kiso Astronomical Observatory on 16 March 1986. The cometary magnetosphere is considered to be changed gradually from the equatorial view to the polar view owing to the gradual change of the IMF from a to b. (As for the fork-type, see also comet Halley on 4 May 1910.)

5 DYNAMIC PRESSURE MODEL FOR THE 31 DECEMBER KNOT EVENT

5.1 Observation of comet Halley and the heliosphere

An especially distinct example of the DE-like knot was photographed on 23 different occasions from at least seven different locations in Japan and China. Fig. 11 displays the 23 photographs taken at the seven sites (Kohchi; Kiso, ϕ 105 cm Schmidt telescope; Beijing, ϕ 60 cm Schmidt; Kagoshima Space Center (KSC) ϕ 50 cm Schmidt; Saitama ϕ 30 cm reflector; Shirakawa ϕ 30 cm reflector; and Sendai, ϕ 20 cm reflector). In the figure, every panel shows an identical portion of the sky, and the knot is indicated by the solid triangles. As is clearly shown in the figure, the comet's orbital motion is upstream of the solar wind, while the knot moved downstream. The knot changed gradually to an arcade and was seemingly about to disconnect from the head at about 1050 UT. The disconnected tail is evident in the last photograph, taken at 1112 UT.

As was reported in our earlier paper (T. Saito *et al.* 1986d), comet Halley was far from the sector boundary on that day. Let us consider that the solar wind started from point P_H on the solar source surface ($R = 2.5\ R_\odot$) and arrived at comet Halley on December 31.4 UT, 1985. The following three conditions played positive roles in shifting the point P_H toward a higher northern heliomagnetic latitude. (1) The north pole of the solar rotational axis was tilted toward comet Halley by 77°.2. (2) The north pole of the equivalent centred dipole for the source-surface magnetic field was tilted toward comet Halley by 15°. (3) Comet Halley was 8°.7 above the ecliptic plane. Hence, P_H was about 18° north of the warped neutral sheet in heliographic latitude. (For a detailed description, see Saito *et al.* 1986c, 1987c, and Oki *et al.* 1987.)

These relations imply that comet Halley was far from the heliospheric neutral sheet when the DE-like knot event took place that night. Therefore, the evidence demands for this DE-like knot event some mechanism other than the dayside crossing model.

5.2 Observed flow of the knot

The observed distance X of the knot from the nucleus is read from the 23 photographs of the knot. The distance means here the component along the Sun–nucleus line, namely,

$$X = \Delta \sin \alpha \cdot \operatorname{cosec} (\beta - \alpha), \qquad (1)$$

where Δ, α and β mean the Earth–nucleus distance, the nucleus–Earth–knot angle, and the Earth–nucleus–Sun angle. The observed values of X are plotted against time, t, as the hollow circles in Fig. 12 (A, C).

The almost linearly arranged distribution of the 23 circles suggests a constant velocity of about 60 km s^{-1}, as Koishikawa & Kasahara (1986) calculated. The straight distribution at such small distances from 4–9×10^5 km seems to imply that a plasma cloud was ejected from the nucleus at about 6.6 h UT on the same day with a constant velocity of ~ 60 km s^{-1}. However, cometary particles are generally considered to leave the nucleus never with a velocity of ~ 60 km s^{-1}, but with an initial thermal velocity of ~ 1 km s^{-1}, and to be accelerated very gradually, as explained in

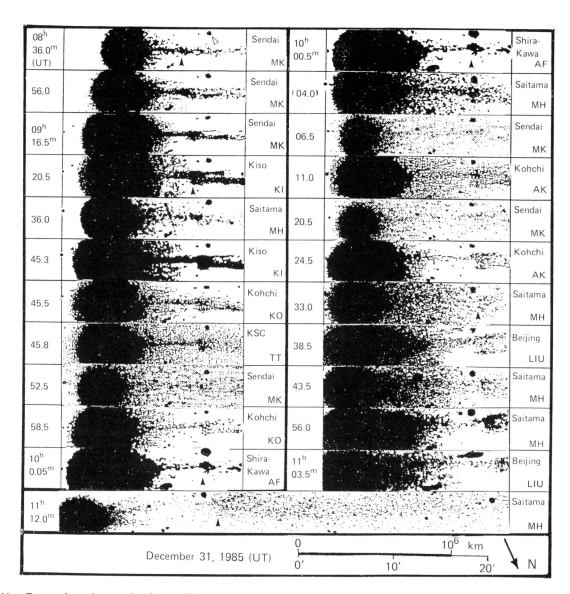

Fig. 11 — Twenty-three photographs of comet Halley taken in Japan and in China on 31 December 1985. The plasmoid indicated by the solid arrow shows the knot structure that was eventually disconnected from the head. Each panel represents the identical portion of the sky. The fixed star indicated by the hollow arrow is BD-2° 5734 (magnitude 9.5). Rightward motion of the knot and leftward orbital motion of the coma are clearly seen. The solid arrows are put on the left edge of the plasmoid, because it has a more distinct boundary owing to the sweeping-plasma effect of the reconnection, while the right edge is vague, owing to the plasma diffusion along the open field structure (cf. Fig. 33A of T. Saito 1988b). The asymmetry of the whole configuration of the plasmoid, which must be different from the more symmetrical plasmoid closed in the O-type field lines in the Earth's magnetotail during a substorm, can be seen more clearly in Fig. 2 of Saito *et al.* (1986c).

the following section. If we call this an ordinary flow, the observed flow velocities of many microstructures in the plasma tails of comet Bennett and comet Halley are very well expressed by this ordinary flow (Celnik 1986, Minami & White 1986, Tomita *et al.* 1987).

In the case of the outstanding kink event of comet Halley on 10–11 January 1986 (see the next section), for example, the ordinary flow showed much smaller velocities (< 1.5 km s^{-1}) than in our case of the 31 December knot event (~ 60 km s^{-1}) over the same

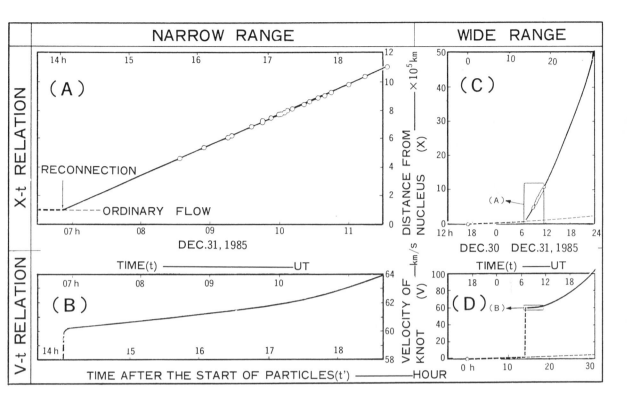

Fig. 12 — (A) X–t relation in narrow and (C) wide time ranges, where X and t are the nucleus–knot distance and Universal Time, respectively. The hollow circles represent the observed values. The broken curves represent the ordinary flow, and the solid curves represent the best curves of the accelerated flow. The acceleration is considered to be generated by the tail field reconnection. (B) V–t relation in narrow and (D) wide time ranges, respectively, where V is the flow velocity of the knot with respect to the nucleus. The curves are obtained by differentiating the curves (A) and (C) against time. Abruptly increased velocity at about 0700 is considered to be due to the tail field reconnection (Saito *et al.* 1987f).

range of distance (~ 4–9×10^5 km).

Consequently, we have to explain, for the knot event, the following three observational peculiarities: (1) the nearly constant and (2) higher velocities (~ 60 km s^{-1}) in comparison with the ordinary flow, and (3) the wide range at small distances (~ 4–9×10^5 km). All three will be explained in section 5.4 by the dynamic pressure model which we proposed earlier (Saito *et al.* 1986c, d). Before the explanation in section 5.4, let us compare quantitatively our observation with the ordinary flow.

5.3 Explanation of the ordinary flow

The ordinary flow of the cometary plasma was proposed by Minami & White (1986) as follows. The plasma in the plasma tail with an externally applied magnetic field perpendicular to the solar wind

velocity, v, is accelerated by the $j \times B$ force, where j is the current flow across the plasma tail.

$$\frac{m_i \, N_c \, V_T \, dV}{V \, dT} = \frac{\sigma_1 \, \sigma_2 (V_o \, V) \, B^2}{\sigma_1 + \sigma_2} \qquad (2)$$

using a molecular mass m_i, comet plasma density in the coma N_c, the initial thermal velocity V_T, the solar wind velocity V_o, conductivity of the solar wind plasma σ_1, and conductivity of the cometary plasma σ_2. By assuming the magnetic field distribution in the tail as,

$$B = B_o[\tanh(X/X_o)]1/2, \qquad (3)$$

equation (2) becomes,

$$V = [V_o\{\cosh(X/X_o) - 1\} + V_T]/\cosh(X/X_o), \quad (4)$$

where X_o is the characteristic length for the acceleration as,

$$X_o = m_i N_c V_T (\sigma_1 + \sigma_2)/B_o^2 \, \sigma_1 \, \sigma_2 \qquad (5)$$

Then the $X-t$ relation becomes:

$$t = \frac{\cosh(X/X_o) \, dX}{V_o\{\cosh(X/X_o) - 1\} + V_T} \qquad (6)$$

The dashed line in Fig. 12(A) indicates a case when the plasma left the nucleus with $V_T = 1$ km s^{-1} at 0150 UT on 30 December 1985. As seen in the figure, the velocity of the ordinarily accelerated case (dashed line) is distinctly less than the velocity expected from the observations (hollow circles).

We shall explain the observed values by means of dynamic pressure model in the next section.

5.4 Dynamic pressure model for the knot event

T. Saito *et al.* (1986c, d) proposed the dynamic pressure model wherein the dynamic pressure of the solar wind increases suddenly, and a field-line reconnection takes place near the triangular mark in the plasma tail in Fig. 9, causing a plasmoid-like the knot on 31 December 1985. It could be compared with the plasmoid in the Earth's magnetotail during the SC-triggered auroral substorm (T. Saito *et al.* 1987c).

The magnetic field lines around the plasmoid are considered to be closed like the letter O in the case of the Earth's magnetotail (Fig. 7.33D of Saito 1988b), while opened like the letter U in this knot event (Fig 7.33A of T. Saito 1988b) as inferred from the arcade-shaped rays, which are evident at 0920.5 UT or 1057.5 UT, for example, in Fig. 11. The marked difference between the O-type and the U-type provides an essential difference between the closed magnetosphere of planets having an intrinsic magnetic field (the Earth, for example) and the open magnetosphere of planets or other bodies having no intrinsic magnetic field (the planet Venus and the cometary nucleus, for example). Hence, this knot event must be important also from the viewpoint of comparative magnetospheres.

Ogino *et al.* (1986) and Ugai (1987) carried out various computer simulations. They showed several cases evidently confirming the dynamic pressure model (T. Saito *et al.* 1987c, f). Russell *et al.* (1986) also proposed that the plasmoid is generated in a cometary

magnetosphere by the sudden pressure increase as well as by an encounter of a shock front and a comet. Since the dynamic pressure model has been confirmed from the viewpoints of both observed characteristics and computer simulation in these ways, let us compare the model quantitatively with the observed flow data in Fig. 12(A).

If reconnection of the tail field lines takes place as is expected from the dynamic pressure model, the tail field lines through the reconnection point at $X = X_R$ move rapidly, so that the sharp angle between the two reconnected lines changes to a more and more blunt angle. The rapid movement of the field lines is associated with a rapid movement of the ambient plasma, forming the sweepings, which must be observed as a plasmoid or a knot.

Then the $V-X$ relation in equation (4) and the $X-t$ relation in equation (5) must be expressed by changing V_T in these equations to V_R as

$$V = [V\{\cosh(X/X_o) - 1\} + V_R]/\cosh(X/X_o), \qquad (7)$$

and,

$$t = \frac{\cosh(X/X_o) \, dX}{V_o\{\cosh(X/X_o) - 1\} + V_R} + t_R \qquad (8)$$

where V_R is the plasma velocity secondarily gained by the reconnection, X_R is the distance of the reconnection point from the nucleus, and t_R is the time of the reconnection as expressed by

$$t = \frac{\cosh(X/X_o) \, dX}{V_o\{\cosh(X/X_o) - 1\} + V_T} \qquad (9)$$

Hence, equations (4) and (6) must be used for the range $0 < X < X_R$, and $0 < t < t_R$, and equations (7) and (8) for the range $X > X_R$, and $t > t_R$.

The solid line in Fig. 12 (A, C) indicates the best-fit curve of (8) for the observation. Note that the solid line is not an exactly straight line, but is slightly curved, and that the expected curve coincides quite well with the observed points.

Hence, we may consider in Fig. 12 from the dynamic pressure model that initially the cometary plasma moved slowly as the ordinary flow along the dashed curve, and secondarily moved rapidly along the solid curve, owing to the field line reconnection. The reconnection point must be located at $X_R \lesssim 3 \times 10^5$ km, because a distinct plasmoid (knot) was already formed at $X \approx 4 \times 10^5$ km. The point must also be located at $X_R \geq 10^5$ km, because the reconnection seems not to occur at such a small distance when the result from the computer simulation (Ugai 1987) is taken into consideration. Therefore the turning point from the dashed curve of the ordinary flow to the solid curve of the accelerated flow may lie within the range of distance $10^5 \leq X \leq 3 \times 10^5$ km.

Figs 12(B) and (D) display the V–t curves for equation (7) expected from Figs 12(A) and (C). The impulsive acceleration due to the reconnection is expressed in Figs 12(B) and (D) with the sudden increase of V at $X = \sim 4$–9×10^5 km. The three observed peculiarities that were pointed out in section 5.2 can be explained by the dynamic pressure model as expressed in Fig. 12.

5.5 Discussion and conclusion

One may wonder if a large-scale sector boundary is not required to detach a tail in the dayside crossing model; a local, 0.5–1.0 day magnetic polarity reversal in the IMF can also suffice, and such small-scale reversals can occur well away from the neutral sheet. Our opinion is as follows.

Generally, we have four main methods to obtain the heliospheric neutral sheet (or the sector boundary, see section 3). The undulation amplitude of the neutral sheet obtained by the potential method tends to be generally larger than those obtained by the other three methods. Apart from possible reasons for the difference (see section 3), let us consider the local polarity reversals by using the potential data, since the potential data are very frequently used by many scientists.

The local reversal of the sector polarity on the source surface forms, in interplanetary space, a subneutral sheet, which has been studied by K. Saito (1982) and T. Saito (1988a) under the designations 'island' (on the solar synoptic chart) or 'subcone neutral sheet' (in interplanetary space). Such small islands appeared actually on the solar source surface as a result of the potential field calculation, e.g., once in 1987, seven times in 1979, five times in 1980, and once in 1981 (see Hoeksema et al. (1983)). It is evident from the data that the islands appear only during the sunspot maximum phase, when the tilt angle of the main neutral sheet is large, and never appear near the

minimum phase, when the tilt angle is small and the aligned dipole component is large (Saito et al. 1987a) as in the case of the 1985/86 apparition of comet Halley.

Next, we examine another case of local polarity reversal supposing two points (P, Q) on the solar source surface: $P(B_P, L_P)$ and $Q(B_Q, L_Q)$, where B and L mean heliographic latitude and Carrington longitude of the two points, respectively. Let us consider that $L_Q = B_Q$, $L_P = L_Q + 10°$, and that the velocities of the solar wind flowing out from P and from Q are V_P and V_Q, respectively. The magnetic fields at both P and Q are assumed to be radially inward (namely, both P and Q are far from the neutral sheet). If $V_P = V_Q$, the sector angles at P′ and Q′ at 1AU, to which the solar wind from P and Q attains, respectively, must be $\phi_P = \phi_Q$ $\sim 315°$, which means the same 'toward' polarity.

However, if the velocity changes linearly from $V_P = 400$ km s^{-1} to $V_Q = 300$ km s^{-1}, the sector angle in the region between P′ and Q′ is $\phi_{PQ} \sim 210°$. According to the definition of the sector polarity, the angle 210°, is 'away', even though the angle is not 135°, the case of the perfect 'away' polarity (see A in Fig. 7). When a comet traverses near this region, it follows the local polarity from 'toward', then 'away' for about one day, and back to 'toward' again. However, any drastic disturbance of the plasma tail would not occur at the two sector boundary crossings, because no heliomagnetospheric neutral sheet is there. Such 'longitudinally' large velocity gradients were in fact frequently observed by both Sakigake and the IPS telescope (Kojima et al. 1987).

The position of the source point, P_H, of the solar wind on the solar source surface to comet Halley on 1987 December 31.4, was estimated to be $B_C = 14.0°$ and $L_C = 159.0°$, where B_C and L_C are heliolatitude and Carrington longitude of P_H, respectively (T. Saito et al. 1986c). A comparison of this point with the synoptic chart of the solar wind velocity obtained by the IPS observation (Kojima et al. 1987) revealed good agreement with the dynamic pressure model.

The positions of the 121 daily source points are compared with the heliospheric neutral sheet from 1 December 1985 to 31 March 1986, where the 'source point' means the position on the source surface from which the solar wind streams out to comet Halley (see Fig. 19 of T. Saito 1988b). Owing to the flattening effect of the neutral sheet (cf. Fig. 6), the comparison is made with the heliospheric neutral sheet derived not from the potential method, but from the IPS method.

We may conclude that the previous dynamic pressure model for the 31 December knot event of comet Halley is substantiated by observed characteristics, computer simulation, and model calculation.

6 WIND INTRUSION MODEL FOR THE JANUARY 10–11 KINK EVENT

6.1 Data and display

Among the eight main kink events which occurred during the 1985/86 apparition, the event on 10–11 January 1986 (Table 1) is concluded to be the most magnificent. Twenty-four photographs of this kink were collected worldwide. The best eleven photographs out of the 24 were exhibited at the 20th ESLAB Symposium in Heidelberg (T. Saito *et al.* 1986e). Analyses of this spectacular tail disturbance were published by many researchers in Heidelberg (described variously as 'kink' or 'DE'): Kubacek *et al.* (1986), Shkodrov *et al.* (1986), Niedner & Schwingenschuh (1986), Brandt & Niedner (1986) and Caron *et al.* (1986). The time range reported for a DE seems to be a little different from that when the kink was observed (for a more detailed discussion on the kink and the DE, see T. Saito 1988b). In the present paper, this plasma tail disturbance will be studied from the viewpoint of the kink event in the following way.

The plasma tail with the kink on the eleven photographs were transcribed onto a copy of the *Palomar Sky Survey Atlas*. The kink transcribed in this way on the eleven panels with the identical portion of the sky is exhibited in Fig. 13 in order of the observation time. The place and the time of the observation are noted on the sides of the panels. Figs. 14(A) and (C) display the movement of the kink on a time (t) versus distance (X) diagram, where X is the distance of the kink from the nucleus, which is located by the Yeomans' calculation. The distance, X, is expressed by the distance along the Sun-comet line, taking into account the relative position relative to the comet, the Sun, and the Earth. The observed distance, X, of the kink against the time, t, in UT is expressed by the hollow circles in Fig. 14(A).

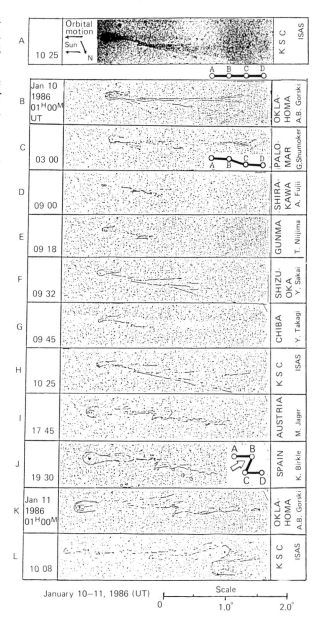

Fig. 13 — The kink event of comet Halley observed for 33 hours from 10 to 11 January 1986 (Tomita *et al.* 1987). The representative eleven out of the 24 photographs were transcribed on the identical area of *Palomar Sky Survey Atlas* by Tomita. A sample photograph taken by a 50 cm Schmidt telescope at the Kagoshima Space Centre is displayed at the top. Note that north is downward as indicated by the arrow N. As in all the figures in this paper, the cometary plasma tail is expressed in the same flow direction of the solar wind from left to right, because all the planetary magnetospheres are so expressed nowadays.

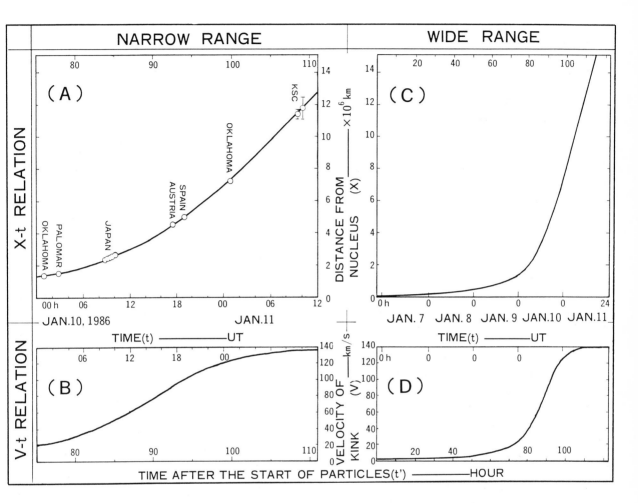

Fig. 14 — (A) X–t relation in narrow and (C) wide time ranges, where X and t are the nucleus–kink distance and Universal Time, respectively. The hollow circles in (A) represent the observed values obtained from Fig. 13. The solid curves represent the best-fit curves from equation (6). (B) V–t relation in narrow and (D) wide time ranges, respectively, where V is the flow velocity of the kink with respect to the nucleus. The curves are obtained by differentiating curves (A) and (C) against time. The curves are considered to represent the ordinary flow of the plasma tail structure, since there are no abrupt increase (Tomita et al. 1987, Saito et al. 1987e).

6.2 Flow velocity of the kink

To explain the observed X-t relation, we considered first the ordinary **j** × **B** force on the cometary plasma. Since the additional acceleration due to the tail field reconnection is not taken into consideration here, let us call this case the ordinary flow. The equation for the ordinary flow is given by equation (6) in section 5.3. The best-fit X-t curve of equation (6) for the observed values is given in the solid curves in Figs. 14(A) and (C). The figure indicates that the observed values are very well expressed by the ordinary flow.

Figs 14(B) and (D) indicate the V–t relation that was obtained by differentiating the equation (6). The name 'ordinary flow' for this case may be appropriate, because there is no abrupt increase of V in the curve in Figs 14(B) and (D), in contrast to the impulsive acceleration due to the tail field reconnection in Fig. 12.

6.3 Formation of the kink

Let us put the four points, A, B, C, and D on a line AD as shown in the inserted illustration in Fig. 13. When the part AB is moved to the upper part of the fixed segment CD, the four points form the letter 'Z'. If we denote $\angle$ ABC (= $\angle$ BCD) as the kink angle, it decreases from 180° to $\sim$40° during the movement of AB.

In the case of the 10–11 January kink event, the kink angle decreased from $\sim$160° at 0h UT on 10 January gradually to $\sim$50° at 0100 on 11 January. In association with the decrease of the kink angle, the segment BC increased from $\sim$5 arc minutes at 09h 00m on 10 January to $\sim$15 arc minutes at 0100. About nine hours later, the Z-shape turned eventually to a large inverted S-shape. These changes of the kink were as if a higher velocity stream intruded toward the corner B of the letter Z (see the arrow in the insert on Fig. 13).

If all the flow around the comet is stable and laminar, the ordinary flow along the tail axis is slowest, because it starts from the nucleus with V_T only $\sim$1 km s^{-1} to V_O. The external (solar wind) flow with the constant velocity of V_O is fastest. The flow velocity in the space between the tail axis and the constant solar wind region must be intermediate, so that the velocity increases with increase of the radial distance from the tail axis (see Fig. 2 of Siscoe et al. 1986). There must be a large velocity shear between the axial flow and the external flow, especially near the coma.

If the velocity of the external flow on the northern side, for example, of the tail increases suddenly, the increased velocity shear makes an intrusion of the solar wind from the northern side into the intermediate region. The sudden deformation of the northern boundary of the cometary magnetosphere would not be observed, because the density of the cometary plasma is extremely low there. However, the intrusion on the boundary would give rise to a sudden intrusion of the intermediate flow from the northern side to the axial ordinary flow. The intrusion to the ordinary flow with high plasma density must be observed as a kinked deformation of the plasma tail, as in the case of the 10–11 January event.

In this case, the external solar wind does not need a large time lag (as in the case of the dayside draping around the cometary ionosphere) to make the intrusion of the intermediate flow to the ordinary one. If we assume 450 km s^{-1} for the high-velocity stream, the source of the stream can be located at B = 17.7° and L = 10.6° on the solar source surface, where B and L are heliographic latitude and Carrington longitude, respectively. The source location seems to be very appropriate, because a source region from where a high-velocity stream with velocity 450 km s^{-1} flows out is identified there by Oki et al. (1987). The point P in Fig. 5(Dg) shows the source identified on the synoptic chart of the solar wind velocity distribution that was obtained by the interplanetary scintillation (IPS) observation (Kojima et al. 1987).

Finally, we may write the following possible scenario for the formation of the 10–11 January kink event. A high-velocity stream flew out from the region at B = 17.7° and L = 10.6° on the source surface and attacked the cometary magnetosphere on 10 January. The attack of the external wind gave rise inside the magnetosphere to the secondary effect of a southward intrusion of the intermediate-velocity flow into the low-velocity ordinary flow along the tail axis, forming the 10–11 January kink event.

6.4 Discussion and conclusion

A possible interpretation was given for the observed flow in section 6.2, and the observed configuration of the kink in section 4. Since the attack of the high-velocity stream was assumed to be moderate, the movement of the kink was explained by a small modification of the ordinary flow, considering neither an extremely large acceleration nor a drastic magnetic field reconnection occurred as in the case of the 31 December kink event discussed in section 5.

The 1986 apparition of comet Halley was predicted to correspond with the aligned phase in the heliomagnetospheric cycle when a low-velocity region near the flat solar sector boundary is generally sandwiched by two high-velocity regions in higher heliographic latitudes, as seen in the coarse axisymmetric distribution of the solar wind velocity on the source surface in Fig. 5(Da). Besides this general latitudinal distribution, there is actually longitudinal inhomogeneity of the velocities, forming several tongues of high-velocity wind source that extend from the higher latitude regions toward the solar equator (Kojima et al. 1987). Our concept of the fine distribution of the solar wind velocity is expressed schematically on the

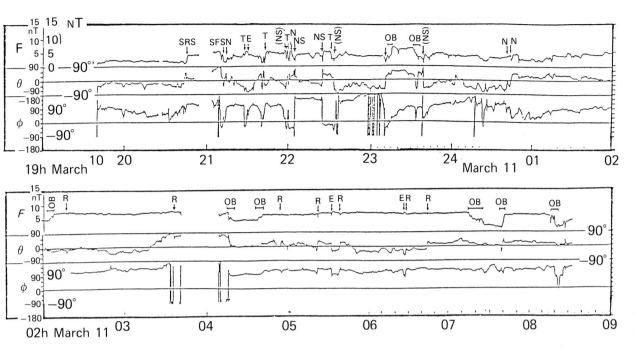

Fig. 15 — Three-component magnetic fields of the solar wind observed by Sakigake on the days of closest approach to comet Halley. Note the multiple switching of the field orientation between the 'away' ($\phi \sim 135°$) and the 'toward' ($\phi \sim -45°$) directions along the Parker spiral, indicating multiple crossings of Sakigake through the nearly horizontal neutral sheet of the heliomagnetosphere with small multiple undulations. Note the synchronized variations of the total field, F, and inclination, θ, at every switching in ϕ. The types of synchronization are indicated by SRS, SF, etc, with arrows. See Saito *et al.* (1986b) for more detailed explanation of these types. The multiple switching that is frequently observed by spacecraft during sunspot minimum phase is the signature of the nearly horizontal neutral sheet during the heliomagnetospheric aligned phase (see Fig. 5(C)).

source surface as viewed from Carrington longitude 10.6° in Fig. 5(Dg), which is based on the IPS observation in Fig. 4 of Kojima *et al.* (1987).

The source region at B = 17.7° and L = 10.6° was one of several tongues of the coronal hole extending from the north to the equator, as shown in Fig. 5 (Dg). This situation agrees with our model, because the kink was observed on the northern side of the plasma tail as seen in Fig. 13. (Note that the solar wind is understood to flow from left to right in all of our figures, as is customary in the case of the Earth's magnetosphere.)

7 QUASI-PARALLEL MODEL FOR THE 11 MARCH EVENT (SAKIGAKE ENCOUNTER)

7.1 Solar wind magnetic field observed by Sakigake

On 10 March 1986 (one day before closest approach) the Sakigake data were telemetered to us for one hour from 0247 to 0347 UT. During this interval, the Sakigake IMF data showed a stable and distinct 'toward' polarity. The next signals from Sakigake were telemetered to us over 13 hours, from 1940 UT on 10 March to 0837 UT on 11 March (see Fig. 15, where the total field F and its θ and ϕ components are shown separately). When the first signals were received at 1940 UT, the IMF direction already showed 'away' polarity, suggesting that some sector boundary crossing(s) occurred from 0347 to 1940 UT on 10 March. At 2049 on the same day, the IMF showed a polarity change (probably of a slow reverse shock), followed by characteristic switches of polarity from 'away' to 'toward', and vice versa, until ∼0200 on 11 March. After this, the total field remained at a fairly stable level of ∼8 nT with away polarity until 0832, except for three intervals (0410–0441, 0713–0740 and 0815–0832 during which F decreased to 2–5 nT, judging from the telemetry signals received in the 8-h interval after 1937 on 11 March, the IMF showed away polarity with some fluctuations, which decreased gradually until 0347 on 12 March. The heliospheric condition inferred from the Sakigake data was confirmed by other solar, interplanetary, and terrestrial observations, including the observations of the Stanford solar magnetogram (T. Hoeksema, personal communication) and interplanetary scintillation data (Kojima *et al.* 1987). Fig. 16 (inset) shows the neutral line, inferred from these data, on the solar source surface. If the neutral line is extended radially and forms the heliospheric neutral sheet, it is considered that between 10 and 11 March Sakigake crossed the neutral sheet (corresponding to the part enclosed by a rectangle in Fig. 16. The IMF disturbances on these days were analyzed and classified into the following categories: (1) The first category represents a group of interplanetary discontinuities including rotational (R), tangential (T), either of these (E), and neither (N). (2) The second category comprises the shock group, which is further classified into slow forward shock (SFS), slow reverse shock (SRS), fast forward shock (FFS), and fast reverse shock (FRS). (3) The third category consists of the heliomagnetospheric neutral sheet (NS) or the current sheet. A tangential discontinuity is regarded as the neutral sheet when it satisfies the following four conditions: (I) switching of the sector polarity occurs

within $180 \pm 30°$ with respect to the Parker spiral, (II) the switching of the sector angle, ψ, corresponds to the switching of the polarity of the θ component, (III) the total field is at a minimum at the moment of switching, and (IV) the interval between two successive discontinuities must be longer than 10 min. Neutral sheets satisfying all four of these criteria occur at 2206 and 2225 UT. (4) The fourth category comprises the peculiar changes in the total field from 2–5 nT to a constant ~ 8 nT, or vice versa. These changes can be considered to be crossings of Sakigake from the solar plasma sheet, through its outer boundary (OB), into a supposed magnetic cloud or a supposed interplanetary lobe, or vice versa. All these groups and subgroups are indicated in Fig. 15 (arrows).

7.2 Geometric relation among the comet, horizontal neutral sheet, and Sakigake

Fig. 16 also shows the positions of Sakigake, the Sun, the supposed neutral sheet, and comet Halley. The neutral sheet aligned along the Parker spiral is conveyed radially by the solar wind with a velocity V_{sw}. If we take the cross-section of the sheet to be a vertical plane incorporating the Sakigake orbit, the cross-section apparently moves rightwards in the plane with $V_N = V_{sw} \tan \chi$, where χ is the angle between the plane and the Parker spiral, and V_N is the apparent velocity component of the neutral sheet along the Sakigake orbit. As the orbital velocity of Sakigake, V_{sk} (~ 70 km s^{-1}), is rightward and is much less than V_N (~ 450 km s^{-1}), the spacecraft crosses the neutral sheet from right to left.

It is certain that Sakigake crossed the neutral sheet at least four times: once (or more) in the interval from 0347 to 1940 on 10 March, at 2206 and 2225 on the same day, and once (or more) from 2225 on 10 March to 1936 on 12 March. Although the calculated inclinations of the tangential discontinuties are taken into consideration in Fig. 16, only those crossing of the neutral sheet that satisfy the four conditions are shown (two open circles linked by a thick solid curve). The average inclination of the neutral sheet may be estimated from points P and Q in Fig. 16. We conclude that Sakigake made multiple crossings of the nearly horizontal neutral sheet.

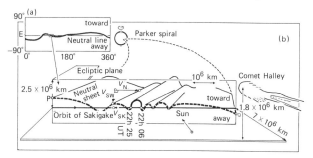

Fig. 16 — (A) Magnetic neutral line on the solar source surface as inferred from the Sakigake/IMF data, Stanford magnetograph data by Hoeksema, and interplanetary scintillation data by Kojima *et al.* (B) Schematic illustration of the relation between comet Halley, Sakigake, and the heliomagnetospheric neutral sheet to explain Fig. 15. Since the rightward co-rotating speed, V_N, of the neutral sheet is higher than the orbital speed; V_{SK}, of Sakigake, the spacecraft is considered to cross relatively the neutral sheet from right to left. The inclination and orientation of each undulation of the neutral sheet are expressed on the basis of minimum variance analysis of the observed IMF data in Fig. 15. The nearly parallel neutral sheet was expected to encounter comet Halley about four hours later, and to cross very slowly, taking nearly one day (Saito *et al.* 1986b).

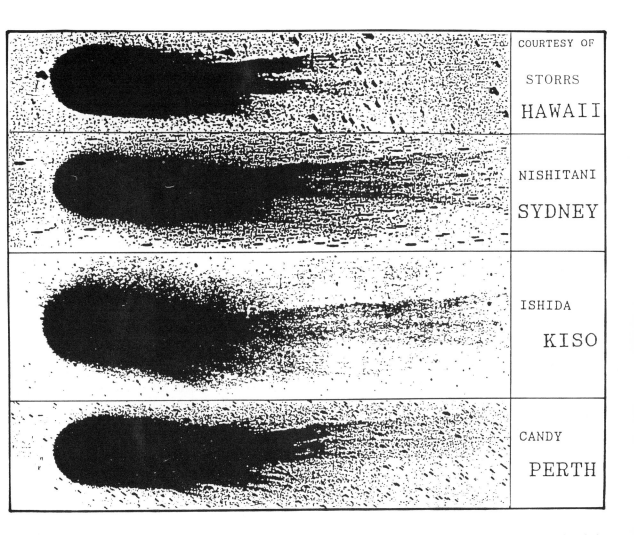

Fig. 17 — Photographs of comet Halley taken on 11 March 1987. Note that no distinct disconnection event was observed, even though the world-first *in situ* observation of the sector boundary was carried out only 1.8 × 10⁶ km upstream of the comet. See also Fig. 3A of Saito *et al.* (1986) for the more detailed structure of the plasma tail of the comet with 4°.5 length taken by the 108 cm Schmidt telescope in the Kiso Astronomical Observatory.

7.3 Response of the comet to the neutral sheet crossing

We now apply the dayside crossing model to our case of the sector boundary crossing between 10 and 11 March 1986. We were fortunate in obtaining precise information concerning the sector boundary located 7×10^6 km upstream of comet Halley on the day of closest approach. If we take 450 km s^{-1} as the solar wind velocity and 0.3 AU for the length of Halley's plasma tail, it takes longer than 1 day (28 h) for the solar wind to pass through the entire plasma tail, and it will be observable from any point on the Earth. However, our conclusion is that no distinct DE can be seen in comet Halley's plasma tail during 11–14 March, in the data from Asia, Australia, and the United States. As an example, Fig. 17 shows photographs of comet Halley taken in Hawaii, Sydney, Kiso, and Perth (Australia).

7.4 Quasi-parallel model for the observation

As the distinct DE did not occur in Halley's plasma tail, we propose the model shown in Fig. 18(A, B) for the approach of a quasi-parallel neutral sheet to the comet. Before the neutral sheet crossing, comet Halley was in the 'toward' hemisphere. Hence, the magnetic field lines of the cometary magnetosphere must be horizontal on average, with a vertical, downward-pointing current sheet (i), as shown schematically in Fig. 18(A). In almost all previous models of the cometary magnetosphere, the vertical (z component) interplanetary magnetic field lines are draped by the cometary nucleus. However, the averaged three-dimensional view must rather comprise horizontal field lines with the vertical tail current sheet, as was almost the case for comet Giacobini–Zinner (Slavin *et al.* 1986).

In our case, a neutral sheet with a small angle of

inclination, I, and with Parker spiral angle χ, was conveyed by the low-speed solar wind as illustrated in Fig.18(A). As $\chi \approx 45°$, the attack angle of the neutral sheet could be as small as the angle I, although the angle could be slightly increased by mass loading upstream of comet Halley. Hence, the southern part of Halley could be dipped first into the away sector, changing the polarity of the field line 'a' from 'toward' to 'away' polarity by means of the dayside crossing process. Associated with the dipping of comet Halley into the apparently upwelling neutral sheet, the polarity of the field lines b, c, d, ... h could be switched one after the other Fig. 18(B) indicates an intermediate stage, in which only the field lines a, b and c have switched to 'away' polarity; this situation gives rise to a neutral line (thick broken line) and a horizontal tail current. The change of the whole field-line polarity from 'toward' to 'away' could be accomplished by a shift of the neutral line and the horizontal tail current from the southern edge to the northern edge. When the dayside-crossing type reconnection is gradually executed from a to h (Fig. 18(B)), there is a drastic DE. For a detailed treatment of the content of this section, see in T. Saito *et al.* (1986b).

8 DISCUSSION AND CONCLUSION

8.1 Summary

The solar wind–cometary tail interaction was studied by concentrating on disconnection events in the plasma tail of comet Halley. The solar wind magnetic field has been observed by our Sakigake/IMF team, as was described in section 2. Our results are summarized as:

(1) The heliospheric neutral sheet was identified during the 1985/86 apparition by the scanning method from the Sakigake data and the minimum velocity method from the IPS data. The neutral sheet obtained by these methods was compared with the sheet obtained by the potential method from the photospheric magnetic field data (section 3).

(2) The plasma tail disturbances of comet Halley were tabulated. The observed DEs were compared with the previous model in which a DE is caused by a reconnection of the field-lines on the dayside of a comet every time that the comet encounters the

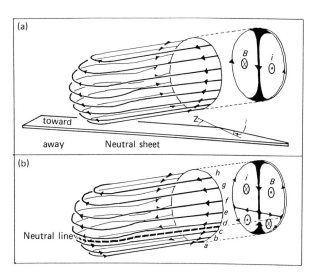

Fig. 18 — Quasi-parallel model to explain the relation between the cometary plasma tail and the heliomagnetospheric neutral sheet on the days of Sakigake encounter with comet Halley. (A) Situation of comet Halley and the neutral sheet before the crossing. The spiral angle and the inclination of the neutral sheet are expressed by χ and I, respectively. Note that the sheet is quasi-parallel to the plasma tail axis. (B) Situation of the cometary magnetosphere during the neutral sheet crossing. Besides the main and nearly vertical current sheet, a sub-nearly horizontal current sheet appears. During the crossing of the quasi-parallel neutral sheet for one day, the sub-sheet moves slowly from the southern edge to the northern edge. Even through the dayside crossing type reconnection takes place, a drastic DE need not occur, because the reconnection must be local and the crossing is gradual (Saito *et al.* 1986b).

heliospheric neutral sheet (the dayside crossing model). The majority of disturbances reported as DEs were found to take place when the comet was far from the heliospheric neutral sheet (section 4).

(3) The knot-type DE that was observed on 31 December 1985 was thought to be explained not by the dayside crossing model, but by the dynamic pressure model, which is related to a sudden increase of the dynamic pressure of the solar wind (section 5).

(4) The 10–11 January kink event was thought to represent the ordinary flow of the tail plasma. The flow was identified by the kink structure that was interpreted as a weak intrusion of the higher-speed plasma tail into the slowest-speed plasma on the tail axis (section 6).

(5) Sakigake provided the very first evidence of the heliomagnetospheric neutral sheet crossing in only 7×10^6 km in front of the comet on 11 March 1986. Nevertheless, no distinct DE was observed. The relation was explained by the quasi-parallel model (section 7).

(6) Comet Halley in this apparition makes it clear that the relation between a DE and the neutral sheet crossing is not unique; DEs occurred without sheet crossing and DEs did not occur at the time of the sheet crossing. In conclusion, we have to consider, as the mechanism of DEs, at least the dynamic pressure process, slipping IMF process, rotating IMF process, Hones' process, and perspective effect, as well as the dayside crossing process (section 4).

8.2 General misunderstanding on comet Halley

Shall we have a good chance to study appropriate comets before the next apparition of comet Halley in 2060/2061? Of course we shall. Let us consider this question and the answer given more closely.

(1) Fresh plasma tail
The comets appropriate for study of the plasma tail–solar wind interaction are not short-period comets like P/Halley or P/Encke, but parabolic comets. This is because the periodic comets have lost much of their plasma during their multiple apparitions.

(2) Tail length
Comet Halley in both its 1910 and 1986 apparitions is estimated to have a tail with length $\sim 8 \times 10^7$ km which was shorter than the tail length of either comet

West or comet Ikeya-Seki. Comet 1843I had a tail with length of 3.2×10^8 km. Two different comets larger than comet Halley appeared in 1811: their tail lengths were 2.1×10^8 and 1.6×10^8 km, respectively.

(3) Maximum brightness
Comet Halley in this apparition ($M_O = 3.9$) was also less bright than either comet West or comet Bennet ($M_O = 3.4$). Daylight comets, like that observed in January 1910, have appeared several times throughout history. Even fainter comets like comet Morehouse, comet Bradfield, etc. have contributed to plasma tail physics, when their observational circumstances were good.

(4) Probability of appearance
Finally, let us list the appearances of some comets in the past, and discuss future observational chances to study plasma tail physics. The comets that have contributed to plasma tail physics recently are: 1957 (Mrkos and Arend–Roland), 1962 (Humason and Seki–Lines), 1963 (Ikeya), 1964 (Tomita–Gerber–Honda), 1965 (Ikeya–Seki), 1969/70 (Tago–Sato–Osaka), 1970 (Bennett), 1973/74 (Kohoutek), 1974 (Bradfield), 1975 (Kobayashi–Berger–Millon), 1975/76 (West), 1980 (West). The probability of appearance is roughly estimated to be 0.6 comet/year or one comet/1.7 years. These values mean that study of the plasma tail–solar wind interaction is not a matter for ~ 70 years' time, but of the present.

8.3 Future observational study on the solar wind–tail interaction

We have concluded that parabolic comets with fresh tails are better than short-period comets, not only as wind vanes and for comparative magnetosphere studies (section 1), but also for the study of the origin of life and primordial material. The observational study of parabolic comets does, however, present difficulties in the forecasting of budgets and in the preparation of spacecraft.

An orbiting or side-running spacecraft is much better for the study of a comet than the instantaneous flyby observation that was carried out by the armada of Halley spacecraft in the last apparition. But study is difficult, because comets have low masses and parabolic trajectories present difficulties.

One of the economical ways for the study of a huge comet like West would be to divert a spacecraft monitoring the solar wind in interplanetary space to the comet. Although ISEE-3 was diverted successfully to observe comet Giacobini–Zinner in this way, the diversion to study a really huge parabolic comet would be difficult in terms of orbital dynamics technique.

One of several possibilities for continual observation from space would be the use of a polar patrol balloon. A combination of a developed high-resolution CCD camera and appropriate software would overcome the swinging of the telescope derived from the floating balloon.

As for observation from the ground, a large Schmidt telescope is powerful, of course. However, such a large telescope is unlikely to be available for long and continuous observation of the wind–tail interaction for large, unexpected comets.

Cooperation between many skilful amateur astronomers worldwide, having moderate-sized Schmidt telescopes, would be fruitful in the study of the dynamics of the plasma tail, as it was for the 31 December 1985 and the 10/11 January 1986 events.

ACKNOWLEDGEMENTS

The author would like to express sincere thanks to the Institute for Space Aeronautical Sciences, especially to Prof. M. Oda and the Planet-A project team for letting him join them as the principal investigator of the magnetic field experiment. Thanks are due to the many coauthors of our 30 or so papers on comet Halley who made it possible for the author to write the present review paper. Thanks are due to the astronomers for sending their valuable photographs. The author's thanks are also due to Dr J.C. Brandt, Dr M. Niedner, and Dr T. Oki for their valuable discussions.

REFERENCES

Alfven, H. (1957). On the theory of comet tails, *Tellus* **9** 92–96

Brandt, J.C. (1982), Observation of dynamics of plasma tails, *Comets*, Wilkenining, L., ed., U. of Arizona Press, 519–537

Brandt, J.C. (1988), The large-scale plasma structure of Halley's Comet–1985–1986, *Halley's Comet 1986*, Ellis Horwood (the present volume)

Brandt, J.C., & Niedner, M.B. (1986), Plasma structures in Comet Halley, *Proc. 20th ESLAB Symp. on the Exploration of Halley's Comet*, Heidelberg, 27–31 October 1986, *ESA SP-250* **1** 47–52

Caron, R., Guerin, P., Koutchmy, S., Sarrazin, M., & Zimmermann, J.P. (1986), The disconnection events of 10 January and 8 April 1986 in the plasma tail of Comet Halley, *Proc. 20th ESLAB Symp. on the Exploration of Halley's Comet*, Heidelberg, 27–31 October 1986 *ESA-250* **3** 143–147

Celnik, W.E. (1986), The acceleration within the plasma tail, the rotational period of the nucleus, and the aberration of the plasma tail of Comet P/Halley 1986, *Proc. 20th ESLAB Symp. on the Exploration of Halley's Comet*, Heidelberg, 27–31 October 1986, *ESA SP-250* **3** 53–58

Hoeksema, J.T., Wilcox, J.M., & Scherrer, P.H. (1983), The structure of the heliospheric current sheet: 1978–1982, *J. Geophys. Res.* **88** 9910

Koishikawa, M., & Kasahara, S. (1986), Node structure in the plasma tail of Comet Halley on 31 December 1985, *Hoshi no TechΔoΔ*, ******, 110–111

Kojima, M., Kakinuma, T., Oyama, K., Mukai, T., Hirao, T., & Miyake, W. (1987), Solar wind structure observed by interplanetary scintillation and spacecraft in 1985 and 1986, *Proc. Res. Inst. Atmosph. Nagoya Univ.* **34** 9–21

Kubacek, D., Pittich, E.M., & Zvolankova J. (1986), Observations of Comet Halley at the Comenius University Observatory, *Proc. 20th ESLAB Symp. on the Exploration of Halley's Comet*, Heidelberg, 27–31 October 1986, *ESA SP-250* **3** 277–280

Marubashi, K., Grebowsky, J.M., & Taylor, H.A. Jr. (1985), Field in the wake of Venus and the formation of ionospheric holes, *J. Geophys. Res.* **90** 1385–1398

Minami, S., & White, R.S. (1986), An acceleration mechanism for cometary plasma tails, *Geophys. Res. Lett.* **13** 849–852

Niedner, M.B. Jr., & Brandt, J.C. (1978), Interplanetary gas XXIII. Plasma tail disconnection events in comets, *Astrophys., J.* **223** 655–670.

Niedner, M.B. Jr., & Schwingenschuh, K. (1986), Plasma-tail activity at the time of the VeGa encounters, *Proc. 20th ESLAB Symp. on the Exploration of Halley's Comet*, Heidelberg, 27–31 October 1986, *ESA SP-250* **3** 419–424

Ogino, T., Walker, R.J., & Ashour-Abdalla, M. (1986), An MHD simulation of the interaction of the solar wind with the out flowing plasma from a comet, *Geophys. Res. Lett.* **13** 929–932

Oki, T., Tomita, K., Saito, K., Kojima, M., Minami, S., Hurukawa, K., & Saito, T. (1987), Effects of the heliomagnetosphere on cometary plasma tail, *Abstract of 1987 Fall Conference of Astronomical Society of Japan*, B90

Russell C.T., Luhmann, J.G., & Elphic, R.C. (1982), Solar wind interaction with comets: lessons from Venus, *Comets*, Wilkenining, L., ed., U. of Arizona Press. 561–587

Russell, C.T., Sanders, M.A., & Phillips, J.L. (1986), Near-tail reconnection as the cause of cometary tail disconnections, *J. Geophys. Res.* **91** 1417–1423

Saito, K. (1982), *Ancient records of stars*, Iwanami Shinsho Series, Iwanami Publ. Co.

Saito, K., & Saito, T. (1986), Effect of the heliospheric neutral sheet to the kinked ion tail of Comet Halley on 13 May 1910, *Proc. 20th ESLAB Symp. on the Exploration of Halley's Comet*, Heidelberg, 27–31 October 1986, *ESA SP-250*, 135–138

Saito, T. (1975), Two-hemisphere model of the three-dimensional magnetic structure of the interplanetary space, *Sci. Rept. Tohoku Univ.*, Ser. 5 **23** 37–54

Saito, T. (1982), *Secrets of the winged sun* — heliomagnetosphere and ancient solar eclipses, ChûΔoΔkΔoΔron-Sha; Tokyo

Saito, T. (1983), A review on 70 recent papers on development of magnetometers for geophysical uses, *Proc. Symp. on Development and Application of Geophysical Magnetometers*, Ed. Saito,

T.; Tohoku Univ. Press, 132–149

Saito, T. (1987), Solar cycle variation of solar, interplanetary, and terrestrial phenomena, *Laboratory and Space Plasmas*, Ed. Kikuchi, H., Springer-Verlag, (in press)

Saito, T. (1988a), Comet, aurora, and magnetic storm, Ky∆o∆ritsu Shuppan Co.

Saito, T. (1988b), Disturbances of cometary plasma tail by the solar wind, Asakura-Shoten, Tokyo

Saito T., Sakurai, T., & Yumoto, K. (1978), The earth's palaeomagnetosphere as the third type of planetary magnetosphere, *Planet. Space Sci.*, **26** 413–422

Saito, T., Yumoto, K., Hirao, K., Aoyama, I., & Smith, E.J. (1986a), Three-dimensional structure of the heliosphere as inferred from observations with a Japanese Halley spacecraft, In: The *Sun and the heliosphere in three dimensions*, Ed. B.G. Marsden, 281–286, D. Reidel Publ. Co.

Saito, T., Yumoto, K., Hirao, K., Nakagawa, T., & Saito, K. (1986b), Interaction between Comet Halley and the interplanetary magnetic field observed by Sakigake, *Nature* **321** 6067, 303–307

Saito, T., Yumoto, K., Hirao, K., Saito, K., Nakagawa, T., & Smith, E.J. (1986c), A disturbance of the ion tail of Comet Halley and the heliospheric structure as observed by Sakigake, *Geophys. Res. Lett.* **13** No. 8 821–824

Saito, T., Takenouchi, T., Saito, K., & Yumoto, K. (1986d), Halley's plasma tail disturbances and solar wind observed by Sakigake, *Proc. Symp. on 7th Solar System Sci.*, ISAS, 6–7

Saito, T., Saito, K., Aoki, T., & Yumoto, K. (1986e), Possible models on disturbances of the ion tail of Comet Halley during the 1985–1986 apparition *Proc. 20th ESLAB Symp. on the Exploration of Halley's Comet*, Heidelberg, 27–31 October 1986, *ESA SP-250*, 155–160

Saito, T., & Akasofu, S.-I. (1987a), On the reversal of the dipolar field of the Sun and its possible implication for the reversal of the Earth's field, *J. Geophys. Res.* **92** A2, 1255–1259

Saito, T., Saito, K., Aoki, T., & Yumoto, K. (1987b), Possible models on disturbances of the plasma tail of Comet P/Halley, *Astronomy and Astrophysics* **187** 201–208

Saito, T., Yumoto, K., Hirao, K., Minami, S., Saito, K., & Smith, E.J. (1987c), Structure and dynamics of the plasma tail of Comet Halley, Part 1. Knot event on 31 December 1985, *Astronomy and Astrophysics* **187** 209–214

Saito, T., & Oki, T. (1987d), Interaction between the heliomagnetosphere and cometary magnetosphere, *Laboratory and Space Plasmas*, Ed. by Kikuchi, H., Springer-Verlag, New York, (in press)

Saito *et al.* (1987e), Structure and dynamics of the plasma tail of Comet Halley, Part 1. Kink event on 10–11 January 1986, *Proc. Chapman Conf. on Plasma Waves and Instabilities in Magnetospheres and at Comets* held in Sendai/Mt. Zao, Japan on 12–16 Oct. 1987, 70–73

Saito *et al.* (1987f), Structure and dynamics of the plasma tail of Comet Halley, Part 2. Knot event on 31 December 1985, *Proc. Chapman Conf. on Plasma Waves and Instabilities in Magnetospheres and at Comets* held in Sendai/Mt. Zao, Japan on 12–16 Oct. 1987, 74–77

Saito, T., Yumoto, K., Hirano, K., Itoh, T., & Nakagawa, T. (1987g), Studies on the interplanetary magnetic field (IMF) observed by Sakigake, *ISAS Bulletin*, Special No. 19, Rept on Studies by Comet-Halley Spacecraft, 85–104

Saito, T., Suzuki, Y., Yumoto, K., Oki, T., Akasofu, S.-I., & Olmsted, C. (1987h), Computer simulation of the heliosphere by the three-dipole model, *Proc. Symp. on 9th Solar System Sci.*,

held in ISAS, in press

Schlosser, W., Schulz, R., & Koczet, P. (1986), the cyan shells of Comet P/Halley, *Proc. 20th ESLAB Symp. on the Exploration of Halley's Comet*, Heidelberg, 27–31 October 1986, *ESA SP-250* **3** 495–498

Schulz, M. (1973), *Astrophys. Space Sci.* **24** 371

Shkodrov, V., Ivanova, V., & Bonev, T. (1986), Observations of Comet P/Halley at the National Astronomical Observatory-Bulgaria *Proc. 20th ESLAB Symp. on the Exploration of Halley's Comet*, Heidelberg, 27–31 October 1986, *ESA SP-250* **3** 195–198

Siscoe, G.L., Slavin, A.J., Smith, E.J., Tsurutani, B.T., Jones, D.E., & Mendis, D.A. (1986), Statics and dynamics of Giacobini–Zinner magnetic tail, Geophys. Res. Lett. **13** 287–290

Slavin, J.A., Smith, E.J., & Tsurutani, B.T. (1986), Giacobini–Zinner magnetotail ice magnetic field observations, *Geophys. Res. Lett* **13** 283–286

Tomita, K.-I., Saito, T., & Minami S. (1987), Structure and dynamics of the ion tail of Comet P/Halley, Part 2. Kink event on 10–11 January 1986, *Astronomy and Astrophysics* **187** 215–219. Table 1. Catalogue of Comet Halley disturbances

Ugai, M. (1987), Computer simulation of a large-scale plasmoid in the long current sheet, *Proc. Chapman Conf. on Plasma Waves and Instabilities in Magnetospheres and at Comets* held in Sendai/Mt. Zao, Japan on 12–16 Oct. 1987, 90–91

Wilcox, J.M., & Ness, N.F. (1965), *J. Geophys. Res.* **70** 5793

Yumoto, K., Saito, T., & Nakagawa, T. (1986), Hydro-magnetic waves near O^+ (or H_2O^+) ion cyclotron frequency observed by Sakigake at the closest approach to Comet Halley, *Geophys. Res. Lett.* **13** No. 8 825–828

Measurements of ion species within the coma of comet Halley from Giotto

H. Balsiger

1 INTRODUCTION

The payload of the Giotto spacecraft included several instruments which were able to measure the mass per charge (m/q) of the ions encountered during the flyby of comet Halley. The instrument, fully dedicated to measurements of the composition and dynamics of ions, was the Ion Mass Spectrometer (IMS) which contained two sensors. The first one, the High Energy Range Spectrometer (HERS), was optimized for the approach phase when solar wind ions and picked-up cometary ions had to be measured over a wide range of angular and energy space; the second one, the High Intensity Spectrometer (HIS), was optimized for the region close to the comet, where cold cometary ions with similar velocity relative to the spacecraft were expected (Balsiger *et al.* 1987a). The Neutral Mass Spectrometer (NMS) was run part-time in a mode where the ion source was switched off. During these periods, ions were measured instead of neutrals (Krankowsky *et al.* 1986a). The NMS had very good mass resolution and hence was able to measure isotopic ratios of some elements in cometary matter. Also, the two plasma analyzers, JPA and RPA, included sensors which were able to measure the mass per charge of the ions. The Implanted Ion Sensor (IIS) of the JPA was designed to study the pick-up of cometary ions in the solar wind (Wilken *et al.* 1987a). The Positive Ion Cluster Composition Analyzer (PICCA) of the RPA with a wide m/q range at modest mass resolution for cold cometary ions (Korth *et al.* 1987) was included in order to identify cometary ion

clusters such as water clusters, etc. The fact that such a large number of mass-discriminating sensors were included in the Giotto payload demonstrates the high priority which was given to plasma composition measurements during mission planning. The reasons are (1) that only mass identification allows unique distinction of solar wind from cometary ions, and (2) that for proper measurement of the plasma properties the mass distribution of the ions must be known.

The aim of ion composition measurements during the Halley flyby was mainly two-fold: (1) To investigate the plasma dynamics of the solar wind interaction with the cometary coma, and (2) to measure the composition of cometary ions as a function of distance to the nucleus in order to make a contribution to the question, 'what is the (volatile) material in the nucleus made of?'

The interaction of a comet with the solar wind is quite complex, but as the results of the recent comet missions have shown, it has been predicted qualitatively quite well. Cometary material first makes its presence felt in the solar wind in the form of neutral gas photo-ionized and accelerated by the solar wind far upstream from the comet nucleus at distances of several million kilometres. Although their density is initially very low, these 'pick-up' ions (e.g. H^+, C^+, O^+, OH^+, H_2O^+, CO^+) 'mass load' the solar wind and begin slowing it. Nearer the comet a bow shock is formed, and at a few thousand kilometres from the nucleus (depending on the gas production rate) the cold cometary ions (inner coma) stand off the solar wind ram pressure (ionopause or contact surface). Between the two major discontinuities, bow shock and ionopause, the so-called cometosheath revealed a complex structure in the plasma and magnetic field during the Giotto flyby which is not yet fully understood. We shall, in this paper, follow the Giotto spacecraft from the first pick-up of cometary ions through the bow shock and the cometosheath to the ionopause.

It was inside the ionopause, only 4 600 km from the nucleus or ~60 seconds before closest approach of Giotto, that the search for original cometary material really began. Thanks to the calculated risk of the Giotto mission to fly close to the nucleus, it was possible to cross this boundary and to measure gas and ions undisturbed by the solar wind. To find original cometary molecules was, of course, the prime task of the Neutral Mass Spectrometer, NMS. However, ion composition measurements are an important complement for understanding the processes in the inner coma. Ions are secondary products of cometary neutral molecules sublimating from the nucleus or from the dust grains. Hence, quite complex model calculations (taking into account photodissociation, photoionization, ion–molecule reactions, ionization, and loss by collisions with electrons, charge exchange with the solar wind, etc.) are needed to eventually reconstruct the composition of the cometary material itself. On the other hand, virtually all material which sublimates from the nucleus or the outflowing dust will, at some distance, be ionized, as molecule or dissociation product. The integral of the ion composition measurements over the full encounter should therefore yield a good estimate of the elemental (not molecular) composition of the volatile material of the comet.

The analyses of ion composition have resulted so far in estimates of the elemental abundance, in several isotopic measurements, and in first attempts of model calculations yielding molecular abundances in the nucleus. All these are relevant to the question of the history of cometary nuclei.

2 COMETARY IONS OUTSIDE AND AT THE BOW SHOCK

The first cometary particles sensed by an approaching spacecraft are high energy particles (see McKenna–

Lawlor, chapter 11 in this volume) and protons resulting from the extensive hydrogen corona. As these H atoms become ionized, either by photoionization or by charge exchange with solar wind ions, the newly formed protons respond to the electric and magnetic fields in the solar wind and move on cycloidal trajectories with gyrocentres moving with the speed of the frozen-in magnetic field lines. These 'pick-up' cometary protons have distinctly different velocity and angular distributions from those of the solar wind, and can hence be readily identified by *in situ* measurement of the proton distribution function. Similarly, one expects to find other cometary ions at distances far away from the nucleus, depending on the outflow velocities of the cometary molecules and their lifetime against ionization. (There is indirect evidence that the energetic ions which have been detected several million kilometres upstream from the shock are cometary water group ions. It is not yet explained how cometary ions — except for H^+ — could be found so far away from the nucleus. For a discussion of energetic ions see McKenna–Lawlor 1989). The narrow pitch angle distribution of pick-up ions is a source of free energy which can stimulate wave activity which in turn can initiate pitch angle diffusion.

Pick-up protons were first detected by the IMS/HERS at 1700 UT on 12 March, the day before encounter at a distance nearly 8 million km from the comet (Balsiger *et al.* 1986, Neugebauer *et al.* 1986, Neugebauer *et al.* 1987a). In Fig. 1 it is shown how these cometary protons have a very special signature which distinguishes them from the solar wind protons. Shown are four polar plots of contours in velocity space of the proton phase space density, $f(v)$, observed with four CEM detectors of the HERS, which covered four 15° sectors between 15° and 75° from the spin axis. CEM A, whose field of view is furthest from the spin axis, contains the solar wind direction. The peak of the solar wind distribution is indicated by 'x'. Three contour levels represent a one-decade change in phase space density. From 'x' the phase space density decreases monotonically over ~5 orders of magnitude. For better identification the outer four contours have been shaded, with a darker shade representing a higher value of $f(v)$. After a minimum (unshaded) the phase-space climbs again (shaded). This outer arc is

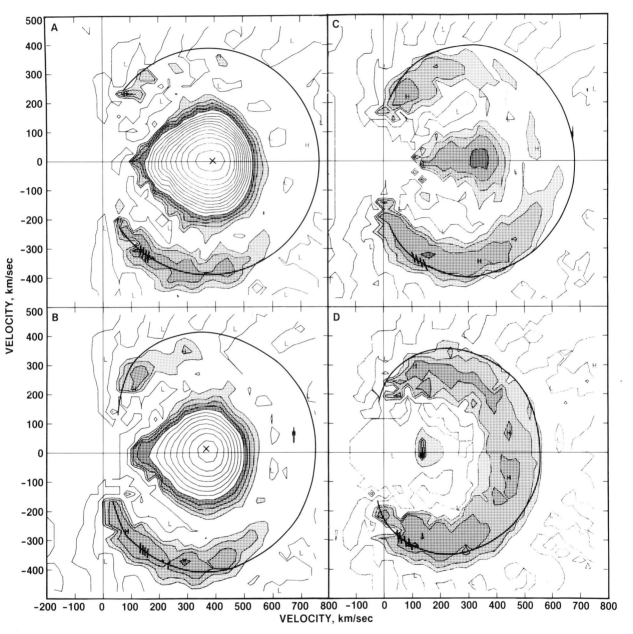

Fig. 1 — Phase space density contour plots of four HERS detectors for the period 0803–0908 UT on 13 March 1986. The plots show the shell of pick-up cometary ions which are — in the solar wind velocity frame — surrounding the main solar wind proton distribution. (Neugebauer *et al.* 1987a).

due to the pick-up protons which are about four orders of magnitude lower in density than the peak of the solar wind proton distribution. With the known relative velocity between spacecraft and comet and a typical hydrogen outflow velocity of 11 km s^{-1} as

measured by Suisei (Kaneda *et al.* 1986) the velocity distribution of cometary protons was computed (a ring in velocity space in the solar wind frame of reference), and the trace of this ring distribution through the instrument field of view is indicated in

Fig. 1 by crosshatching. The fact that in Fig. 1 the pick-up protons include but extend beyond the hatched ring distribution is evidence for pitch angle scattering leading to a spherical shell distribution rather than a ring. The curves showing the intersection of such a shell with the instrument field of view are also given. The population of picked-up protons is spread out along these curves, indicating a nonuniformly filled shell distribution with the shell centre and radius determined by the velocities of the solar wind and the comet. Fig. 1 is typical for distances between several million km from the comet to quite close to the shock, although the degree of filling of the shell changes. A thickening of the shell was observed as the bow shock was approached by the spacecraft. This could be due to energy diffusion and/or increased turbulence of the solar wind flow close to the shock. However, other effects also have to be considered. Firstly, at a given time, the instrument detected cometary protons which were ionized over a range of distances upstream of the spacecraft. 'Older' protons from further upstream where the solar wind speed was presumably higher, should be on a larger-radius shell than locally picked-up ones. Secondly, the protons in the shell would be adiabatically heated during deceleration and compression of the solar wind (Ip & Axford 1986). For a more detailed analysis of pick-up

protons we refer to Neugebauer *et al.* (1989a,b).

A similar study has been performed on pick-up of water group ions with data from the IIS by Wilken *et al.* (1987b). IIS had lower mass-resolution than IMS but higher sensitivity, and it was able to identify water-group ions from the comet as far as 1.5×10^6 km from the nucleus, hence clearly outside the bow shock. Here too the shell distributions are quite wide in energy with distinct tails on the high-energy side, which are associated by the authors with the presence of 'older' ions that were incorporated further upstream at higher solar wind speed. The IIS and IMS teams have also derived the radial dependence of picked-up ions. Neugebauer *et al.* (1987a) determine for the protons a drop-off proportional to $R^{-1.75}$ (least squares fit) between the shock front and 10^7 km, and Wilken *et al.* (1987b) find for the water group ions a dependence somewhat steeper than R^{-2} between the shock front and 2×10^6 km. They also computed the relative mass loading, i.e. the number density of the pick-up water group ions divided by the solar wind proton number density. This mass loading increased between 1.5×10^6 km and the shock front (1.16×10^6 km) by almost an order of magnitude, reaching a value of $\sim 1\%$ at the shock.

At the bow shock the pick-up ions and solar wind

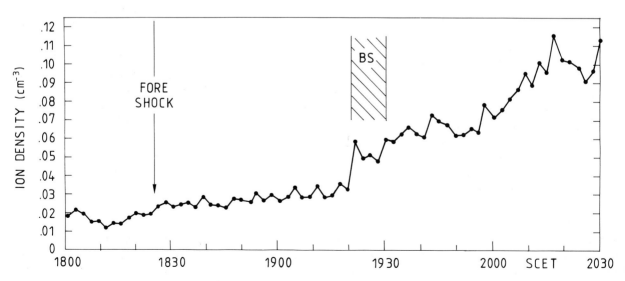

Fig. 2 — Radial profile of water group ion density in the region of the bow shock (BS) from the IIS instrument. At the bow shock, the density of heavy pick-up ions increases by a factor of two. (Wilken *et al.* 1987b).

ions were heated, and the cometary protons could no longer be distinguished from the solar wind protons. The density of the oxygen showed a sudden increase at the bow shock by a factor of two (Fig. 2). Inside the shock, high-energy tails of the proton distribution functions and oxygen shell distributions at energies well above the local pick-up acceleration limit, are evidence for ions picked-up at higher speed, but possibly also of additional energization processes. The increased velocity spread and the rather homogeneous shell structure can be explained by the enhanced levels of turbulence in the shock region. On the other hand, the variable slopes of the oxygen radial profile (Fig. 2) and complex energy spectra with peaks clearly below the energy which is expected for local pick-up ions (Wilken *et al.* 1987b, Neugebauer *et al.* 1987a,b) indicate that the simple pick-up mechanism (integrated over a wide source region) will probably not explain all the features observed behind the shock and in the cometosheath (cf. Rème *et al.* 1987, B. Goldstein *et al.* 1987).

3 THE COMETOSHEATH

The region between bow shock and ionopause (contact surface) has proved to be a very complex one which is far from being fully understood. Mainly based on electron data, it can be divided up into three outer regions and an inner region (Rème *et al.* 1987). In Table 1 we summarize the characteristics of these regions, and give the locations of discontinuities as observed during the Giotto encounter. In addition to the discontinuities mentioned by Rème *et al.* (1987), we list here three more tentative discontinuities X, Y, Z, which will be further discussed below.

Even though it is not yet clear if the features listed in Table 1 are temporal or fundamental, we use them in the following as orientation guides in our description of the cometary plasma environment; in turn the ion composition measurements in the different subregions may shed light on the cause of these boundaries.

Unfortunately, continuous and comparable data sets on composition in all these regions do not yet exist. None of the different instruments mentioned in the introduction (HERS, HIS, IIS, PICCA), which shared the task of composition measurements, was able by itself to cover the wide range of phase space

which would be needed for the cometosheath ions. Whereas HERS and IIS have blind spots in the ram direction and hence do not see really cold cometary ions, HIS and PICCA look only into the ram direction. The first mentioned sensors start losing the cometary ions from their fields of view around 80 000 km when the second group of sensors has to take over. There is a certain amount of overlap between the two groups, but further analyses are needed to match them in the critical region around 10^5 km in order to get a complete coverage of all ions (hot and cold) and to investigate what processes govern the transition between solar wind and comet dominated plasma. It will also be very important to include the magnetic field data in this investigation because the magnetic field shows large fluctuations in magnitude and direction in the pile-up region (Neubauer *et al.* 1986, Neubauer 1987). Hence, before a more thorough analysis is available, we prefer to remain very cautious in deciding which of the observed discontinuities (other than shock front and ionopause) are fundamental boundaries of the solar wind/comet interaction and which are temporal effects caused by either the solar wind or a variable outflow of cometary gas and plasma.

For an overview of the whole cometosheath see Figs. 3, 4, and 9 giving radial profiles of solar wind alpha particles (Fig. 3 and 4) and of some major cometary ions measured by HERS between 3.5×10^5 and 6×10^4 km (Fig. 4) and by HIS between 1.5×10^5 and about 1000 km (Fig. 9). The discontinuities mentioned in Table 1 are marked with arrows and the respective acronyms. Roughly speaking, the region between BS and PB is dominated by solar wind, 'contaminated' by pick-up cometary ions, whereas the region between X and C is clearly dominated by cometary ions. Between PB and X we have a transition region with, in our opinion, as yet undetermined contributions of the two sources.

3.1 The outer regions

The region between the bow shock (BS) and the pile-up boundary (PB) is clearly dominated by solar wind plasma. We show in Fig. 3 density, velocity, and thermal speed of the second most abundant solar wind ion He^{++} (mass per charge $m/q = 2$). Also given is the

Table 1 — Description of discontinuities and regions in Halley's 'cometosheath', the turbulent region between bow shock and ionopause

Distance from nucleus (km)	Encounter time by Giotto (GRT)	Name	Acronym	Description of particle and field characteristics	References
1.16×10^6	1930	Bow shock	BS		Neubauer et al. (1986)
1.16×10^6 – 8.5×10^5		Outer region 1		Compared to solar wind: — increased electron density — increased p,α densities, velocities, temperatures, direction change of solar wind ions — increased density of heavy pick-up ions	Rème et al. (1987) Balsiger et al. (1986) Neugebauer et al. (1987b) Wilken et al. (1987b)
8.5×10^5	2044		1/2		Rème et al. (1987)
8.5×10^5 – 5.5×10^5		Outer region 2 Mystery region		Hot (keV) electrons Increased α densities	Rème et al. (1987) Balsiger et al. (1986)
5.5×10^5	2159		2/3	Sudden slowing of bulk flow, leading to bi-modal ion distribution Sudden decrease of hot electrons	Thomsen et al. (1987) d'Uston et al. (1987)
5.5×10^5 – 1.35×10^5		Outer region 3		n_e, v_e, T_e decrease, hot electrons disappear Increase of heavy cometary pick-up ions	Rème et al. (1987) Amata et al. (1986) Balsiger et al. (1987b) Hodges et al. (1986) Shelley et al. (1987)
				Charge exchange of He^{++} into He^{+}	
1.35×10^5	2338	Pile-up boundary Cometopause	PB	Start of magnetic field pile-up Sudden (10^4 km) decrease in electron density, change of electron distribution function from isotropic to cigar shaped along field lines	Neubauer et al. (1986) d'Uston et al. (1987)
1.6×10^5		Cometopause		Sudden drop of proton density (factor 3) On Vega: sudden (10^4 km) appearance of cometary ions in ram direction	B. Goldstein et al. (1987) Gringauz et al. (1986)
1.35×10^5 – 4.6×10^3		Inner region			Rème et al. (1987)
8.6×10^4	2350	'Discontinuity' X	X	Gradual loss of energetic solar wind ions (probably by charge exchange) Second rapid drop of proton density (factor 3)	Johnstone et al. (1986) B. Goldstein et al. (1987) Krankowsky et al. (1986b)
2.7×10^4	0004	'Discontinuity' Y	Y	Start of sharp decrease in ion temperature	Schwenn et al. (1987) Hodges et al. (1986)
1.6×10^4	0007	'Discontinuity' Z	Z	Transition from radially outflowing plasma to stagnant plasma	Balsiger et al. (1986) Schwenn et al. (1987)
4.6×10^3	0010	Ionopause Contact surface	C	Magnetic field goes to zero Strong drop of ion temperature by ~ 1000 K	Neubauer et al. (1986) Balsiger et al. (1986) Krankowsky et al. (1986b) Schwenn et al. (1987)
605	0011	Closest approach	CA		

Fig. 3 (Reinhard, Chapter 1, p. 6) — The Japanese spacecraft Suisei, launched on 19 August 1985, which passed the nucleus of comet Halley at a distance of 151 000km on the sunward side on 8 March 1986. (Photograph courtesy of ISAS).

Fig. 7 (Reinhard, Chapter 1, p. 16) — The ultraviolet spectrometer aboard the Pioneer Venus Orbiter (PVO) spacecraft made systematic observations of comet Halley while in orbit around the planet Venus. It was particularly well placed to monitor the comet during perihelion passage in early February 1986 when observation from Earth was difficult. (Photograph courtesy of NASA).

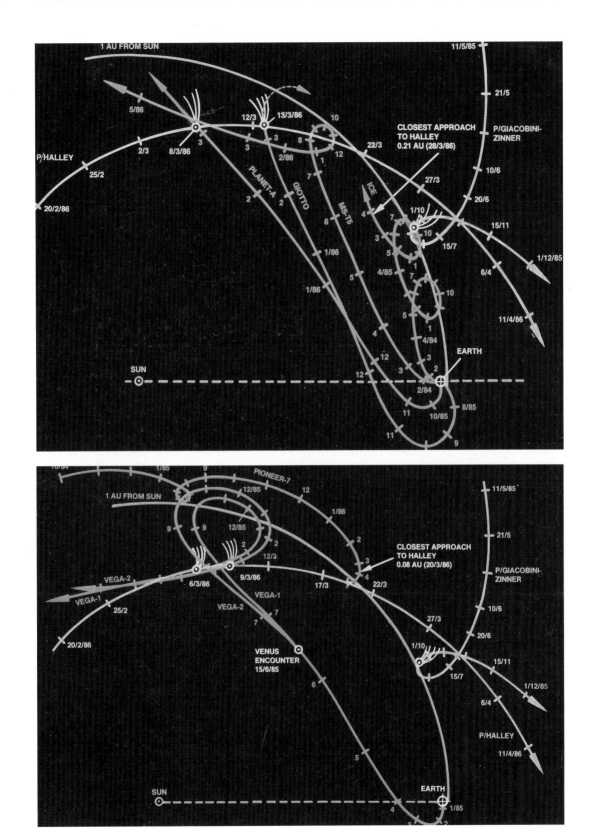

Fig. 8 (Reinhard, Chapter 1, p. 9) — Trajectories of the various Halley spacecraft relative to a fixed Sun-Earth line (courtesy of R. Farquhar).

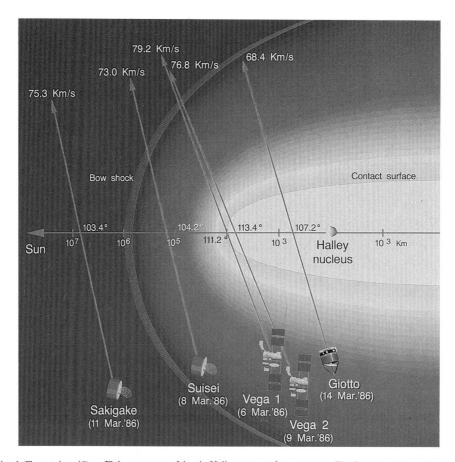

Fig. 9 (Reinhard, Chapter 1, p. 18) — Flyby geometry of the six Halley spacecraft at encounter. The Sun is to the left; the distance scale is logarithm[...]
For each mission the flyby dates are given at the bottom, flyby phase angles in the centre, and flyby speeds at the top.

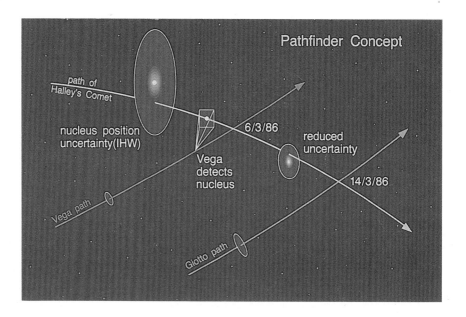

Fig. 10 (Reinhard, Chapter 1, p. 19) — Principle of the Pathfinder Concept (courtesy of J.F. Jordan).

Fig. 5 (Brandt, Chapter 3, p. 35) — Halley's comet on 1986 March 8.47 showing plasma tail (blue) and the dust tail (pale yellow). (Photograph by W. Liller, Large-Scale Phenomena Network of the IHW, from Easter Island).

Fig. 10 (Brandt, Chapter 3, p. 36) — Halley's comet on 1986 April 14.51 showing a wide dust tail (pale yellow), a narrow plasma tail (blue), and the globular cluster ω Centauri. Image taken on Easter Island by W. Liller as part of the IHW/L-SPN's Island Network. See Fig. 11 for additional detail on the comet.

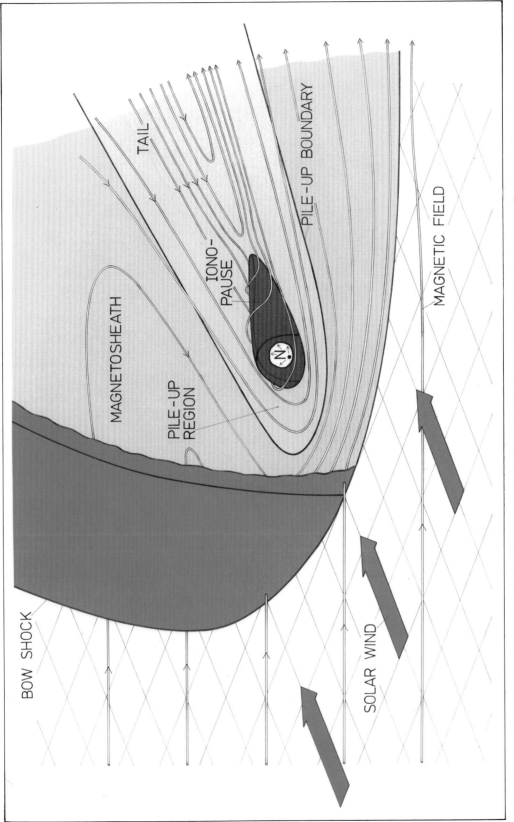

Fig. 4 (Neubauer, Chapter 5, p. 84) — Three dimensional view of the comet solar wind interaction. The blue bow shock shape has been cut out to allow the viewer to look on the symmetry plane with the magnetosheath and pile-up region. The three-dimensional cavity shape is shown in red as a closed volume in which ionization is essentially balanced by recombination globally. The figure thereby combines the cavity observations of Giotto with the observations of ICE where no tailward extension of the cavity as such was observed (Slavin *et al.* 1987). The regions are not drawn to scale because of the widely differing dimensions. The magnetic field and boundary topology should, however, be correct.

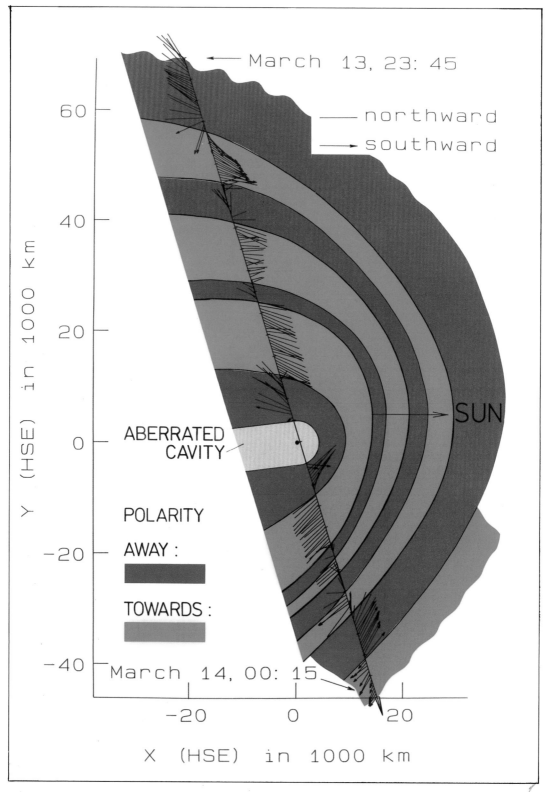

Fig. 5 (Neubauer, Chapter 5, p. 84) — Projections on the *X-Y* plane on the unit vectors **B**/|**B**| along the Giotto encounter trajectory. The unit vectors are based on 4-second averages of magnetic field vectors divided by two. The slabs of opposite interplanetary polarity are shown in red and blue, where red denotes field lines coming from the Sun and vice versa (From Raeder *et al.* 1987).

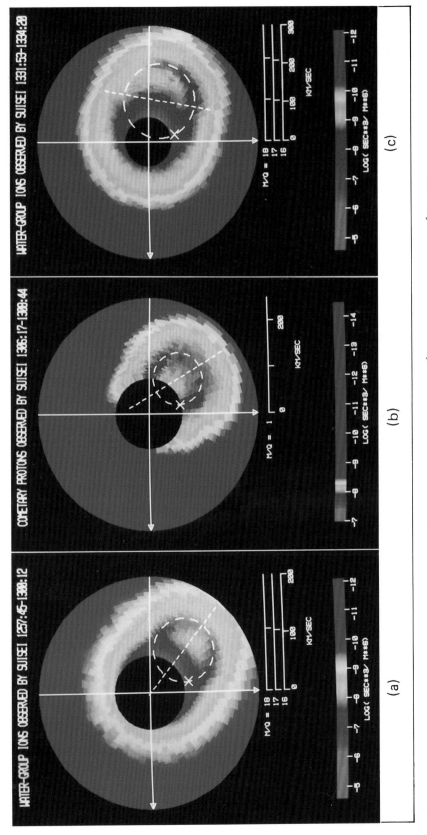

Fig. 6 (Miyake *et al.*, Chapter 12, p. 196 & 197) — Phase space distributions of ions (a:left) around 1.53×10^5km, (b:middle) around 1.51×10^5km, and (c:right) 1.92×10^5km from the nucleus. The colour scale is made by assuming that all ions were 0^+ for (a) and (c), or protons for (b). Horizontal and vertical arrows give the directions of $+V_x$ and $+V_y$, respectively. The CSE frame is used except that the origin of the phase is shifted to the spacecraft frame. The velocity scales are for H_2O^+, OH^+, and O^+ from the top in panels (a) and (c), and for protons in (b). Circles indicate the fitting positions of pick-up shells for O^+ (a) and (c) and for protons (b). Dashed lines show symmetry axes of the observed shell structure.

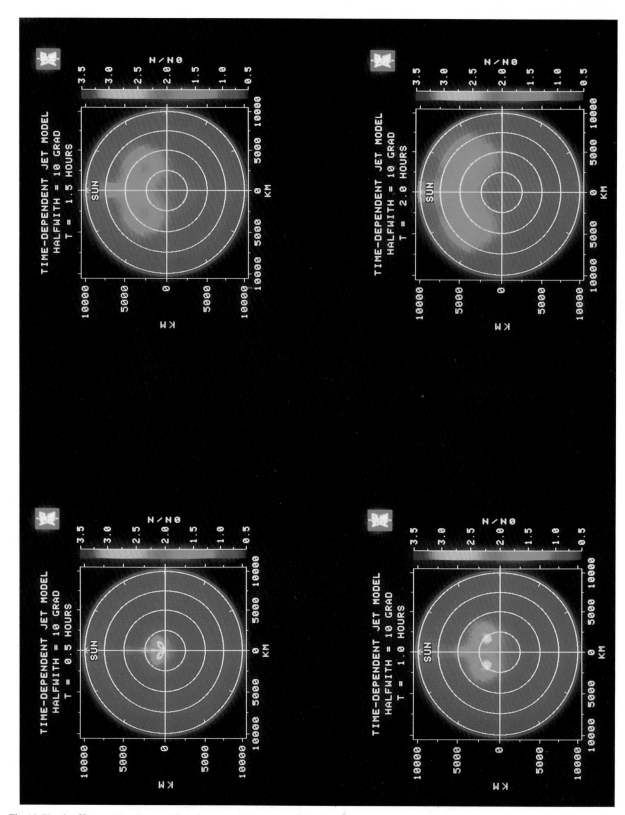

Fig. 15 (Kömle, Chapter 15, p. 242) — Time-dependent hydrodynamic simulation illustrating the development of a gas outburst on the cometary surface lasting 30 minutes. It is assumed that the source of the outburst is a region of 10° extent around the subsolar point on the cometary nucleus.

Fig. 5 (Feldman, Chapter 17, p. 265) — False-colour hydrogen Lyman-α image of comet Halley from the 13 March 1986 sounding rocket experiment of McCoy *et al.* (1986). (Naval Research Laboratory photograph, courtesy of R.P. McCoy)

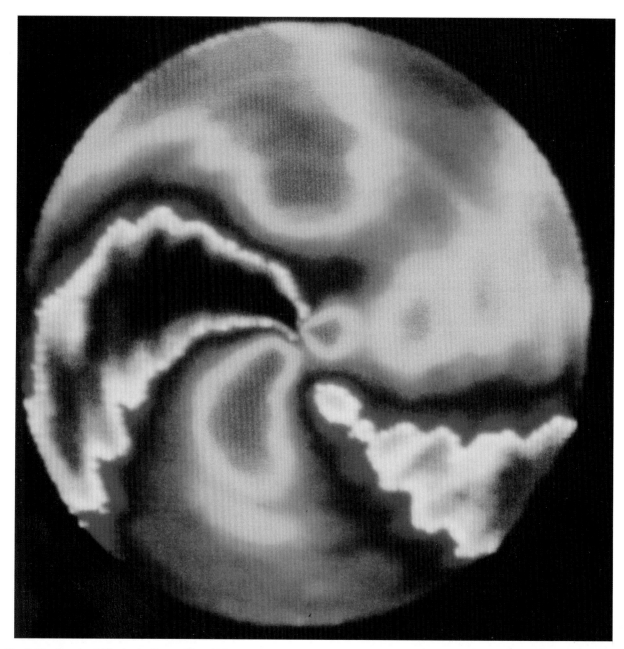

Fig. 8 (Krankowsky & Eberhardt, Chapter 18, p. 281) — Image of comet Halley obtained on 23 April 1986, through a CN filter. The false colour representation of the intensities shows in red the three prominent CN jets. The photograph was taken at Perth Observatory, Western Australia, by a team of astronomers from the University of Maryland (Michael A'Hearn and Susan Hoban) and from Perth Observatory (Peter Birch, Craig Bowers, Ralph Martin) and enhanced at NASA-Goddard Space Flight Center by Daniel Klinglesmith (A'Hearn *et al.* 1986a, 1986b).

elevation angle which demonstrates how these ions change direction relative to the Sun–comet–line as they have to move around the cometary obstacle. Here the main discontinuities BS, 1/2, 2/3, and PB are clearly visible. The so-called 'mystery region' (outer region 2) is particularly distinguished by the increase of the alpha particle density by a factor of three.

The pick-up oxygen (coming from dissociation of cometary water) has already been observed by IIS outside the bow shock, as discussed in the previous section. After its initial density jump at the shock by a factor of two its absolute density steadily increases further, and, of course, its abundance relative to the solar wind protons goes up from the 1% value measured at the shock. The transition between outer regions 2 and 3 is marked by a sudden change in the energy spectrum of the heavy ion population (Wilken et al. 1987b). As the spacecraft left the mystery region (boundary 2/3) the heavy pick-up ions split into a low-

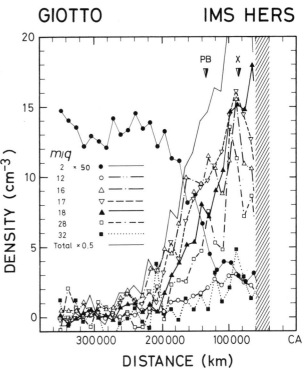

Fig. 4 — Radial profiles of ion densities for the main cometary ion species from HERS. For comparison, the profile of solar wind alpha particles ($m/q = 2$) is given. Beginning with the shaded area, ions start to disappear from the field of view of the HERS sensor. Discontinuities PB and X (Table 1) are marked with arrows. (Balsiger et al. 1987b).

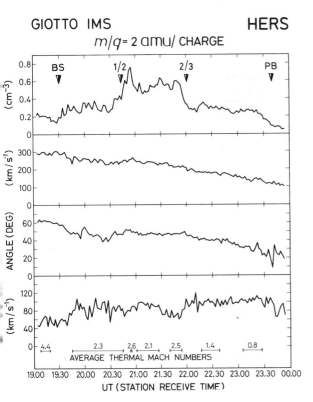

Fig. 3 — Plasma parameters for the $m/q = 2$ ions (mainly solar wind alpha particles) during Giotto encounter. Several discontinuities (see Table 1) are marked with arrows. (Balsiger et al. 1986).

and high-energy peak. According to Thomsen et al. (1987), this splitting is due to a sudden decrease of the bulk flow of the (solar wind) ions which tends to strongly separate the 'old' pick-up ions from the 'new' ones generated locally at lower speed. Before boundary 2/3 the low-energy O+ component had also been present, but was part of the low-energy tail of the O+ distribution which will always develop when new pick-up ions are added to the old ones (see also our discussion in the bow shock section). From around 2.5 × 10⁵ km IIS, PICCA, HERS, and NMS measure quite consistently ion intensities which are strongly increasing with decreasing distance from the comet (Amata et al. 1986, Rème et al. 1987, Balsiger et al. 1986, Hodges et al. 1986). The radial dependence is estimated to be R^{-x}, with $x \geq 2$; no dramatic discontinuity for the ions is observed at the pile-up boundary PB in contrast to the protons (B. Goldstein

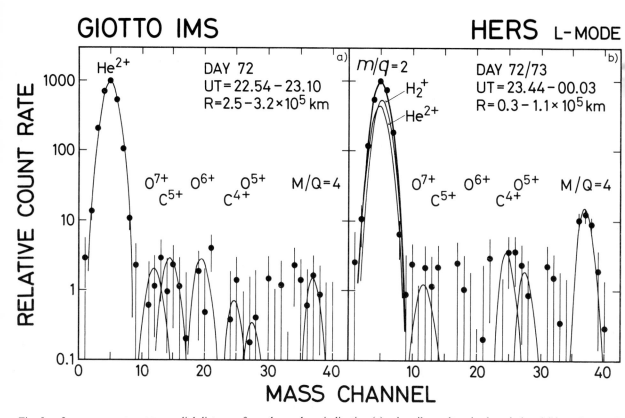

Fig. 5 — Ion mass spectra at two radial distances from the nucleus, indicating (a) primarily unaltered solar wind and (b) a mixture of partially charge exchanged solar wind ions (in particular $^4He^+$ and cometary hydrogen H_2^+). (Balsiger *et al.* 1986).

et al. 1987) and to the electrons (d'Uston *et al.* 1987). This observation we shall discuss further in section 3.2.

In the outer region 3, charge exchange of the solar wind ions with the outflowing cometary gas has been recognized to become increasingly important. Fig. 5 includes two mass spectra from the HERS L-mode (m/q-range 2–4 amu/e). Whereas spectrum (a) from outside 2.5×10^5 km reveals a typical solar wind spectrum with the multiply charged oxygen and carbon ions, spectrum (b) from inside 1.1×10^5 km has a strongly increased signal at $m/q = 4$. This is interpreted as being $^4He^+$ resulting from charge exchange of the $^4He^{2+}$ with cometary gas. Also shown in this figure is the appearance of the cometary hydrogen molecule at $m/q = 2$. Fig. 6 is from a study of Shelley *et al.* (1987). It demonstrates how the charge exchange, as expected, becomes more and more important when the comet is approached. $^4He^+$ starts

rising relative to the total helium ($m/q = 2 + 4$) around the inner mystery region boundary (2/3), and the slope of the curve becomes very steep at the cometopause (PB). (This latter feature is, of course, consistent with Fig. 4 which shows $m/q = 2$ dropping steeply at PB). Whereas the qualitative picture of the increasing importance of charge exchange toward the comet is as expected, the observed quantity of He^{++} being transformed into He^+ at distances below $\sim 10^5$ km is much higher than expected (by a factor of 5 to 10; Shelley *et al.* 1987, Fuselier *et al.* 1988).

3.2 The pile-up boundary/cometopause

The region around the pile-up boundary (PB) is of special interest because of the transition from a solar wind dominated hot plasma to one that is dominated by cold cometary ions. However, the status of the available data analysis does not yet allow an assess-

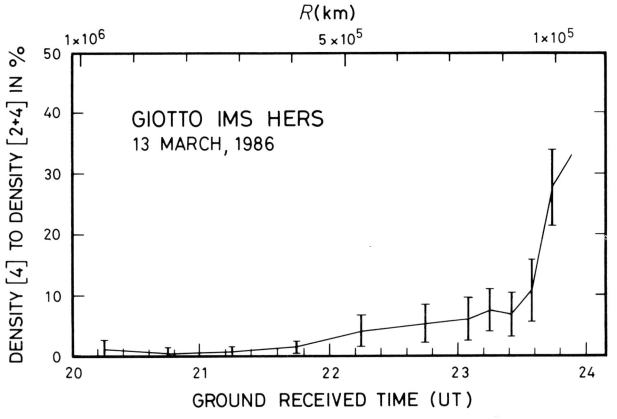

Fig. 6 — Ratio of the density of $m/q = 4$ ions to the combined densities of $m/q = 2$ and 4 ions versus distance to the nucleus along the inbound trajectory. The strong increase of charge exchange around the pile-up boundary/cometopause is evident. (Shelley *et al.* 1987).

ment of the full plasma populations; as we discussed earlier, none of the instruments could fully cover the transition between hot and cold plasma. The highly variable magnitude and direction of the magnetic field add to the difficulty in analyzing the particle data between discontinuities PB and X.

Fig. 7 shows that the pile-up boundary which has been defined by the magnetic field (Neubauer *et al.* 1986) and the electron data (Rème *et al.* 1987) coincides with a strong gradient in the proton density. A similar gradient is observed between 2349 and 2350, a location which we tentatively marked X. It is at X, 8.6×10^4 km from the nucleus, where Johnstone *et al.* (1986) had noted the gradual loss of energetic solar wind ions, probably due to charge exchange. In Fig. 4 we gave radial profiles of the major individual ions as detected by the HERS: $m/q = 2$ (He^{++}), 12 (C$^+$), 16 (mainly O$^+$), 17 (mainly OH$^+$), 18 (mainly H$_2$O$^+$), 28

(mainly CO$^+$), and 32 (mainly S$^+$). (For a discussion of the identification of the ions as given in parenthesis, see Balsiger *et al.* 1986). As expected, the most abundant cometary ions are water and ions resulting from the dissociation of water, OH$^+$ and O$^+$ (and H$^+$, which is not shown in the profile). The relative density values can be obtained from Table 2.

Fig. 4 demonstrates the steep but continuous increase of cometary ions toward the comet while in parallel the density of solar wind He^{++} decreases. Roughly at the cometopause, cometary ions become dominant over solar wind protons and alpha particles, mainly because of a steep decrease of the protons at this location (Fig. 7). The rather gradual increase of ion density through the pile-up boundary (PB) disagrees with a thin chemical boundary ($\sim 10^4$ km) postulated from the PLASMAG data on Vega (Gringauz *et al.* 1986); the HERS observation is, how-

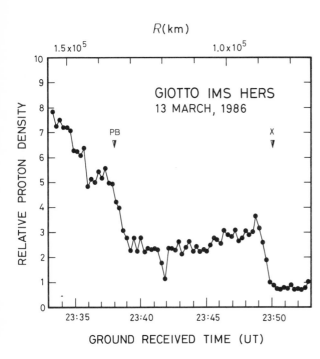

$R(\mathrm{km})$

GIOTTO IMS HERS
13 MARCH, 1986

RELATIVE PROTON DENSITY

PB

X

23:35   23:40   23:45   23:50

GROUND RECEIVED TIME (UT)

Fig. 7 — Radial profile of the relative proton density (arbitrary units). The proton density strongly decreases at about the pile-up boundary/cometopause (PB) and at 'discontinuity' X. (B. Goldstein *et al.* 1987).

able if displayed in higher time resolution than it is indicated in Fig. 4 (see Balsiger *et al.* 1987b). The shaded area in Fig. 4 marks the region where cometary ions have been observed to disappear from the field of view of the HERS (Balsiger *et al.* 1987b). This apparent drop in ion densities starting around 90 000 km is due to the fact that the cometary ion population had become so cold and stagnant that it could no longer be detected by HERS. A complete analysis of HIS, which was measuring in the ram direction, will be needed in order to obtain a profile of the total ion density (hot and cold) and to decide if the cometary ion profile reveals a yet unexplained discontinuity X (similar to the one in the proton profile).

In addition to the major ions discussed above, HERS could also identify several minor ions if the data were integrated over longer time periods. Fig. 8 shows a mass-per-charge spectrum averaged over the region of highest densities ($213–82 \times 10^3$ km). The displayed M- and H-modes cover different mass-per-charge regions with an overlap at the water group (see Balsiger *et al.* 1987a for a description of the modes). The minor ions which are clearly identified are $m/q = 13$ (mainly CH^+), 14 (CH_2^+ and N^+), and 15 (CH_3^+ and NH^+); more marginal is the identification of 19 (H_3O^+) and 23–24 (Na^+ and/or C_2^+). These are important constituents for modelling the cometary ionosphere. For this region and for two adjacent regions we also give the ion number densities for the major ions (Table 2). Except for the innermost region they have been corrected for the population lying outside the field of view of the instrument. Comparing the ions resulting from dissociation of water, 16, 17, and 18, it is evident how the relative importance of these constituents is changing with distance from the nucleus. Whereas at large distances water is strongly

ever, in agreement with Amata *et al.* (1986) and Hodges *et al.* (1986). If such a chemical boundary existed during the Giotto flyby, it had to be located at a different location (e.g. at X) and not at the pile-up boundary (PB). Further analysis of this very dynamic and complex region with inclusion of magnetic field data is needed to shed light on this question. As a matter of fact, the magnetic field changes direction several times inside the pile-up boundary (Neubauer 1986), and also the plasma density is much more vari-

Table 2 — HERS ion number densities (cm^{-3})

Range ($\times 10^3$ km)	$m/q = 2$	12	16	17	18	28	32
344–213	0.26 ±0.01	0.23 ±0.04	1.3 ±0.2	0.7 ±0.2	0.3 ±0.1	0.2 ±0.1	0.15 ±0.1
213–82*	0.23 ±0.02	2.5 ±0.4	14.5 ±2.2	14.3 ±2.3	10.2 ±1.7	7.7 ±1.4	2.5 ±0.6
82–16	(0.03) ±0.01	(1.2) ±0.2	(5.2) ±0.8	(7.0) ±1.1	(7.0) ±1.1	(3.8) ±0.7	(1.5) ±0.3

* Densities have been corrected for ions outside field of view.

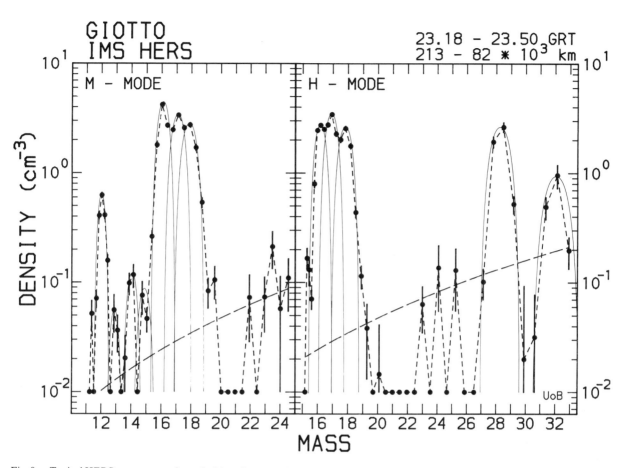

Fig. 8 — Typical HERS mass spectra from the M- and H-mode (see text). The dashed line represents a limit of confidence. Only data points with 2σ above this limit should be considered trustworthy. The solid lines represent theoretical mass peaks which have been fitted to the data. Errors are 1σ statistical errors. (Balsiger *et al.* 1987b).

dissociated into H^+, O^+, and OH^+, hence making O^+ the most abundant of the heavy ions, the H_2O^+ (18) dominates at smaller distances.

3.3 The inner region

As Giotto moved 'inward' from $\sim 10^5$ km the plasma environment became less turbulent, and solar wind protons and alpha particles became negligible. Fig. 9 gives an overview on the major (here rather cold) ions measured by the HIS sensor (only count rates are available at this time). The dependence of the main ions on radial distance outside $\sim 10^4$ km goes roughly with R^{-2}. Outside 6×10^4 km the dissociation products of water OH^+ and O^+ are more abundant than the ion of the mother product H_2O^+. Below $2 \times$

10^4 km the H_3O^+ ion becomes more abundant than H_2O^+, demonstrating the importance of ion-molecule reactions at high total densities. Between here ('discontinuity' Z) and the ionopause, the plasma was found to be stagnant (Fig. 10). This stagnant plasma has a local density peak around 10^4 km, termed the plasma pile-up region (Balsiger *et al.* 1986). Actually, both particles and magnetic field (Neubauer 1987) show a maximum here. Whereas the stagnation of the plasma in front of the ionopause was predicted by a model (Ip & Axford 1982), the observed ion density profile with a maximum at 10^4 km and a minimum just outside the ionopause was quite unexpected (cf. Fig. 13, points A and B). Candidate processes which have been mentioned to produce such a feature are electron impact ionization via the so-called critical

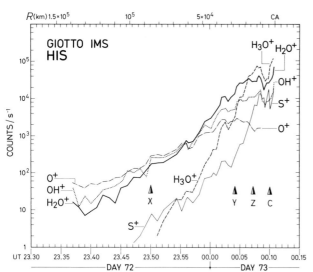

Fig. 9 — Radial profiles of the HIS count rates for major cometary ions. Time is ground station received time (GRT) in UT. 'Discontinuities' X, Y, Z, and the ionopause (C) are marked with arrows (cf. Table 1). (Balsiger *et al.* 1986).

velocity ionization effect (Alfvén 1954), local rapid loss of plasma along the draped magnetic field lines in the region of minimal density, and a temporal feature due to change in the solar wind condition or/and the outgassing rate. Ip *et al.* (1987) ruled out these processes and suggested that the minimum would be produced by enhanced electron dissociative recombination, a process which is efficient in a regime of low plasma velocity.

At this point it is interesting to note that in the stagnation region the plasma is decoupled from the neutral gas. The outflow of neutral water vapour was even accelerated here. The ion temperature, on the other hand (Fig. 10), increases from the ionopause to reach about 30 000 K at 2.7×10^4 km (discontinuity Y), while the gas temperature remains below 500 K over this range (Lämmerzahl *et al.* 1987, Schwenn *et al.* 1987).

4 THE INNER COMA

At 4 600 km from closest approach, Giotto crossed on the inbound pass the ionopause (contact surface) of Halley's comet (Neubauer *et al.* 1986, Balsiger *et al.* 1986, Krankowsky *et al.* 1986b). This very clear discontinuity, marked C in Fig. 10, was characterized

by a steep temperature drop of the ions by ~ 1000 K, by an increase of the ion outflow velocity from zero to about 800 m s^{-1} and by a drop of the magnetic field to zero (Balsiger *et al.* 1986, Schwenn *et al.* 1987, Lämmerzahl *et al.* 1987, Neubauer *et al.* 1986). Inside C, the Giotto spacecraft was basically immersed in a cold plasma of purely cometary origin, Halley's ionosphere. (For a discussion of energetic ions also observed inside the ionopause see Goldstein *et al.* 1987; Eviatar *et al.* 1989). In this cold plasma it was possible to detect more complex molecular ions and even to measure several isotopic ratios, (a) because several compounds were not yet broken up by solar UV so close to the nucleus, and (b) sensors like NMS (in the ion mode) and PICCA had better mass resolution for cold ions.

Fig. 11 is a comparison of two IMS mass-per-charge spectra, one outside (6 000 km) and one inside (1 500 km) the ionopause. The most striking differences in (b) with respect to (a) are: the increased $m/q = 19$ (H_3O^+) demonstrating the increase in ion–molecule reactions at higher density; the higher variety of peaks in the region 24–34 possibly including HCN^+, CO^+, N_2^+, H_2CO^+, $C_2H_n^+$ hydrocarbons, S^+, H_2S^+;

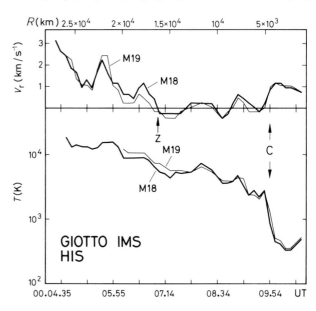

Fig. 10 — Radial profiles of ionospheric flow speed relative to the comet, and temperature derived from HIS mass peaks 18 (H_2O^+) and 19 (H_3O^+). The ionopause (C) and the 'discontinuity' Z (Table 1) are marked with arrows. (Balsiger *et al.* 1986).

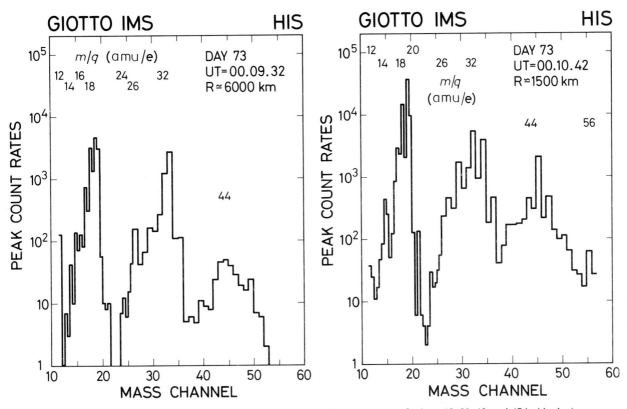

Fig. 11 — HIS mass spectra outside (a) and inside (b) the ionopause. The appearance of m/q = 19, 30, 45, and 47 inside the ionopause demonstrates the importance of ion–molecule reactions (see text). (Balsiger *et al.* 1986).

several peaks around 44 (CO_2^+) which according to Krankowsky *et al.* (1986b) are probably CS^+ (44), HCS^+ and HCO_2^+ (45) and CH_2SH^+ (47); and the presence of iron-ions indicated by the peak observed at m/q = 56.

The PICCA sensor, an electrostatic analyzer, was designed to measure water clusters and other heavy molecules in the cold plasma region where all ions have virtually the same (ram) velocity. Fig. 12 gives an example of such a measurement at 11 000 km, hence from outside the ionopause, but it should be quite representative also for the ionosphere. Whereas the broad peak to the left comes from the masses around 45 (see discussion above), the mass peaks in the right frame were tentatively identified as m/q = 64, 76, ~86, ~100 (Korth *et al.* 1986). The authors first suggested that a probable identification is S_2^+ and CS_2^+ for peaks 64 and 76; later a case was made that these heavy peaks were evidence for the presence of

polymerized formaldehyde (Mitchell *et al.* 1987). Here, as in the case of the HIS results, more careful analyses and, in particular, involvement of chemical models are needed to make proper identifications.

In the inner coma, the NMS in the ion mode had high enough mass resolution to determine isotopic ratios of hydrogen, oxygen, and sulphur (Krankowsky *et al.* 1986b, Eberhardt *et al.* 1987b). They are given in Table 3. The hydrogen and oxygen isotopic ratios are derived from the water group ions (m/q = 18, 19, 20, 21). As these mass locations are not only occupied by water, the calculations involve substantial corrections and are dependent on some assumptions. For example, one has to assume how much NH_4^+ contributes to m/q = 18. Therefore, the errors are still quite large but will be reduced with more progress in the model calculations. The ratios given in Table 3 are consistent with Solar System isotopic abundances, and the D/H ratio, in particular, is in the range which is observed

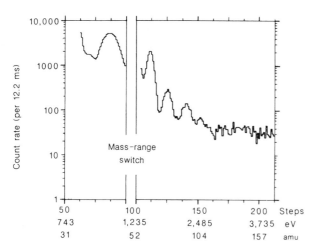

Fig. 12 — Count rates for heavy ions ∼ 11 000 km from the nucleus from the PICCA instrument, indicating the presence of complex molecules above 50 amu/e in the coma. (Korth *et al.* 1986).

for those Solar System objects that acquired their hydrogen as part of volatile molecules, e.g. as ices (Eberhardt *et al.* 1987b).

Table 3.   Isotopic ratios of ions in Halley's coma

$$0.6 \times 10^{-4} \lesssim D/H \lesssim 4.8 \times 10^{-4}$$

$${}^{18}O/{}^{16}O = 0.0023 \pm 0.0006$$

$${}^{34}S/{}^{32}S = 0.0450 \pm 0.010$$

Important clues for the proper understanding of the size of the ionosphere and of the complex source and loss processes are the location of the ionopause (at given gas production rates and given solar wind conditions) and the radial dependences of ion density, velocity, and temperature. It has become clear that in the pressure equilibrium between the solar wind outside and the particle pressure inside the ionopause the drag force which the neutral particles exert on the solar wind ions is an important factor. Without this drag the ionopause would be much closer to the nucleus. For a more thorough discussion see Cravens (1986) and Haerendel (1987).

As regards the ion source and loss processes, numerical modelling will have to be used after proper data analysis has led to radial dependence of the different ion and neutral species. However, a clue is already available from the HIS data, which have

revealed an R^{-1} radial dependence of the ions inside the ionopause, compared with R^{-2} outside (Fig. 13). Such an R^{-1} dependence would be expected for local photochemical equilibrium (i.e. when the photoionization of the neutral gas is locally balanced by recombination of ions with electrons). A more detailed analysis will probably reveal a more complex picture, in particular as it is now known that not only the nucleus but also the outstreaming dust has to be considered as a source for neutral gases (Eberhardt *et al.* 1987a).

4.1 Elemental and molecular abundance derived from ion measurements

As already discussed, it is a disadvantage of ion investigations that one deals mainly with secondary

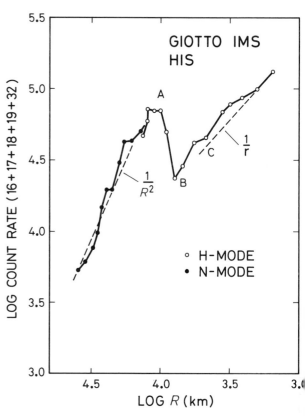

Fig. 13 — Radial profile of the sum of the HIS count rates for the major cometary ions (16, 17, 18, 19, and 32). Inside the ionopause (C), the count rate follows an R^{-1} dependence; outside an R^{-2} dependence fits the data quite well. Ions are either piled-up in the region of the magnetic field maximum (A) or depleted in the region of stagnating plasma (B). (Balsiger *et al.* 1986).

products (after dissociation, ionization, ion molecule reactions, charge exchange, etc.). On the other hand, all cometary molecules will eventually end up as (mainly elemental) ions. An integral of the ion composition over the full encounter should therefore give a good estimate of the elemental (not molecular) abundance of the volatile cometary material. The IMS team has given such a preliminary estimate, and has compared it to solar abundance (Table 4). One concludes that within the (quite large) error limits the major elemental abundances agree with solar abundances, except for the nitrogen which (in molecular form) is very volatile and is depleted in comets (Balsiger *et al.* 1986, Geiss 1987).

The fact that the H_3O^+/H_2O^+ ratio in the inner coma is very sensitive to the presence of NH_3, and, furthermore, the fact that CH_3^+ does not react with water and hence dominates the $m/q = 15$ peak allowed Allen *et al.* (1987) to compute from the IMS spectra (see Fig. 11b) with an Eulerian model of chemical and physical processes the relative gas production rates of the two mother molecules. With a water abundance of 80% (Krankowsky *et al.* 1986b) abundances for methane and ammonia of $\sim 2.5\%$ and $\sim 2\%$, respectively, are deduced. After assessment of the assumptions used in the model and possible errors we would consider these values as upper limits.

Table 4 — Element abundances in the gas phase of P/Halley

	Gas phase Halley	'Solar System'
C/O	0.33	0.61
N/O	≤0.01	0.13
S/O	0.03	0.026

Preliminary estimate from IMS ion spectra (Balsinger *et al.* 1986)

5 SUMMARY AND DISCUSSION

We have reviewed the ion composition data from the experiments on the Giotto spacecraft. Although the data analysis is not yet complete, it can be stated that the measurements have greatly enhanced our knowledge of the interaction of the solar wind with comet Halley and of the composition of the volatile component of the nucleus. The most difficult region to analyze is that around the pile-up boundary/cometopause where the transition between solar wind

dominated and comet dominated plasma happens, and where the plasma properties change dramatically; hence, several sensors have to be used for the analysis of the full plasma population, and only after this analysis is available can a final assessment of the interaction phenomena in the cometosheath be made.

The interaction between solar wind and comets had been predicted qualitatively rather well. The Halley missions have confirmed the existence of two main discontinuities, of a bow shock and of an ionopause (contact surface), and with the now available data the missing parameters can be determined and quantitative models established. The transition between solar wind dominated and comet dominated plasma happens around the pile-up boundary/cometopause, a discontinuity which is still somewhat controversial because not all investigators deduce the same thickness for it. In addition to the quite clear discontinuities — shock front and ionopause — and the above mentioned pile-up boundary/cometopause, several other discontinuities and regions have been observed in the cometosheath during the Giotto encounter (see Table 1). It remains to be seen which of those are more temporal and which are fundamental discontinuities of the solar wind–comet interaction.

Composition measurements contribute to the understanding of a major interaction mechanism, the mass loading of the solar wind with cometary ions by the pick-up process. Except for the very energetic ions where the second order Fermi mechanism is suggested as an acceleration process (see McKenna–Lawlor 1989, for a discussion of energetic ions), a superposition of picked-up cometary ions can probably explain most of the observed distributions of the lower energy ions. The locally picked-up ions in the slowed down solar wind have less energy than the old ones originating further upstream. Turbulence inside the shock front may explain the observed smearing of the energy distributions of ions which have been well separated outside the bow shock (cf. Neugebauer *et al.* 1987b), and wave–particle interaction may be responsible for the pitch angle scattering of ions into shell distributions (compared with the expected ring distributions for a pure pick-up process).

Starting around 5×10^5 km from the nucleus, charge exchange with the neutrals becomes an increasingly important loss mechanism for the solar wind ions. This is directly observed for He^{++} decaying into He$^+$ (Shelley *et al.* 1987), but there is indirect evidence that this is also an important loss mechanism for protons (R. Goldstein *et al.* 1987). Whereas the qualitative picture — increasing importance of charge exchange with decreasing distance to the comet — is as expected, the observed quantity of He^{++} being transformed into He$^+$ at distances below $\sim 10^5$ km is by a factor of 5 to 10 higher than expected. Hence, charge exchange may be a more important loss mechanism for solar wind ions (including protons) than previously thought.

Ion composition measurements have, of course, also contributed to our knowledge of the composition of the volatile component of the nucleus. We distinguish here between molecular, elemental, and isotopic composition. Because ions are secondary and tertiary products of sublimating gas molecules, rather complex models are needed to deduce from ion abundances the molecular abundances. By means of such model calculations, the abundances of two molecules, methane (2.5%) and ammonia (2%), have been shown to be amazingly low. In particular, the low abundance of methane relative to CO ($\sim 10\%$; Eberhardt *et al.* 1987a) will have implications for models of the early Solar System. In a solar nebula with thermodynamical equilibrium, methane should be more abundant.

Preliminary estimates of the elemental abundance in cometary volatile matter from ion measurements give good agreement with solar abundances for carbon, oxygen, and sulphur, whereas nitrogen, which in molecular form is very volatile, is depleted (Table 4). The isotopic ratios of sulphur, oxygen, and hydrogen determined from ion measurements (Table 3) also agree with Solar System values. Hence, isotopic and elemental abundances are well in agreement with a Solar System origin of comets.

On the other hand, there are several observations which give evidence that interstellar molecules may have survived the Solar System formation unharmed, because comets were formed at low temperatures in far out regions (20–30 AU) of the Solar System. It came as a surprise that carbon and nitrogen were to a large extent not found in their reduced form as methane and ammonia but as CO and probably N$_2$ (which has been lost due to its volatility). This

resembles what might be expected in interstellar molecular clouds where low temperatures lead to unusual chemistry. Also the D/H ratio in cometary water could be consistent with such an origin (Geiss 1987). Hence, comets may provide us with a link between the early Solar System and its ancestor, an unnamed interstellar molecular cloud.

The ionopause and the region of stagnant plasma in front of it have been predicted by models (cf. Ip & Axford 1982) in principle. However, the location of the ionopause was unpredictable, and again the *in situ* measurements were needed for the proper determination of parameters (e.g. ion–neutral drag). Also unpredicted was the pile-up of ions at the same location as the magnetic field pile-up (Fig. 13, point A) or, in other words, the minimum of the ion density in the stagnation region (Fig. 13, point B). At this time, the most likely mechanism to explain this phenomenon seems to be enhanced electron dissociative recombination due to a change in electron temperature (Ip *et al.* 1987).

Ion–molecule reactions in the coma clearly manifest themselves inside 20 000 km when H$_3$O$^+$ becomes the dominant ion (Fig. 9). Inside the ionopause, more complex reaction products like H$_2$CO$^+$, C$_2$H$_n^+$ hydrocarbons, CS$^+$, HCS$^+$, HCO$_2^+$, and CH$_2$SH$^+$ are found, and must eventually be explained by models of the inner coma. There is also evidence for heavier cometary ions possibly including polymerized formaldehyde in the coma (Fig. 12). Modelling of the PICCA sensor response as well as modelling of the ion chemistry in the coma will be needed to unfold these quite crude mass peaks which each cover several masses.

ACKNOWLEDGEMENTS

I acknowledge all the stimulating discussions within the whole IMS team, and in particular those with J. Geiss who has also critically reviewed this paper. This paper would not have been possible without a Giotto mission and without the experiment teams. I therefore acknowledge the dedicated support of the many technicians within ESA and at the many laboratories which provided the experiments. I thank I. Peter and G. Troxler for typing and E. Sennhauser for graphics. This work was supported by the Swiss National Science Foundation and by the Bern Government.

REFERENCES

Allen, M., Delitsky, M., Huntress, W., Yung, Y., Ip, W.-H., Schwenn, R., Rosenbauer, H., Shelley, E., Balsiger, H., & Geiss, J. (1987). Evidence for methane and ammonia in the coma of Comet Halley; *Astron. Astrophys.* **187** 502

Alfvén, H. (1954) *On the origin of the solar system*, Oxford, Clareton Press

Amata, E., Formisano, V., Cerulli-Irelli, R., Torrente, P., Johnstone, A.D., Coates, A., Wilken, B., Jockers, K., Winningham, J.D., Bryant, D., Borg, H., & Thomsen, M. (1986) The cometopause region at Comet Halley; *Proc. 20th ESLAB Symposium on the Exploration of Halley's Comet*, M., Heidelberg, *ESA SP-250* **1** 213

Balsiger, K. Altwegg, H., Bühler, F., Geiss, J., Ghielmetti, A.G., Goldstein, B.E., Goldstein, R., Huntress, W.T., Ip, W.-H., Lazarus, A.J., Meier, A., Neugebauer, M., Rettenmund, U., Rosenbauer, H., Schwenn, R., Sharp, R.D., Shelley, E.G., Ungstrup, E., & Young, D.T. (1986) Ion composition and dynamics at Comet Halley, *Nature* **321** 330

Balsiger, H., Altwegg, K., Benson, J., Bühler, F., Fischer, J., Geiss, J., Goldstein, B.E., Goldstein, R., Hemmerich, P., Kulzer, G., Lazarus, A.J., Meier, A., Neugebauer, M., Rettenmund, U., Rosenbauer, H., Säger, K., Sanders, T., Schwenn, R., Shelley, E.G., Simpson, D., & Young, D.T. (1987a) The ion mass spectrometer on Giotto, *J. Phys. E: Sci. Instrum.* **20** 759

Balsiger, H., Altwegg, K., Bühler, F., Fuselier, S.A., Geiss, J., Goldstein, B.E., Goldstein, R., Huntress, W.T., Ip, W.-H., Lazarus, A.J., Meier, A., Neugebauer, M., Rettenmund, U., Rosenbauer, H., Schwenn, R., Shelley, E.G., Ungstrup, E., & Young, D.T. (1987b) The composition and dynamics of cometary ions in the outer coma of comet P/Halley, *Astron. Astrophys.* **187** 163

Cravens, T.E. (1986) The physics of the cometary contact surface, *Proc. 20th ESLAB Symposium on the Exploration of Halley's Comet*, Heidelberg, *ESA SP-250* **1** 241

Eberhardt, P., Krankowsky, D., Schulte, W., Dolder, U., Lämmerzahl, P., Berthelier, J.J., Woweries, J., Stubbemann, U., Hodges, R.R., Hoffman, J.H., & Illiano, J.M. (1987a) On the CO and N_2 abundance in Comet Halley, *Astron. Astrophys.* **187** 481

Eberhardt, P., Dolder, U., Schulte, W., Krankowsky, D., Lämmerzahl, P., Hoffmann, J.H., Hodges, R.R., Berthelier, J.J., & Illiano, J.M. (1987b) The D/H ratio in water from Halley, *Astron. Astrophys.* **187** 435

Eviatar, A., Goldstein, R., Young, D.T., Balsiger, H., Rosenbauer, H., & Fuselier, S.A. (1989) Energetic ion fluxes in the inner coma of comet P/Halley, *Astrophys. J.* **339** 545–557

Fuselier, S.A., Shelley, E.G., Balsiger, H., Geiss, J., Goldstein, B.E., Goldstein, R. & Ip, W.-H. (1988) Cometary H_2^+ and solar wind He^{2+} dynamics across the Halley cometopause, *Geophys. Res. Lett.* **15** 549–552

Geiss, J. (1987) Composition measurements and the history of cometary matter, *Astron. Astrophys.* **187** 859

Goldstein, B.E., Neugebauer, M., Balsiger, H., Drake, J., Fuselier, S.A., Goldstein, R., Ip, W.-H., Rettenmund, U., Rosenbauer, H., Schwenn, R., & Shelley, E.G. (1987) Giotto-IMS observations of ion flow velocities and temperatures outside the contact surface of Comet Halley, *Astron. Astrophys.* **187** 174

Goldstein, R., Young, D.T., Balsiger, H., Bühler, F., Goldstein, B.E., Neugebauer, M., Rosenbauer, H., Schwenn, R., & Shelley, E.G. (1987) Hot ions observed by the Giotto ion mass spectrometer at the Comet Halley contact surface, *Astron. Astrophys.* **187** 220

Gringauz, K.I., Gombosi, T.I., Tátrallyay, M., Verigin, M.I., Remizov, A.P., Richter, A.K., Apáthy, I., Szemerey, I., Dyachkov, A.V., Balakina, O.V., & Nagy, A.F. (1986) Detection of a new 'chemical' boundary at Comet Halley, *Geophys. Res. Lett.* **13** 613

Haerendel, G. (1987) Plasma transport near the magnetic cavity surrounding Comet Halley; *Geophys. Res. Lett.* **14** 673

Hodges, R.R., Illiano, J.M., Berthelier, J.J., Krankowsky, D., Lämmerzahl, P., Woweries, J., Stubbemann, U., Hoffman, J.H., Eberhardt, P., Dolder, U., & Schulte, W. (1986) Measurements of thermal ion energy spectra from the Giotto encounter with Comet Halley; *Proc. 20th ESLAB Symposium on the Exploration of Halley's Comet*, Heidelberg, *ESA SP-250* **3** 415

Ip, W.-H. & Axford, W.I. (1982) Theories of physical processes in the cometary comae and ion tails, In: *Comets* (ed. L.L. Wilkening), 588

Ip, W.-H. & Axford, W.I. (1986) The acceleration of particles in the vicinity of comets, *Planet. Space Sci.* **34** 1061

Ip, W.-H., Schwenn, R., Rosenbauer, H., Balsiger, H., Neugebauer, M., & Shelley, E.G. (1987) An interpretation of the ion pile-up region outside the ionospheric contact surface; *Astron. Astrophys.* **187** 132

Johnstone, A., Coates, A., Kellock, S., Wilken, B., Jockers, K., Rosenbauer, H., Stüdemann, W., Weiss, W., Formisano, V., Amata, E., Cerulli-Irelli, R., Dobrowolny, M., Terenzi, R., Egidi, A., Borg, H., Hultquist, B., Winningham, J., Gurgiolo, C., Bryant, D., Edwards, T., Feldman, W., Thomsen, M., Wallis, M.K., Biermann, L., Schmidt, H., Lust, R., Haerendel G., & Paschmann, G. (1986) Ion flow at Comet Halley; *Nature* **321** 344

Kaneda, E., Hirao, K., Shimizu, M., & Ashihara, O. (1986) Activity of Comet Halley observed in the ultraviolet, *Geophys. Res. Lett.* **13** 833

Korth, A., Richter, A.K., Loidl, A., Anderson, K.A., Carlson, C.W., Curtis, D.W., Lin, R.P., Rème, H., Sauvaud, J.A., d'Uston, C., Cotin, F., Cros, A., & Mendis, D.A. (1986) Mass spectra of heavy ions near Comet Halley, *Nature* **321** 335

Korth, A., Richter, A.K., Loidl, A., Güttler, W., Anderson, K.A., Carlson, C.W., Curtis, D.W., Lin, R.P., Rème, H., Cotin, F., Cros, A., Médale, J.L., Sauvaud, J.A., d'Uston, C., & Mendis, D.A. (1987) The heavy ion analyser PICCA for the Comet Halley flyby with Giotto, *J. Phys. E: Sci. Instrum* **20** 787

Krankowsky, D., Lämmerzahl, P., Dörflinger, D., Herrwerth, I., Stubbemann, U., Woweries, J., Eberhardt, P., Dolder, U., Fischer, J., Herrmann, U., Hofstetter, H., Jungck, M., Meier, F.O., Schulte, W., Berthelier, J.J., Illiano, J.M., Godefroy, M., Gogly, G., Thévenet, P., Hoffman, J.H., Hodges, R.R., & Wright, W.W. (1986a) The Giotto neutral mass spectrometer, *ESA SP-1077* **109**

Krankowsky, D., Lämmerzahl, P., Herrwerth, I., Woweries, J., Eberhardt, P., Dolder, U., Herrmann, U., Schulte, W., Berthelier, J.J., Illiano, J.M., Hodges, R.R., & Hoffman, J.H. (1986b) *In situ* gas and ion measurements at Comet Halley, *Nature* **321** 326

Lämmerzahl, P., Krankowsky, D., Hodges, R.R., Stubbemann, U., Woweries, J., Herrwerth, I., Berthelier, J.J., Illiano, J.M.,

Eberhardt, P., Dolder, U., Schulte, W., & Hoffman, J.H. (1987) Expansion velocity and temperatures of gas and ions measured in the coma of Comet Halley, *Astron. Astrophys.* **187** 169

McKenna–Lawlor, S.; Energetic ion species in the Comet Halley plasma, the present volume, 1989

Mitchell, D.L., Lin, R.P., Anderson, K.A., Carlson, C.W., Curtis, D.W., Korth, A., Rème, H., Sauvaud, J.A., d'Uston, C., & Mendis, D.A. (1987) Evidence for chain molecules enriched in carbon, hydrogen, and oxygen in Comet Halley; *Science* **237** 626

Neubauer, F.M. (1987) Giotto magnetic field results on the pile-up region and cavity boundaries, *Astron. Astrophys.* **187** 73

Neubauer, F.M., Glassmeier, K.H., Pohl, M., Raeder, J., Acuña, M.H., Burlaga, L.F., Ness, N.F., Musmann, G., Mariani, F., Wallis, M.K., Ungstrup, E., & Schmidt, H.U. (1986) First results from the Giotto magnetometer experiment at Comet Halley; *Nature* **321** 352

Neugebauer, M., Lazarus, A.J., Altwegg, K., Balsiger, H., Goldstein, B.E., Goldstein, R., Neubauer, F.M., Rosenbauer, H., Schwenn, R., Shelley, E.G., & Ungstrup, E. (1986) The pick-up of cometary protons by the solar wind, *Proc. 20th ESLAB Symposium on the Exploration of Halley's Comet*, Heidelberg, *ESA SP-250* **1** 19

Neugebauer, M., Lazarus, A.J., Altwegg, K., Balsiger, H., Goldstein, B.E., Goldstein, R., Neubauer, F.M., Rosenbauer, H., Schwenn, R., Shelley, E.G., & Ungstrup, E. (1987a) The pick-up of cometary protons by the solar wind, *Astron. Astrophys.* **187** 21

Neugebauer, M., Neubauer, M., Balsiger, H., Fuselier, S.A., Goldstein, B.E., Goldstein, R., Mariani, F., Rosenbauer, H., Schwenn, R., & Shelley, E.G. (1987b) The variation of protons, alpha particles, and the magnetic field across the bow shock of Comet Halley, *Geophys. Res. Lett.* **14** 995

Neugebauer, M., Goldstein, B.E., Balsiger, H., Neubauer, F.M., Schwenn, R., & Shelley, E.G. The Density of Cometary Protons Upstream of Comet Halley's Bow Shock; *J. Geophys. Res.*, Vol. **945**, No. A2, 1261–1269, 1989a.

Neugebauer, M., Lazarus, A.J., Balsiger, H., Fuselier, S.A., Neubauer, F.M., & Rosenbauer, H. The velocity distributions of cometary protons picked up by the solar wind; *J. Geophys. Res.*, Vol. **94**, No. A5, 5227–5239, 1989b.

Rème, H., Sauvaud, J.A., d'Uston, C., Cros, A., Anderson, K.A., Carlson, C.W., Curtis, D.W., Lin, R.P., Korth, A., Richter, A.K., & Mendis, D.A. (1987) General features of the Comet Halley — solar wind interaction from plasma measurements, *Astron. Astrophys.* **187** 33

Schwenn, R., Ip, W.-H., Rosenbauer, H., Balsiger, H., Bühler, F., Goldstein, R., Meier, A., & Shelley, E.G. (1987) Ion temperature and flow profiles in Comet Halley's close environment; *Astron. Astrophys.* **187** 160

Shelley, E.G., Fuselier, S.A., Balsiger, H., Drake, J.F., Geiss, J., Goldstein, B.E., Goldstein, R., Ip, W.-H., Lazarus, A.J., & Neugebauer, M. (1987) Charge exchange of solar wind ions in the Comet Halley coma; *Astron. Astrophys.* **187** 304

Thomsen, M.F., Feldman, W.C., Wilken, B., Jockers, K., Stüdemann, W., Johnstone, A.D., Coates, A., Formisano, V., Amata, E., Winningham, J.D., Borg, H., Bryant, D., & Wallis, M., (1987) *In situ* observations of a bi-modal ion distribution in the outer coma of Comet Halley; *Astron. Astrophys.* **187** 141

d'Uston, C., Rème, H., Sauvaud, J.A., Cros, A., Anderson, K.A. Carlson, C.W., Curtis, D., Lin, R.P., Korth, A., Richter, A.K. & Mendis, A. (1987) Description of the main boundaries seen by the Giotto electron experiment inside the Comet Halley — solar wind interaction regions, *Astron. Astrophys.* **187** 137

Wilken, B., Weiss, W., Stüdemann, W., & Hasebe, N. (1987a) The Giotto implanted ion spectrometer (IIS): physics and technique of detection, *J. Phys. E. Sci. Instrum* **20**, 778–785

Wilken, B., Johnstone, A., Coates, A., Borg, H., Amata, E. Formisano, V., Jockers, K., Rosenbauer, H., Stüdemann, W. Thomson, M.F., & Winningham, J.D. (1987b) Pick-up ions at Comet Halley's bow shock: observations with the IIS spectrometer on Giotto; *Astron. Astrophys.* **187** 153

9

Some results of neutral and charged particle measurements in the vicinity of comet Halley by Vega-1, 2 spacecraft

K. I. Gringauz and M. I. Verigin

1 INTRODUCTION

Both Vega-1 and Vega-2 spacecraft carried plasma instrument scientific packages PLASMAG-1 which included neutral particle flux sensors and ion and electron energy spectrometers†

Two hemispherical electrostatic analyzers measured the energy/charge spectra of ions arriving from the spacecraft–comet relative velocity direction (the Cometary Ram Analyzer CRA) and from the direction of the Sun (the Solar Direction Analyzer SDA). Electrons were measured by a cylindrical electrostatic analyzer (EA) which was orientated perpendicular to the ecliptic plane. PLASMAG-1 also included two Faraday cups. The Solar Direction Faraday Cup (SDFC) measured the solar wind ion fluxes, while the Ram Faraday Cup (RFC), in addition to measuring the total flux of ions arriving from the ram direction, provided information on the neutral particle flux from the comet by detecting the secondary electrons and ions produced by the neutrals striking a metallic collector.

A description of the scientific payload and the detailed results of the plasma measurements have been published elsewhere (Apáthy et al. 1986, Balebanov et al. 1987, Galeev et al. 1986a, b, 1987, Gringauz et al. 1983, 1985, 1986a b, c, d, e, 1987a, b, Remizov et al., Verigin et al. 1986, 1987a, b). The present paper, which is mainly a compilation of the

above mentioned papers, contains the main observational results of the PLASMAG-1 scientific packages and our present understanding of these results.

2 THE NEUTRAL DENSITY DISTRIBUTION

The neutral particle density was estimated in situ aboard the Vega-1, 2 spacecraft by means of a rather primitive multi-electrode Ram Faraday Cup in a special mode of operation (Apáthy et al. 1986,

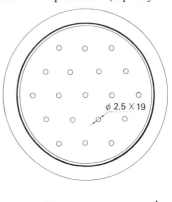

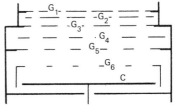

Fig. 1 — Schematics of the PLASMAG-1 Ram Faraday Cup (RFC) for measuring the neutral gas distribution in the vicinity of Halley's comet. The 19 holes in G_1 are of 2.5 mm diameter.

† The package was developed in the Space Research Institute of the USSR Academy of Sciences and in the Central Research Institute for Physics of the Hungarian Academy of Sciences, with participation of the Max-Planck-Institut für Aeronomie (FRG).

Gringauz *et al.* 1983, 1985, 1986a, b, Remizov *et al.* 1986). Fig. 1 (Remizov *et al.* 1986) is a schematic of this sensor which is different from the Faraday cups traditionally used on space probes. To avoid damage to the grid by cometary dust, the analyzing grid system was substituted by a flat disk electrode system G_1–G_6. There were 19 holes on each disk and the diameter of the holes increased toward the collector C.

When neutral particle fluxes were measured, voltages of -40 V and $+3500$ V were applied to electrodes G_2 and G_4, respectively. Photo- and secondary electrons of the surrounding plasma with energies $E_e < 40$ eV, and cometary and solar wind ions of energies $E_i < 3500$ eV were thereby diverted from the collector by these potentials.

In this way the collector current I_c of the RFC is given by the following, in the vicinity of the comet:

$$I_c \approx I_{se} - I_{si} + I_{ph} - I_e \qquad (1)$$

where I_{se} and I_{si} are the secondary electron and sputtered ion currents, which are proportional to the flow of neutral particles (the individual short bursts in I_c are associated with the registration of dust particles), I_{ph} is the photoelectron current, and I_e is the current of energetic electrons reaching the collector. To register I_{se} and I_{si} separately, a potential of $+40$ V and -60 V was applied to electrode G_6, respectively. In the first case, when the current of secondary electrons is measured, I_{si} is suppressed, and I_c is determined by I_{se}, I_{ph}, and I_e, thus:

$$I_{ce} \approx I_{se} + I_{ph} - I_e \qquad (2)$$

In the second case, when the sputtered ions current is registered, I_{se} is suppressed, I_{ph} decreases significantly and changes its sign (this current is now not produced by photoelectrons originating from the collector, but by the photoelectrons originating from electrode G_6), and the current I_e slightly decreases since in this case only the electrons with energy $E_e > 60$ eV can reach the collector. Thus, the collector current (ions) is:

$$I_{ci} \approx -I_{si} - I_{ph} - I_e \qquad (3)$$

The following equation was used to determine the neutral gas density on the basis of the measured $I_{ce,i}$ values:

$$n_n = \frac{I_{ce,i} - I_o}{q V_{sc} S Y_{e,i}} \qquad (4)$$

Here I_o is the sum of currents I_{ph} and I_e, q is the electron charge, $V_{sc} = 79.2$ km s^{-1} (76.8 km s^{-1}) is the velocity of Vega-1 (Vega-2) relative to the comet, $S = 0.93$ cm^2 is the total surface area of the holes of electrode G_1 (see Fig. 1), and $Y_{e,i}$ is the yield of secondary electrons and ions provided by neutral particles impacting the collector. In general, I_o in equation (4) is not a constant, as it will depend on time and on the distance from the cometary nucleus. However, for the estimation of n_n we assumed $I_o =$ constant and used the value as measured by RFC at large distances from the comet, where $I_{se,i} \approx 0$.

The accuracy of n_n estimations based on RFC data in the regions where $I_{ce,i} \gg I_o$ depends mainly on our present knowledge of the secondary electron Y_e and sputtered ion Y_i yields from the Ni emitter (collector) used. The secondary electron emission from the metal surface may be caused in two ways (Kaminsky 1965). The potential emission occurs when the impinging particle is an ion or an excited neutral. There is no minimum kinetic energy to start this process, but the influence of it on the I_{ce} value is small, as the number of ions with $E_i > 3500$ eV and the number of excited neutrals is small relative to the number of neutrals in the ground state. The kinetic electron emission occurs as a result of the energies of impinging particles exceeding the threshold energy. For the water group neutrals this energy is close to their energy relative to the spacecraft. Finally, the secondary ion emission results from the thermal sputtering of metal by a localized heating process.

There are no direct experimental data on $Y_{e,i}$ for Ni targets bombarded with water group molecules with $V_{sc} \approx 80$ km s^{-1}. Fig. 2 presents the results of measurements of $Y_{e,i}$ dependence on energy for Al and Au targets bombarded by water molecules (Shmidt & Azens 1987). These measurements were conducted with energies greater than ~ 700eV while the energy of water group molecules relative to the spacecraft is ~ 500 eV. The $Y_e \approx 0.3$ and $Y_i \approx 0.005$ values which were used during the preliminary processing of the RFC data are marked in Fig. 2 by asterisks.

The yield values presented in Fig. 2 were chosen by the PLASMAG-1 team just after receiving the first RFC cometary flyby data on 6 March 1986, and Fig. 3 (Gringauz *et al.* 1986b) (computed from expression (4) and $Y_{e,i}$) was presented to the scientific community

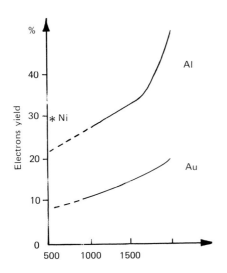
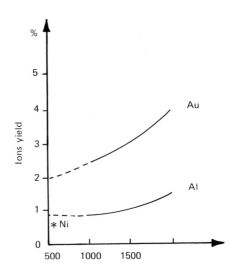

H$_2$O energy, eV

Fig. 2 — Yields of secondary electrons and sputtered ions from Al and Au targets (Shmidt & Azens 1987) compared with the values used for RFC measurements processing (asterisks).

meeting on the next day. Both $Y_{e,i}$ values were chosen as providing a reasonable agreement with the experimental data available (Fig. 2), and simultaneously providing the continuity of the neutral gas density profile n_n (R) (Fig. 3) as a function of R at the cometocentric distances $R \approx 10^5$ km, where the secondary electron registration mode is replaced by the sputtered ion registration mode. The optimization analysis of $Y_{e,i}$ selection is not yet completed, so the profile presented should be considered as preliminary; we evaluate its uncertainty to be a factor of ≈ 2.

In addition to the n_n (R) profile, the dashed line in Fig. 3 represents a simple fit to the data, assuming R^{-2} $\exp(-R/\lambda)$ density dependence. The ionization scale length was estimated to be $\lambda \approx 2 \times 10^6$ km, and a value of $\approx 1.3 \times 10^{30}$ molecules s^{-1} was obtained for the total gas production rate Q_o.

The data shown in Fig. 3 were obtained during the inbound pass of the flyby. On the inbound leg of Vega-2, the estimated n_n values were half those for Vega-1 at distances 1.5×10^4 km $< R < 10^5$ km, although the n_n values estimated from the data on both spacecraft are significantly closer to each other further away from the nucleus. Moreover, the values of n_n estimated from the I_{ce} measurements on the outbound leg of both Vega-1 and Vega-2 at distances 1.5×10^5 km $< R < 3 \times 10^5$ km are about half the corresponding n_n values on the inbound leg (see Figs. 4, 5 in Remizov et al. 1986).

These and other deviations might be caused by jets of neutral gas originating from the rotating nucleus.

VEGA-1 PLASMAG-1

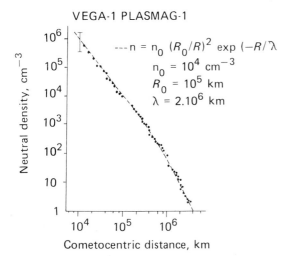

$$--- n = n_0 \, (R_0/R)^2 \exp (-R/\lambda)$$
$$n_0 = 10^4 \text{ cm}^{-3}$$
$$R_0 = 10^5 \text{ km}$$
$$\lambda = 2.10^6 \text{ km}$$

Fig. 3 — Radial profile of the overall neutral gas density, as measured inbound by the Vega-1 Ram Faraday Cup. Comparison with the distribution predicted by theory (dotted line) yields an ionization scale length of $\lambda \sim 2 \times 10^6$ km and an overall production rate of $Q_o \sim 1.3 \times 10^{30}$ molecules s^{-1}.

Also, (1) the solar UV radiation scattered by the comet may have some influence on the instrument performance, and (2) the solar radiation pressure can influence the asymmetry of the neutral gas distribution along the spacecraft trajectory. Although the neutral gas density estimations presented above were produced on the basis of a very simple instrument, the gas production rate of Halley's comet estimated from the RFC observations on board Vega-1 and Vega-2 is in reasonable agreement with the estimates provided by other instruments on board Vega-1, Vega-2, and Giotto. The production rates for the H_2O and OH molecules were determined as $Q_{H_2O} \approx 4 \times 10^{29}$ s^{-1} and $Q_{OH} \approx 1.7 \times 10^{30}$ s^{-1} from the spectroscopic observations by the three-channel spectrometer (TKS) on board Vega-2 in the visible and infrared spectral ranges (Krasnopolsky *et al.* 1986), and as $Q_{OH} \approx 9 \times 10^{29}$ s^{-1} in the near ultraviolet range (Moreels *et al.* 1986). The value of Q_o estimated from the data measured by PID onboard Vega-1 varies between $10^{30} < Q_o < 4 \times 10^{30}$ s^{-1} at cometocentric distances $R < 1.5 \times 10^5$ km.

Finally, a preliminary estimate of the gas production rate of $Q_o \approx 6.9 \times 10^{29}$ s^{-1} was provided by the instrument NMS on board Giotto (Krankowsky *et al.* 1986). Besides, at cometocentric distances 2×10^5 km $< R < 3 \times 10^6$, the RFC was the only sensor on spacecraft that was able to measure the distribution of neutral gas along the trajectory.

3 COMETARY PLASMA PHENOMENA

The main feature of near cometary space is the increase, by orders of magnitude, of the population of neutral particles with approach to the nucleus of the comet, as considered above. This phenomenon determines the peculiarities of plasma flow in this region.

The plasma sensors of the PLASMAG-1 scientific package are shown schematically in Fig. 4 (Apáthy *et al.* 1986). The essential feature of the CRA and SDA analyzers, is the quadrupole electrostatic lenses in front of curved analyzing plates. These lenses increase the fields of view of the CRA and SDA up to 14° × 32° and 30° × 38°, respectively. The number of grids in the SDFC is the same as the number of flat electrodes in the RFC (Fig. 1). For the ion flux measurements, the voltages −40 V and −60 V were applied to the electrodes G_2 and G_6. Then the subtraction of

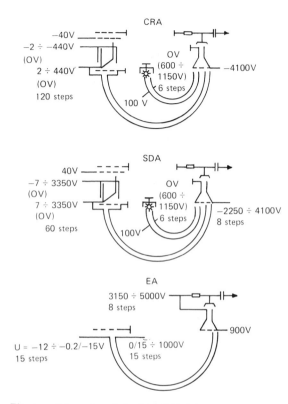

Fig. 4 — Schematics of the PLASMAG instrument.

collector C currents, measured with voltages 0 V and 3500 V applied to the G_4 electrode, permits the evaluation of the net ion flux fed to the SDFC and RFC. A more detailed description of the instrumentation is given in Apáthy (1986) and Gringauz *et al.* (1983, 1986a).

For convenience of further discussion of near-cometary plasma phenomena we present first, in Fig. 5, a general overview of the different plasma formations, as identified from the PLASMAG-1 plasma observations on board Vega-1 and -2 during their encounter with comet Halley (Gringauz *et al.* 1985). The terms 'cometosheath' and 'cometopause', were introduced by the PLASMAG-1 team (Gringauz *et al.* 1986b) to indicate the principal difference with respect to the physical processes occurring over this cometary regions as compared to those observed in the magnetosheath of planets with strong intrinsic magnetic fields, or in the ionosheath of planets with intense gravitational fields. These terms are now generally accepted. Let us consider the peculiarities of

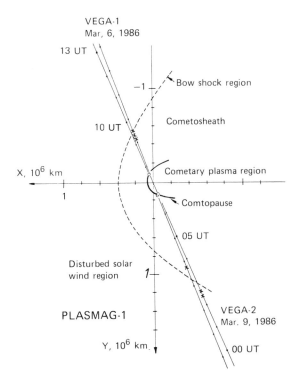

Fig. 5 — General overview of the inbound and outbound locations of the 'bow shock' and of the 'cometopause' as well as of the 'cometosheath' and of the 'cometary plasma region', as identified from PLASMAG-1 plasma observations on board Vega-1 and -2 during their encounter with comet Halley.

the plasma flow within the different plasma formations, beginning from the outermost ones.

3.1 Solar wind region disturbed by picked-up ions and the bow shock

In Fig. 6 the decrease in the solar wind proton velocities with cometocentric distance R is shown, as measured by the SDA on both Vega-1 and Vega-2 during their inbound legs. (Verigin *et al.* 1986, 1987a). The location of the bow shock, which was determined from simultaneous measurements of the plasma, the plasma waves, and of the magnetic field (Galeev *et al.* 1986a) is marked by S. As one can see, the decrease of the solar wind velocity due to mass loading by heavy cometary ions had already started at a distance of 2–3 $\times$ 10^6 km from the nucleus, i.e. 1–2 $\times$ 10^6 km upstream of the bow shock.

The process of mass loading and deceleration of the solar wind by new ions, originating from the cometary

neutrals, is well known in principle, and has already been considered in a number of hydrodynamic models of solar wind interaction with comets (see, e.g. Shmidt & Wegmann 1982). These ions first form a ring distribution in velocity space. The Alfvén wave turbulence being excited by the ion-cyclotron instability of such a distribution isotropizes the newly formed ions in the coordinate system moving with the solar wind (Sagdeev *et al.* 1986).

On the basis of the kinetic theory developed in that paper, a complete system of equations for the variation of the isotropic part of cometary ion distribution function and the hydrodynamical parameters of the flow along the **X** axis (parallel to the solar wind direction) was presented and solved in Galeev (1986, 1987) and Galeev *et al.* (1986a). In a three-dimensional case, if the main contribution to the plasma pressure P comes from the cometary ions with average mass $m_i \approx 17\, m_p$, the hydrodynamic part of this system can be written in the following form:

$$\left. \begin{aligned} \mathrm{div}(\rho V) &= Q(R) \\ \rho(v, \nabla)V &= -\nabla P - Q(R)V \\ \mathrm{div}(PV) &= QV^2/2, \end{aligned} \right\} \quad (5)$$

where ρ and **V** are the plasma flow density and velocity, and $Q = m_i n_n v_n/\lambda$ is the mass loading rate.

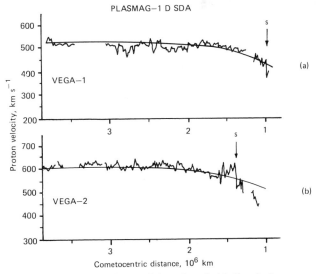

Fig. 6 — Radial dependence of the solar wind bulk velocity, as observed by Vega-1 and Vega-2 on their inbound trajectories upstream of the bow shock of Comet Halley (marked by S).

From equations (5) one can obtain the expected variation of the flow parameters along the Vega-1, -2 trajectories at large cometocentric distances, where the initial velocity u_o and density ρ_o are slightly perturbed by the cometary ions:

$$u = u_o - \frac{4}{3\rho_o} L_o(R, \varphi),$$

$$v = \frac{\cos\varphi.\sin\varphi}{3\rho_o} L_2(R, \varphi),$$

$$\rho = \rho_o + \frac{1}{u_o}\left(2L_o + \frac{3\cos^2\varphi - 1}{6} L_2 + \frac{\cos^2\varphi.\sin^2\varphi}{2} L_4\right),$$

$$P = \frac{u_o}{3} L_o(R, \varphi) \tag{6}$$

where,

$$L_n(R, \varphi) = R \int_{\cos\varphi}^{\infty} Q(R\sqrt{\zeta^2 + \sin^2\varphi})\zeta^n \, d\zeta,$$

u and v are parallel and perpendicular to X axis components of the plasma flow velocity vector, and $\phi \approx 110°$ is the angle between this axis and the Vega-1, 2 trajectory.

The cometocentric dependence of $V(R) \approx u(R)$ computed in accordance with (6) along the Vega-1, 2 trajectories, is shown in Fig. 6a ($u_o = 540$ km s^{-1}) and in Fig. 6b ($u_o = 620$ km s^{-1}) by smooth curves. The neutral density profile $n_n(R)$ was used in accordance with the RFC data described in the previous section; the initial ρ_o values were based on SDA measurements onboard Vega-1 ($\rho_o/m_p \approx 12$ cm^{-3}) and Vega-2 ($\rho_o/m_p \approx 11$ cm^{-3}). The reasonable agreement between the computed and measured velocity profiles is obvious.

The plasma flow deceleration while approaching the cometary nucleus is not continuous, and the bow shock is formed in front of the comet as a result of increasing mass loading by heavy cometary ions. The bow shock positions observed by Vega-1,2 are shown by crosses in Fig. 5. Fig. 7 (Verigin et al. 1986, 1987a) presents the energy spectra measured by the SDA onboard Vega-1 when crossing the bow shock on the inbound leg and on the outbound leg. Also, the gradual slowing down of protons, and a gradual widening of the energy spectra, can be observed when approaching the comet, i.e. the ion temperature is increasing several hours before reaching the bow

shock. After a data gap of 20 minutes around 0300 UT, the ion temperature was already so high that the proton distribution overlaps the α particle peak in the spectrum. The gradient of the velocity is increasing significantly when crossing the bow shock at a distance of 1.02×10^5 km from the nucleus (~ 0346 UT).

The variation of the plasma parameters seems to be more complicated when crossing the bow shock on the outbound leg of Vega-1. The highest gradient in the plasma velocity was observed between 0900 and 0930 UT at a distance of about 5.5×10^5 km from the nucleus. At the same time the ion temperature stayed high, so that the α peak could not be distinguished from the proton distribution until 1130 UT ($R = 1.2 \times 10^6$ km). Thus, on the outbound leg the bow shock was crossed at a distance of $5.5 \pm 1 \times 10^5$ km (Galeev et al. 1986a), and the high ion temperature which can be observed until 1220 UT for about another 8×10^5 km is associated with the high lever of magnetohydrodynamic (MHD) activity in the foreshock region.

The latter distance seems to be determined only by the distribution of the neutral gas density n_n around Halley's comet. The increase of n_n to 30 cm^{-3} at a distance of 1.3×10^6 km from the nucleus (Gringauz et al. 1985, 1986b, Remizov et al. 1986) seems to be enough to ensure turbulent heating of the solar wind ions caused by the unstable beam-like distribution of ions of cometary origin. The location of the bow shock

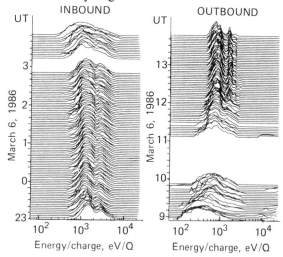

Fig. 7 — Ion energy spectra measured by Vega-1 on the inbound leg and on the outbound leg around the bow shock of Halley's comet.

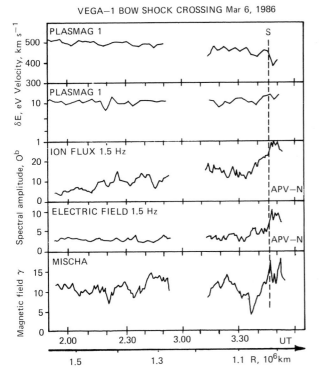

Fig. 8 — The behaviour of solar wind parameters during the inbound bow shock crossing by the Vega-1 spacecraft on 6 March 1986. From top to bottom: solar wind velocity and effective temperature measured in solar direction; spectral amplitude of ion flux fluctuations and electric field oscillations with the frequency $f = 1.5$ Hz; magnetic field strength.

and the gradient of the ion velocity related to the shock is naturally different on the inbound leg as compared to the outbound leg of the Vega-1 trajectory as seen in Fig. 5, since the bow shock was crossed around the flanks on the inbound leg but closer to the subsolar point on the outbound leg. This is the reason for the difference between the dimensions of the regions of hot solar wind observed inbound and outbound around the bow shock.

The behaviour of other plasma parameters across the bow shocks can be judged from the data presented in Fig. 8 (Galeev et al. 1986a). The solar wind velocity and the effective temperature δE on this figure are the first and the second moments of SDA ion energy spectra, respectively. The crossing of this quasiperpendicular bow shock is identified most easily by the sharp increase in intensity of the plasma waves. These lower hybrid plasma waves are excited by the beam of

cometary ions leaking forward from behind the subshock front, similar to the excitation of these waves at the quasiperpendicular Earth bow shock by the beam of protons reflected from the shock (Galeev 1987). The jump of the magnetic field at the shock front is masked in Fig. 8 by the strong MHD turbulence of a loaded solar wind.

The more complicated phenomena were seen at the quasi-parallel cometary shocks observed by Vega-1 on the outbound leg (Fig. 7) and by Vega-2 on the inbound leg of the trajectory (Galeev et al. 1986a). Further study and comparison with computer simulations of these formations (Galeev 1987) is required.

3.2 Cometosheath region and the cometopause

Downstream of the cometary bow shock both Vega spacecraft encountered a region which was called the cometosheath (Gringauz et al. 1986b). The principal difference between the plasma flow within this region and that within the regions downstream of bow

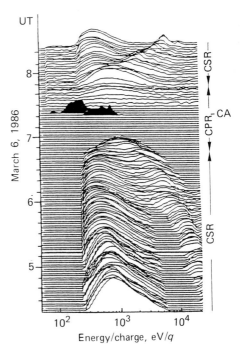

Fig. 9 — Sequence of energy-per-charge spectra observed by the solar-direction ion electrostatic analyzer SDA on board Vega-1: (a) on the inbound and outbound leg in the cometosheath regions (CSR) and in the cometary plasma region (CPR).

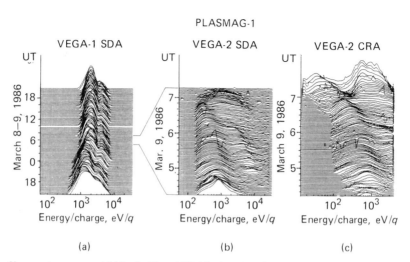

Fig. 10 — Sequences of ion spectra measured (a) by the Vega-1 SDA in the solar wind during the time of the encounter of Vega-2; (b) by the Vega-2 SDA downstream of the cometary bow shock and up to a cometocentric distance of $\sim 1.5 \times 10^4$ km; and (c) by the Vega-2 Cometary Ram Analyzer (CRA) over the same distance range.

shocks near other celestial bodies is readily seen from the ion spectra as measured by the Vega-1 SDA (Fig. 9) and by the Vega-2 SDA (Fig. 10b) and CRA (Fig. 10c) over the cometosheath. It is worth noting that the comparison of the ion spectra observed by the Vega-1 SDA (Fig. 10a) in the undisturbed solar wind and, simultaneously, by the Vega-2 SDA (Fig. 10b) in the cometosheath region certainly proves that the plasma disturbances in this region are not of solar wind origin.

At a cometocentric distance of $\sim 8 \times 10^5$ km the SDA post-bow shock ion spectra are characterized by a wide, single maximum distribution of decelerated and heated solar wind ions (see the lower parts of Figs 9 and 10b). The proton bulk velocity is 350–400 km s^{-1}. Approaching the nucleus, the following are observed: (1) the direction of the solar wind bulk flow changes smoothly from the solar toward the ram direction, as the maximum intensities at the same energies gradually decrease (increase) in the SDA (CRA), respectively (Fig. 10b, c); (2) the bulk velocity of the solar wind and its temperature gradually decrease (Fig. 9, 10b, c); (3) a second maximum starts to appear in the SDA spectra at higher energies (Fig. 9), indicating that the corresponding ions may initially have had energies larger than 25 keV, which are not observable by the SDA. Certainly, these latter ions must be of cometary origin. Comparing the energies at the two peaks in the ion spectra, it is

reasonable to deduce that these ions originate from the water-group (O$^+$, OH$^+$, H$_2$O$^+$, H$_3$O$^+$).

As was noted above, after a few cyclotron periods newly created cometary ions are isotropized in the coordinate system moving with the solar wind (Sagdeev *et al.* 1986). In this way the newly created ions are arriving from the solar direction with a maximum energy $4m_i/m_p$ times higher than the proton energy, independently of the angle between the solar wind flow and the magnetic field direction.

In the solar wind upstream of the bow shock, the energy of these heavy ions exceeds the upper limit of the energy range of the SDA ($E_i/q < 25$ keV). But in the outer regions of the cometosheath, where the solar wind flow is already decelerated by the shock and by the mass-loading process, the SDA measurements are in reasonable accordance with the abovementioned process of implantation of cometary ions into the plasma flow at large distances from the comet (Verigin *et al.* 1986, 1987a).

With deeper penetration into the cometosheath, the proton energy gradually decreases. The energy of the cometary ions observed in the SDA decreases faster, so that the ratio of the energy of heavy ions to the energy of protons is also decreasing (Fig. 9). At distances of $3-4 \times 10^5$ km from the nucleus (0600–0615 UT) the velocity of cometary ions from the solar direction decreases to the value of the proton velocity, owing to some collective process, which is not fully understood. Also, the fluxes of these two populations become comparable. Later, the rate of energy decrease increases further for the heavy ions, while the energy of the protons practically ceases to change.

Around the cometopause (~ 0645 UT), which separates the cometosheath from the cometary plasma region, solar wind protons disappear from the acceptance angle of the SDA, and cometary ions produce a peak around 1 keV in the energy spectrum (Figs 9, 10b). When the velocity of Vega-1,2 relative to the comet is taken into account, the velocity of the heavy ions relative to the comet can be estimated to be a few tens of kilometres per second in the vicinity of the cometopause, while the proton velocity is still around 200 km s^{-1} in this region.

After closest approach (CA in Fig. 9) these characteristic changes in the cometosheath plasma occur again on the outbound leg, but in reverse order.

The energy spectra of cometary ions as measured by PLASMAG-1 in the inner region of the cometosheath might be interpreted in two different ways. Here, the ratio of the energy of cometary ions E^1_i to the proton energy E_p measured by the SDA is smaller than $4m_i/m_p$ (~ 60–80), and it gradually decreases with decreasing cometocentric distances. The IIS (implanted ion sensor) of the JPA instrument on board Giotto was able to measure ions up to energies of $E_i/q < 90$ keV and to estimate the mass of the ions (Johnstone *et al.* 1986). According to the measurements of IIS there is another more energetic branch in the energy spectrum corresponding to water group ions with an energy of $E^2_i \approx 4m_i/m_p E_p$ as well as the branch which was also observed by the SDA. If the ratio $E^2_i/E_p \approx 4m_i/m_p$ is constant in the cometosheath, the ions with energies E^2_i might have been locally ionized and picked up by the process discussed above. In this case, the ions observed by the SDA with energies $E^1_i < E^2_i$ might have been ionized further upstream of the population with energies E^2_i, and they might have lost their energy because of some later collective processes (Gringauz *et al.* 1985).

However, there is a preferable explanation for the existence of these two different cometary ion populations: the ions with the energy E^2_i were actually created far upstream from the point of observation (Johnstone *et al.* 1986, Verigin *et al.* 1987a), but the proton energy slightly changes in the cometosheath as

observed by the SDA. Hence the ratio E^2_i/E_p does not change much, or stays almost constant, as in the first case. The ions registered with energies E^1_i were created in the vicinity of the spacecraft. First, these ions are only partly taken into the solar wind flow; however, when approaching the cometopause, these ions are no longer a minor population. At the start of this process there is no complete isotropization in the coordinate system moving with the solar wind. But when approaching the cometopause, the density of these ions (Fig. 9) is increasing, thus the energy of the solar wind flow will not be large enough to increase the velocity of all the newly created heavy ions. In the vicinity of the cometopause these ions are accelerated only to a velocity of a few tens of kilometres per second. Since the spacecraft has a velocity of ~ 80 km s^{-1} relative to the comet, the slowly moving newly created cometary ions will be observed by the SDA with an energy of about 1 keV around the cometopause. There is an additional fact in favour of the possibility that the cometary ions belonging to the less energetic branch E^1_i were created not very far from the spacecraft: the flux of these ions increases with decreasing R (see Fig. 9), corresponding to the increase in the neutral gas density. The fluxes of heavy ions, however, belonging to the high-energy branch do not increase significantly when approaching the nucleus (see Fig.2 of Johnstone *et al.* (1986).

The characteristic feature of the electron plasma component in the cometosheath is a decrease of electron temperature T_e with approach to the cometopause. From the data presented in Fig. 11 (Gringauz

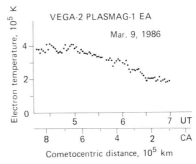

Fig. 11 — Distribution of the electron temperature in the cometosheath of comet Halley between 8×10^5 and 1.6×10^5 km (cometopause). CA = closest approach.

et al. 1986e, 1987a) it readily follows (1) that there is an overall decrease of T_e from about 4×10^5 K to about 2×10^5 K within this region; (2) that T_e decreases fastest within the region of about $5.5–6 \times 10^5$ km, but stays approximately constant outside; and (3) that there are significant fluctuations in T_e in the region outside about 6×10^5 km. It should be mentioned that at a cometocentric distance of about $5–6 \times 10^6$ km, i.e. in the undisturbed solar wind, the electron temperature was about 2×10^5 K, and that it had already increased to its maximum value of about $3.5–4 \times 10^5$ K, at a distance of about 3×10^5 km upstream of the cometary bow shock.

During the encounter of the Giotto spacecraft with comet Halley, electrons were measured by the electron electrostatic analyzer EESA of the RPA experiment (Rème *et al.* 1986). In the undisturbed solar wind the electron temperature was about 2.5×10^5K and it occasionally increased to about 3.5×10^5K beyond 10^5 km upstream of the bow shock. Downstream, in the region from 1.15×10^6 km to 5.5×10^5 km, T_e stayed practically constant (3.5×10^5 K) but with large fluctuations superimposed, while afterwards, when approaching the cometopause, T_e decreased to about 2×10^5 K. Thus, outside the cometopause both the Vega-2 and Giotto electron measurements are in excellent agreement, not only with respect to the actual values of T_e but also with respect to its overall radial dependence.

We propose that the decrease in electron temperature from about 4×10^5 K near the bow shock to about 2×10^5 K at the cometopause, as observed by Vega-2 and Giotto independently, is due to inelastic collisions between the thermal electrons and the cometary neutral gas (Gringauz *et al.* 1986e, 1987a). Near the cometopause at $R \approx 1.6 \times 10^5$ km the neutral density is about $n_o \approx 5 \times 10^3$ cm^{-3} (Remizov *et al.* 1986). The energy loss of electrons caused by inelastic collisions with neutrals of the water group is $N_o L_o \approx 2 \times 10^{-11}$ eV cm^{-1}, where $L_o \approx 4 \times 10^{-15}$ eV cm^2 is the energy loss function for electrons with an energy of about 40 eV (Olivero *et al.* 1972), and N_o is the number density of neutrals of the water group. The plasma velocity near the cometopause is $V \approx 200$ km s^{-1} (Galeev *et al.* 1987). The characteristic time for the plasma flow through the cometopause is therefore $\tau \approx 2R/V \approx 1.5 \times 10^3$ s (taking into account that $\phi \approx$

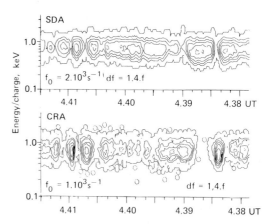

VEGA-2 PLASMAG-1 Mar. 9, 1986 R = 7,4.10⁵ km

Fig. 12 — Spectrograms of the ion fluxes as observed by the Sun-directed (SDA) and ram-directed (CRA) ion electrostatic analyzer on board Vega-2 simultaneously in the cometosheath. Note the variations in the direction of the plasma flow of cometary ions.

110°). During this time the thermal electrons, with an average velocity of about 4 000 km s^{-1}, have traversed a distance of about 6×10^6 km. Thus, according to the above estimate, their energy loss is about 12 eV.

As well as the large-scale changes in the plasma flow in the cometosheath, which determine the global picture of the solar wind flowing around the comet, a high level of MHD turbulence with a wide frequency range is characteristic for this region. An example of the variations in the direction of the plasma flow within the cometosheath is presented in Fig. 12 (Verigin *et al.* 1986, 1987a). In these spectrograms, registered by the CRA and the SDA at a distance of about $R \approx 7.4 \times 10^5$ km from the nucleus, the characteristic duration of the variations is 20–30 s. The outermost isolines correspond to a counting rate of f_o, and each inner isoline represents a counting rate which is 1.4 times greater than the corresponding value of the adjacent outer line. The variations in the flow intensity observed by both analyzers are caused by variations in the direction of the ion flow. When the intensity of the ion flow decreases in the direction of the CRA analyzer, the counting rates observed by the SDA increase simultaneously, and vice versa. The approximate value of this declination of the proton flow from the original direction can be estimated as

$$\delta\alpha \approx \frac{2\kappa T}{m_i V^2} \cdot \ln\left(\frac{N_{max}}{N_{min}}\right) \approx 5° \qquad (7)$$

where the proton temperature is $T \approx 3 \times 10^5$ K and their velocity $V = 350$ km s^{-1}, and where the ratio of the maximum to the minimum counting rates is $N_{max}/N_{min} = 3$–5.

But the most spectacular variation of the cometo-sheath plasma flow characteristics occurs at the cometopause, which separates the cometosheath itself, controlled by the solar wind proton flow, from the inner cometary plasma region dominated by slowly moving cometary ions (Gringauz *et al.* 1985, 1986b, c, d, 1987b). This boundary was not theoreti-cally predicted but was discovered on the basis of PLASMAG-1 data from both Vega spacecraft. The existence of the cometopause was confirmed later by the plasma wave experiment APV-N onboard the Vega spacecraft (Savin *et al.* 1986) and by the instruments PICCA and JPA onboard Giotto (Korth *et al.* 1985, Amata *et al.* 1986).

The top panel of Fig. 13 (Galeev *et al.* 1987) shows the ion spectrogramm measured by the CRA of the PLASMAG-1 package in the vicinity of the cometo-pause. The outermost isolines correspond to a count rate of 10^3 s^{-1}, and the ratio between count rates represented by adjacent isolines is two. The two vertical dashed lines indicate the time interval 0643–0645 UT when Vega-2 crossed the cometopause (Gringauz *et al.* 1986c).

As shown by the spectrogram of Fig. 13, the typical energy/charge ratio of ions detected by the CRA significantly increases at the cometopause from ~ 170 eV to ~ 900 eV. This feature is due to changes in the distribution function and ion composition of the plasma; protons ($m_p = 1$ amu) are most abundant in the cometosheath, while water group ions ($m_i \approx 16$–18 amu) dominate inside the cometopause. Proton fluxes detected by the CRA significantly decrease after crossing the cometopause, though their energy spectra become wider (Gringauz *et al.* 1985, 1986c) and their typical energy increases to ~ 250 eV.

Before crossing the cometopause, the ion energy spectra observed by the SDA show two maxima (see Figs 9, 10b). The first maximum, which is essentially due to protons is typically ~ 300 eV; the heavy water group ions produce a second maximum at ~ 900 eV. After crossing the cometopause, protons practically disappear from the acceptance angle of the SDA, while the energy of heavy ions is hardly changed

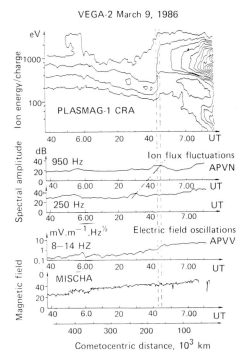

Fig. 13 — Plasma and field data collected by four different instruments during the last 100 min before closest approach. From top to bottom: spectrogram of ion flux in the ram direction, plasma wave activity in three different frequency ranges, total magnetic field. The cometopause is indicated by dashed lines.

(Gringauz *et al.* 1985, 1986c).

Electron spectra measured by the Electron Analyz-er, and hence the density and temperature of the electrons, do not show any characteristic variation when the cometopause is crossed (Gringauz *et al.* 1985, 1986c).

Magnetic field measurements performed by the magnetometer MISCHA on board Vega-2 are also consistent with the conclusion that the plasma density does not change significantly at the cometopause. As seen from the bottom panel of Fig. 13, the magnitude of the total magnetic field is practically constant in this region. Only minor changes are observed in the B_y and B_z components (Reidler *et al.* 1985).

The middle part of Fig. 13 shows the plasma wave activity measured by the instruments APV-N and APV-V on board Vega-2. In general, the filter channels shown in this figure (and also other channels not shown here) are characterized by an increase of the average amplitude of plasma and electric field oscillations from a cometocentric distance of $1.5-2 \times 10^5$ km when the Vega-2 spacecraft approaches the nucleus. In the vicinity of the cometopause (around 0630–0650 UT), plasma wave oscillations are observed in the whistler frequency range (0.2–1 kHz) and the amplitude of the electric field suddenly increases in the lower hybrid frequency range (8–14 Hz) during a 2 min interval when the spacecraft crosses the cometopause.

The wave activity in the lower frequency range can be seen in Fig. 14 (Galeev *et al.* 1987) where more detailed measurements of plasma, magnetic field, and waves are presented. The top panel shows the ion spectrogram measured by the CRA. Here the difference between count rates represented by adjacent isolines is 440 s^{-1}, and the outermost isolines correspond to a count rate of 10^3 s^{-1}. Dots on the spectrogram mark the maxima of the ion flux in an interval of 10 min around the cometopause. A comparison between the spectrograms simultaneously measured by the CRA and SDA (see Fig. 1 in Gringauz 1986c, where the spectrograms are colour

coded) shows that the fluctuations of ion fluxes measured by the two sensors are in anticorrelation. This is an indication of the existence of large-scale MHD variations in the direction and/or in the velocity of the plasma flow with a characteristic period $T \approx 1$ min.

These large-scale MHD waves at the cometopause are reflected in the electric field by oscillations at the lower hybrid frequency (2–32 Hz) and in the B_z component of the magnetic field by fluctuations with the same characteristic period ($T \approx 1$ min). The correspondence between the maxima of electric field, magnetic field, and ion fluxes is indicated by arrows in Fig. 14.

The sudden decrease of proton fluxes within a ≈ 2 min interval (corresponding to $\Delta \approx 10^4$ km along the trajectory of Vega-2) in the ram and solar direction cannot be explained without taking into account collisionless deceleration processes and isotropization of the proton distribution function; this phenomenon may be caused by an instability due to the relative motion of solar wind protons and cometary ions. We shall therefore estimate the velocity of the protons and ions observed by PLASMAG-1.

Outside the cometopause, the typical energy of solar wind protons is ~ 170 eV in the ram direction and ~ 300 eV in the solar direction. Their estimated bulk velocity relative to the spacecraft is $v_{pr} \approx 250$ km s^{-1}. The velocity of the protons relative to the comet $v_p \approx 200$ km s^{-1} is given by a possible vector diagram shown by Fig. 15 (Galeev *et al.* 1987), where the spacecraft velocity relative to the comet $v_{sc} \approx 80$ km s^{-1} taken into account. Fig. 15 also shows the acceptance angles of the CRA and SDA sensors and the regions of velocity space in which protons can be observed by these analyzers (areas shaded with vertical lines). In a similar way, the velocity of heavy cometary ions can be estimated; $v_{ir} \approx 120$ km s^{-1} relative to the spacecraft, and $v_i \approx 60$ km s^{-1} relative to the comet.

The measured magnetic field direction is not far from parallel to the plasma flow, and a 'fire-hose' instability might develop. Outside the cometopause, when the proton number density is $n_p \approx 10-20$ cm^{-3} and the magnitude of the magnetic field is $B \approx 40\gamma$, the condition for a fire-hose instability caused by the solar wind flow through the cometary plasma is not

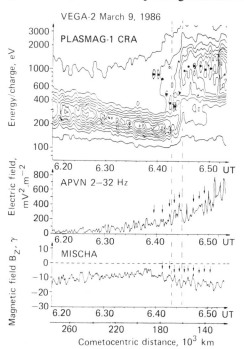

Fig. 14 — Fluctuations of ion flux, electric field and B_z component (pointing toward the north pole of the ecliptic) of the magnetic field around the cometopause (indicated by dashed lines). Maxima are shown by dots and arrows.

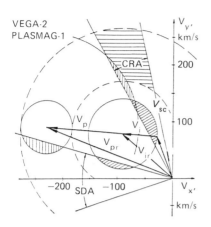

Fig. 15 — Possible proton and ion velocity vector diagrams and the acceptance angles of the SDA and CRA ion analyzers in the spacecraft reference frame. Particles from the shaded areas can be observed by the analyzers (sloping lines: cometary ions, vertical lines: protons outside the cometopause, horizontal lines: protons inside the cometopause).

fulfilled. On the other side of this boundary, when the number density of heavy ions is $n_i \approx 10$ cm^{-3} and the velocity is $v_i \approx 60$ km s^{-1} relative to the comet, the increasing ionization of cometary neutrals leads to an instability caused by the flow of solar wind protons and picked-up cometary ions relative to the newly ionized cometary gas. As a consequence, protons are decelerated, and pitch angle scattering takes place; the intensity of proton fluxes detected by both CRA and SDA is therefore decreasing.

A clear indication of the instability which develops near the cometopause is the large-scale variation of the plasma flow correlated with the oscillations of the perpendicular magnetic field component (relative to the main field direction), as seen in Fig. 14. The characteristic scale of these oscillations is $v_{sc}T \approx 5000$ km along the spacecraft trajectory which is comparable with the thickness of the cometopause Δ, but it is much larger than the Larmor radius of cometary ions $Q_{ci} = v_i/\omega_{ci} \approx 300$ km ($\omega_{ci} \approx 0.2$ s^{-1} is the cyclotron frequency of water group ions). The plasma velocity vector was not measured by the Vega spacecraft, and it is not possible to determine the mode of oscillation excited by the instability. Since the magnitude of the magnetic field is almost constant there, the oscillations seem to be perpendicular. The amplitude of the velocity perturbation δv can be estimated from the oscillation of the perpendicular magnetic field component δB,

$$\delta v \approx (\delta B/B)v_A \approx 10 \text{ km s}^{-1} \tag{8}$$

where $v_A \approx 60$ km s^{-1} is the Alfvén velocity. This effect can induce strong modulations in the ion flows observed by both analyzers. The large scale of these oscillations compared to the ion Larmor radius indicates that these waves are certainly not caused by cyclotron resonance. In other words, we possibly observe the development of a fire-hose instability with a significantly larger growth rate than in the case of a resonance instability (Galeev et al. 1987). The source of energy of such an instability is the kinetic energy of the newly born cometary ions relative to the plasma flow; the density of this flow is much larger than the density of the original solar wind proton flow.

The significant decrease of the proton flux at the cometopause and in the cometary plasma region shows the importance of the charge exchange between protons and cometary gas for producing cometary ions here. The characteristic time for charge exchange is,

$$\tau_{ct} \approx (\sigma_{ct} v_p n_n)^{-1} \approx 6 \times 10^3 \text{ s} \tag{9}$$

where $\sigma_{ct} \approx 2 \times 10^{-15}$ cm^2 is the cross-section for charge exchange, $n_n \approx 4 \times 10^3$ cm^{-3} is the number density of neutral gas in the vicinity of the cometopause at a cometocentric distance $R \approx 1.6 \times 10^5$ km (Remizov et al. 1986), and $v_p \approx 200$ km s^{-1} is the velocity of the proton flow outside the cometopause, which is of the same order of magnitude as the proton gyrovelocity inside the cometopause after pitch angle scattering. Since the above estimated τ_{ct} is comparable to the characteristic time of the plasma flow around the comet (of the order of $2R/v_i \approx 5 \times 10^3$ s in the vicinity of the cometopause for a flow velocity $v_i \approx 60$ km s^{-1}), charge exchange is effective in this region.

As a consequence of the existence of a cometary ion beam in the plasma flow, the intensity of plasma waves is increasing in the lower hybrid frequency range at the cometopause (see Figs 13, 14). The growth of this wave is limited by the quasilinear relaxation of the ion beam to a steady state (Formisano et al. 1982), and the wave intensity in the lower hybrid range measured by the instruments APV-V (Fig. 13) and APV-N (Fig. 14) is in reasonable agreement with the theoretical estimate (Galeev et al. 1987).

The excited lower hybrid waves accelerate the suprathermal electrons which are in Čerenkov resonance (Galeev & Khabibzachmanov 1985). If the lifetime of the suprathermal electrons was significantly larger than the time of acceleration, their maximum density would be determined by the condition that the electron Landau damping be small compared to the growth rate of the instability. The acceleration of suprathermal electrons along the magnetic field lines leads to the excitation of oblique Langmuir waves (whistlers in high β plasma) due to the growing anisotropy in the velocity distribution of electrons. The frequency of high-frequency Langmuir waves increases as the plasma is decelerating in the vicinity of the cometopause because the energy of suprathermal electrons is decreasing. This effect is marked by a dashed-dotted line in Fig. 13. The increase in the level of lower hybrid and whistler mode plasma oscillations is a consequence of the rapid mass-loading and deceleration of the solar wind by cometary ions in the vicinity of the cometopause. It is not responsible for the mass-loading process (Galeev et al. 1987).

3.3 Cometary plasma region and unsteady phenomena

After crossing the cometopause, during the further approach to the cometary nucleus, the CRA sensor observed an increase of the counting rates of heavy cometary ions by an order of magnitude (see, for example, Figs. 13, 14). However, the increase of cometary plasma density was neither monotonic nor stationary. This is shown in Fig. 16 (Gringauz et al. 1986d, 1987b) exhibiting two high-resolution spectrograms, as observed during 0700–0704 UT and during 0706–0710 UT, or at distances of about 8×10^4 km and 5×10^4 km, respectively (time runs from right to left). The counting rates corresponding to the outermost lines are given by $f_o = 4 \times 10^3$ s^{-1} and $f_o = 10^4$ s^{-1}, respectively. The increase in the counting rates is characterized in steps of df between two adjacent isolines with $df = 4 \times 10^2$ s^{-1} and $df = 10^3$ s^{-1} respectively. From these figures it readily follows that a quasiperiodic modulation in the intensity of cometary ions exists. This modulation is most pronounced in the energy-over-charges range around 500–600 eV, which corresponds to ions of the water group. The typical normalized amplitude (A) in the modulation of

these latter ions is about,

$$A = (f_{max} - f_{min})/(f_{max} + f_{min}) \approx 0.05\text{--}0.1$$

Moreover, we find that the modulation period τ_M decreases when approaching the nucleus from about $\tau_M \approx 10$ s at 8×10^4 km to about $\tau_M \approx 8$ s at 5×10^4 km.

It should be noted that this type of modulation differs from the sporadic, burst-type density enhancements which have been observed by Vega-2 at about 4×10^4 km and 3×10^4 km (see Fig. 4 in Gringauz et al. 1985). These bursts were associated with enhancements in the plasma wave intensity around the lower hybrid frequency, and their occurrence was possibly connected with the critical ionization velocity effect of Alfvén (Galeev et al. 1986).

The observed quasiperiodic modulation in the heavy ion density in the cometary plasma region could be explained by the development of the large-scale instability caused by the anisotropic velocity distributions of cometary ions. This instability is connected with the cyclotron resonance of ions by Alfvén oscillations, therefore the wavelength of these oscillations can be estimated as $\lambda \approx v_\parallel/f_{ci}$ where f_{ci} is the cyclotron frequency of the ions and $v_\parallel$ is their velocity along the magnetic field. Taking the magnetic field of $B \approx 50\gamma$ and the spatial dimensions of the plasma flow fluctuations as ~ 8 s $\times$ 80 km s$^{-1} \approx 640$ km into account, the field-aligned velocity can be estimated as $v_\parallel \approx 30$ km s^{-1} which seems to be reasonable.

Well inside the cometary plasma region both the thermal and the bulk velocities of ions become small compared to the spacecraft relative velocity of ≈ 80 km s^{-1}. The energy-over-charge spectra break up in several well-defined sub-distributions with their maximum intensity located at distinct E_i/q values. This situation is depicted in Fig. 17 (Gringauz et al. 1986d, 1987b), showing a sequence of 4 s averaged ion energy spectra as obtained by the CRA sensor at a distance of 1.4–1.7×10^4 km from the nucleus. From the left-hand distribution it readily follows that the bulk velocity of the protons is close to the spacecraft relative velocity. Thus, the energy/charge E_i/q spectra can be transformed into mass/charge m_i/q spectra.

In this way Fig. 17 suggests the presence of H$^+$, C$^+$, CO$^+_2$, and Fe$^+$ ions in the cometary plasma region. The peak at $14 < m_i/q < 20$ may originate from the

VEGA-2 PLASMAG-1 March 9, 1986

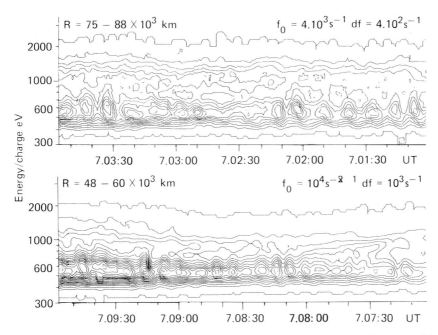

Fig. 16 — Contour plots of the ion intensity vs. energy/charge and time as measured by the CRA on board Vega-2 in the cometary plasma region at distances of 8×10^4 km and 5×10^4 km from the nucleus.

H_2O parent molecules, such as O^+, OH^+, H_2O^+, and H_3O^+, while the peak at $24 < m_i/q < 34$ may be due to parent molecules of CO/CO_2 or to molecules containing N/S, such as CO^+, N_2^+, H_2CO^+, HCO^+, CN^+, HCN^+, or O_2^+ or to atomic ions like Mg^+, Al^+, Si^+, P^+, or S^+. Finally, the minor peaks at $m_i/q = 2, 8, 70$, and 85 might stem from H^+_2 and O^{++} and from some heavier, not yet identified organic ions. We have marked these different groups of cometary ions at the top of Fig. 17.

For these groups we have determined their densities at different cometocentric radial distances by summing the respective count rates. The result is shown in Fig. 18 (Gringauz *et al.* 1986d, 1987b), together with the total sum Σ of all ions and with the expected R^{-2} slope. Straight lines correspond to the regions where the densities could be determined more accurately. From Fig. 18 it readily follows that, in principle, really none of the curves, except may be the curve of mass-group 70^+, can be approximated by a pure R^{-2} dependence. For the total sum we find a dependence of $R^{-2.4}$ for $3 \times 10^4 < R < 1.5 \times 10^4$ km.

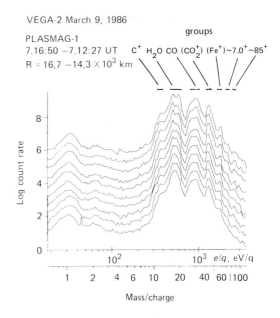

Fig. 17 — Time sequence of 4s averaged ion energy/charge (mass/charge) spectra, as observed by the CRA sensor on board Vega-2 at a cometocentric distance of $(1.4–1.7) \times 10^4$ km.

A somewhat unexpected result is the observed well defined peak at $m_i/q = 56$ in the CRA ion spectra. This peak was also identified in the IMS (Balsiger *et al.* 1986) and NMS (Krankowsky *et al.* 1986) measurements onboard Giotto. However, it has not been observed by the Giotto-PICCA instrument (Korth *et al.* 1986), although otherwise this sensor is quite similar to the CRA sensor onboard Vega-2. It seems quite natural to identify this peak with the occurrence of Fe^+ ions in the innermost part of the cometary plasma region, although it should be noticed that, owing to the energy resolution of the CRA sensor of 5.5%, there is an uncertainty of about 3 amu in the mass resolution. From the observed mass composition of cometary dust particles, iron has been identified in most of the mass spectra analyzed (Kissel *et al.* 1986). Thus, sputtering of dust grains with large metallic cores and no layer of ice could account for the release of metallic ions in the cometary coma, (Ip & Axford 1986) although other processes discussed in (Geiss *et al.* 1986) could also account for the occurrence of metal ions.

As indicated in Fig. 18, the density of cometary ions decreases somewhat faster than R^{-2}. This result is consistent with the observations of the IMS (Balsiger *et al.* 1986) and the PICCA instrument (Korth *et al.* 1987) onboard Giotto. The neutral gas density, however, decreases as R^{-2}, as observed both by Vega (Gringauz *et al.* 1985, 1986b, Remizov *et al.* 1986) and by Giotto (Krankowsky *et al.* 1986). This somewhat faster decrease in the ion density could be associated with the fact that the loss rate of ions increases with increasing distances from the nucleus, owing to an increasing convective outflow of the cometary ions. Vega-2 observed a neutral gas density of $n_o \approx 1.5 \times 10^5$ cm^{-3} at 2×10^4 km (Remizov *et al.* 1986). For a characteristic time of ionization of $\tau_i \approx 10^6$ s, the local production rate for ions is $Q_i \approx n_n/\tau_i \approx 0.15$ cm^{-3} s^{-1}. To obtain an ion density of $n_i \approx 10^3$ cm^{-3} at these distances, a value for the velocity of the convective outflow of $V_i \approx Q_i R/n_i \approx 3$ km s^{-1} would be sufficient. This value for V_i is in reasonable agreement with the velocity determined for the water group ions from the IMS measurement onboard Giotto (Balsiger *et al.* 1986).

On the other hand, if the loss of cometary ions is mainly caused by the convective outflow, then the ratios of the densities of the ions and of their parent molecules must be close to each other. In Krankowsky *et al.* (1986) a value of 5% for the CO_2 content relative to the content of H_2O was estimated. From our CRA measurements we find that 70–80% of the total ion content belongs to the water group ions and 2–5% to the ions with $m_i/q \approx 44$ (CO_2^+. In turn, as we find 15–20% for the content of the CO/CO_2 group ions we can estimate the content of the parent molecules H_2O and CO/CO_2 based on our CRA data to be 70–80% and 15–20%, respectively. This agrees with the production rates of H_2O and CO as determined from the UV measurements onboard the IUE spacecraft (Festou *et al.* 1986).

As an example of the large-scale unsteady cometary phenomena, the inhomogenuity formed in the cometary plasma region by the solar wind IMF parameters change can be considered (Verigin *et al.* 1987b). The general feature of the plasma measurements onboard both Vega craft is the absence of any plasma registration by SDA analysers inside the cometopause. However, in case of Vega-1 this sensor

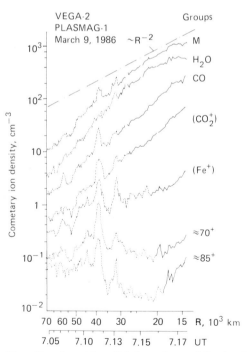

Fig. 18 — Cometocentric radial dependence of the densities of certain groups of cometary ions, as defined in Fig. 17 and of the total sum Σ of all ions. For comparison the R slope is inserted.

observed a burst, lasting for about 5 min, of ions with energies $E_i = 100–1000$ eV approximately, near the closest approach to comet Halley, marked in black in Fig. 9. For roughly the same time interval as in Fig. 9, the SDFC sensor observed ion flux coming from the solar direction, which was evaluated as $(5–8) \times 10^9$ cm^{-2} s^{-1} (Verigin *et al.* 1987b). The density of cometary ions by RFC data could only be estimated to a few thousands per cm^3.

Inspecting the evolution of the magnetic field direction along the Vega-1 trajectory, the authors of magnetic measurements concluded that in addition to the global draping of the magnetic field lines around the comet, the portion of the field observed near closest approach is a remnant of the IMF of the previous polarity (Riedler *et al.* 1986, Schwingenschuh *et al.* 1986).

When interacting near closest approach, two portions of magnetic fields of opposite polarities will lead to an X-type reconnection pattern of the magnetic field lines (Fig. 19). The region in which the SDA and SDEC sensors observed the enhanced flux of ions is indicated by dots in this diagram. It seems

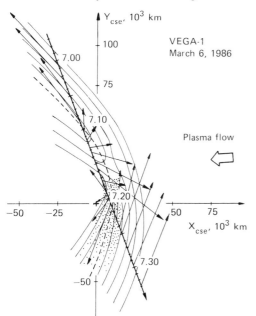

Fig. 19 — Overall topology of the magnetic field around closest approach as deduced from the measurements of the magnetic field direction (arrows) along the Vega-1 trajectory. The dotted areas represents the region in which the burst of accelerated ions (see Fig. 6) was observed.

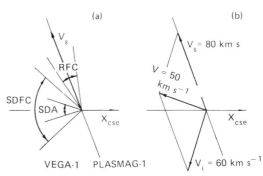

Fig. 20(a) — Schematic representation of the look-directions and the fields of view of the SDA, SDFC, and RFC sensors on board Vega-1 relative to the cometary nucleus and to the spacecraft velocity direction V.

Fig. 20(b) — Vector diagram of the spacecraft velocity V_s and of the ion velocities V_i and V in the cometary frame and in the spacecraft frame of reference, respectively. V_i is presumed to be roughly parallel to the magnetic field lines. Only for large values of $V_i \gtrsim 35$ km^{-1} will ions be detected by the SDFC/SDA sensors, as in this case V will fall into the field-of-view of the SDA/SDFC sensors.

quite reasonable to suggest that these ions were accelerated by the non-stationary process of magnetic field reconnection near closest approach (Verigin *et al.* 1987b).

Why are these ions then observed by the SDA/SDFC sensors only on the outbound leg of the Vega-1 trajectory, not on the inbound leg? The answer is provided by Fig. 20a,b (Verigin *et al.* 1987b). Here, together with look-directions and the angular fields of view of the SDA, SDFC, and RFC sensors (Fig. 20a), the possible diagram of the velocity vectors V_s, V_i, and V is shown. Here, V_s is the velocity of ions at rest in the cometary frame of reference, V_i is the velocity of accelerated ions in the same frame, and $V = V_i + V_s$ is their velocity in the Vega-1 frame of reference. This diagram holds for the outbound situation (V_i is supposed to be parallel to the surface separating magnetic fields of opposite direction which are marked by the dashed curve in Fig. 19). For the inbound case the direction of V_i is close to the direction of V_s, hence the direction of V is close to the same direction (and out of the field of view of SDA/SDFC).

From ion composition measurements (Gringauz *et al.* 1986d, 1987b, Balsiger *et al.* 1986) we know that cometary ions are predominantly the water-group ions with $m_i = 16–18$. At their velocity relative to Vega-1 of about 50 km s^{-1} (see Fig. 20), they will be observed at energies $E_i > 200$ eV by the SDA analyzer. This is in agreement with the actual measurements shown in Fig. 6. From the SDFC measurements we estimated the total ion flux as $(5–8) \times 10^9$ cm^{-2} s^{-1}. Taking now their velocity V relative to the spacecraft into account, we are able to estimate their density as $n_i > (1–2) \times 10^3$ cm^{-3}. This value agrees roughly with the number density estimated from the RFC observations.

Based on these different, yet self-consistent, observations we may, in summary, conclude that the burst of cometary ions observed by the SDA and SDFC sensors onboard Vega-1 near closest approach to comet Halley was produced by the motion of cometary ions of the water-group accelerated up to some tens of km s^{-1} (Verigin et $al.$ 1987b). This acceleration could be caused by merging of interplanetary magnetic field lines of opposite polarity retarded by the presence of cometary plasma and neutrals.

3.4 The aurora-like phenomena in the vicinity of cometary nucleus

The only essential difference between Vega and Giotto plasma observations within the cometary plasma region is the observation of keV electrons onboard Vega-2 during its closest approach to Comet Halley. Fig. 21 (Gringauz et $al.$ 1985, 1986b) presents two typical electron spectra as measured by the EA in the solar wind and near the closest approach of Vega-2 to the cometary nucleus. The main difference between these spectra is the appearance of the energetic

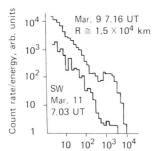

VEGA-2 PLASMAG-1

Fig. 21 — Two electron spectra, as measured by the Vega-2 electron analyzer in the solar wind and near closest approach to the cometary nucleus at $\sim 1.5 \times 10^4$ km respectively. Notice the occurrence of keV electrons close to the nucleus.

cometary electron component, which could serve as an additional source of ionization of cometary neutrals.

The presence of ~ 1 keV electrons during the Vega-2 flyby and the absence of similar electrons during the Giotto flyby could be due to either an instrumental effect, or to the sporadic occurrence of these particles. The possible instrumental effect was discussed in Gringauz et $al.$ (1986e, 1987a). The electrons with energies of about 1 keV could be generated by a 'magnetospheric substorm', similar to the auroral events occurring in the terrestrial magnetosphere. The occurrence of such events in the cometary magnetotail has been proposed by Ip (1976) and Ip & Mendis (1976). A schematic representation of a 'cometary substorm', as suggested by Ip & Axford (1982) and Mendis et $al.$ (1985) is shown in Fig. 22 (Gringauz et $al.$ 1986e, 1987a). In the steady state the cometary tail electric current adopts the usual Θ-shape configuration. If the cross-trail current becomes partly disrupted, the induced tail-aligned currents can discharge through the cometary coma (Fig. 22). Such a process can accelerate electrons to energies of a few keV, as in the case of auroral electrons in the terrestrial upper atmosphere. Vega-2 might have observed such substorm generated energetic electrons, while Giotto might not.

We now summarize other observations that point also toward the possibility that energetic electrons may occur sporadically in the coma of Comet Halley. In Feldmann et $al.$ (1986) the two UV spectra measured by IUE on 18–19 March 1986, on the tailward side of the nucleus at a distance of about 40 000 km and about 30 min apart, were compared. It was found that the CO_2^+ emission had decreased to less than a quarter, while the OH^+ brightness had remained nearly constant. The authors proposed that this difference could be explained by the existence of localized currents of energetic electrons close to the nucleus.

The far-UV observations of Comet Halley in February 1986 with a sounding rocket experiment showed the presence of $\lambda = 153.6$ nm line tentatively identified as oxygen emission (Woods et $al.$ 1986a). This intercombination line is not excited by solar fluorescence as the other emissions are, and it suggests the presence of a region of electron excitation within

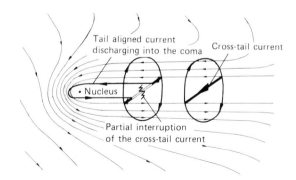

Fig. 22 — Schematics of a cometary cross-tail disruptive event, which could have produced the keV energetic electrons observed by Vega-2 during closest approach (see Fig. 21).

the inner coma (Woods *et al.* 1986a). The consideration of a collisional processes with the electrons helps toward the solution of the 'carbon puzzle' (Woods *et al.* 1986), which is connected with the presence of unexpectedly large amount of neutral and ionized carbon in the vicinity of P/Halley and other comets (Balsiger *et al.* 1986, Festou 1984).

4 CONCLUSIONS

The results of *in situ* neutral and plasma measurements during the Vega-1, 2 encounters with P/Halley which have been presented above, generally confirmed the MHD description of the solar wind interaction with cometary plasma (see, for example, Ip & Axford 1982), yet led to the discovery of a number of phenomena that were not envisaged by the existing theories.

In accordance with the predictions:

- the deceleration of the solar wind flow due to mass loading starts a few million kilometres from the nucleus;
- the bow shock is formed near $R \approx 10^6$ km, although the possibility of the absence of this structure is also discussed;
- the additional solar wind mass loading by cometary ions and the associated increases of plasma pressure and density leads to the very low velocities of the flow at $R \lesssim 10^5$ km;
- some evidences of the 'substorm' phenomenon in the cometary magnetosphere were obtained.

On the other hand:

- the distribution function of the multicomponent

plasma within the cometosheath and the cometary plasma region turned out to be a stable but essentially unequilibrium function — there are two branches of the picked-up cometary ions in the cometosheath. The protons and the heavy ions in the cometary plasma region seem to be decoupled;

- the existence of the cometopause — the relatively sharp, yet unpredicted boundary within the cometary coma at cometocentric distances of about 1.6 $\times 10^5$ km — was revealed;
- the continuous cooling of the electron plasma component through the cometosheath, due to inelastic collisions with neutrals, was observed;
- it was shown that the increase of heavy ion densities within the cometary plasma region is not monotonic but lumpy, quasiperiodic, with a 'period' decreasing while approaching the nucleus;
- the unexpected presence of Fe^+ ions in the P/Halley plasma was demonstrated.

These and a number of other peculiarities of near-cometary plasma are not yet finally and quantitively analyzed. There are new possibilities for such studies on the basis of the qualitative analyses of the data obtained.

REFERENCES

Amata, E., Formisano, V., Cerulli-Irelli, R., Torrente, P., Johnstone, A.D., Wilken, B., Jockers, K., Winningham, J.D., Bryant, D., Borg, H., & Thomsen, M. (1986) The cometopause region at Comet Halley, In: *Exploration of Halley's Comet, ESA SP-250* **1** 213–218

Apáthy, I., Remizov, A.P., Gringauz, K.I., Balebanov, V.M., Szemerey, I., Szendrö, S., Gombosi, T., Klimenko, I.N., Verigin, M.I., Keppler, E., & Richter, A.K. (1986) Plasmag-1 experiment: solar wind measurements during the closest approach to Comet Giacobini–Zinner by the ICE probe and to Comet Halley by Giotto and Suisei spacecraft, In: *Exploration of Halley's Comet, ESA SP-250* **1** 65–70

Balsiger, H., Altwegg, K., Bühler, F., Geiss, J., Ghielmetti, A.G., Goldstein, B.E., Goldstein, R., Huntress, W.T., Ip W.-H., Lazarus, A.J., Meier, A., Neugebauer, M., Rettenmund, U., Rosenbauer, H., Schwenn, R., Sharp, R.D., Shelley, E.G., Ungstrup, E., & Young, D.T. (1986) Ion composition and dynamics at Comet Halley, *Nature* **321** 330–334

Balebanov, V.M., Gringauz, K.I., Verigin, M.I. (1987) Plasma phenomena in the vicinity of the closest approach of VeGa-1, -2 spacecraft to the Halley Comet nucleus, In: *Symposium on diversity and similarity of comets, ESA SP-278*, 119–127

Feldman P.D., A'Hearn, M.F., Festou M.C., McFadden L.A., Weaver H.A., & Woods T.N. (1986) Is CO_2 responsible for the outburst of Comet Halley?, *Nature* **324** 443–436

Festou, M.C. (1984) Aeronomical processes in cometary atmospheres: the carbon compounds puzzle, Adv. Space Res. **4**, No. 9, 165–175

Festou, M.C., Feldman, P.D., A'Hearn, M.F., Arpigny, C., Cosmovici, C.W., Danks, A.C., McFadden, L.A., Gilmozzi, R., Patriarchi, P., Tozzi, G.P., Wallis, M.K., & Weaver, H.A. (1986) IUE observations of Comet Halley during the Vega and Giotto encounters, *Nature* **321** 361–363

Formisano, V., Galeev, A.A., & Sagdeev, R.Z. (1982) The role of critical ionization velocity phenomena in the production of inner coma cometary plasma, *Planet. Space Sci* **30** 791–797

Galeev, A.A. (1985) Solar wind interaction with Comet Halley, *Adv. Space Res* **5** No. 12, 155–163

Galeev, A.A. (1986) Theory and observations of solar wind/cometary plasma interaction processes, In: *Exploration of Halley's Comet, ESA SP-250* **1** 3–17

Galeev, A.A. (1987) Encounters with comets: Discoveries and puzzles in cometary plasma physics, *Astron. Astrophys* **187** 12–20

Galeev, A.A. & Khabibzachmanov, I.Kh. (1985) On the nature of plasma waves in the Io torus, *Letters Astron. J.*, (in Russian) **11** 292–297

Galeev, A.A., Gribov, B.E., Gombosi, T., Gringauz, K.I., Klimov, S.I., Oberz, P., Remizov, A.P., Riedler, W., Sagdeev, R.Z., Savin, S.P., Sokolov, A.Yu., Shapiro, V.D., Shevchenko, V.I., Szegö, K., Verigin, M.I., & Yeroshenko, Ye.G. (1986a) Position and structure of the Comet Halley bow shock: Vega-1 and Vega-2 measurements, *Geophys. Res. Lett.* **13** 841–844

Galeev, A.A., Gringauz, K.I., Klimov, S.I., Remizov, A.P., Sagdeev, R.Z., Savin, S.P., Sokolov, A.Yu., Verigin, M.I., & Szegö, K. (1986b) Critical ionization velocity effects in the inner coma of Comet Halley: measurements by Vega-2, *Geophys. Res. Lett.* **13** 845–849

Galeev, A.A., Gringauz, K.I., Klimov, S.I., Remizov, A.P., Sagdeev, R.Z., Savin, S.P. Sokolov, A.Yu., Verigin, M.I., Szegö, K., Tátrallyay, M., F Grard, R., Yeroshenko, Ye.G., Mogilevsky, M.J., Riedler, W., & Schwingenschuh, K. (1987) Physical processes in the vicinity of the cometopause interpreted on the basis of plasma, magnetic field and plasma wave data measured on board the Vega-2 spacecraft, In: *Symposium on diversity and similarity of comets. ESA SP-278.* 83–87

Geiss, T., Boshler, P., Ogilvie, K.W., & Caplan, M.A. (1986) Origin of metal ions in the coma of P/Giacobini–Zinner, *Astron. Astrophys* **166** L1–L4

Gringauz, K.I., Apáthy, I., Denshikova, L.I., Gombosi, T., Keppler E., Klimenko, I.N., Remizov, A.P., Richter, A.K., Skuridin, G.A., Somogyi, A., Szabó, L., Szemerey, I., Szendrö, S., Verigin, M.I., Vladimirova, G.A., & Volkov, G.I. (1983) The Vega probe instrument package for measuring charged particles with energies less than 25 keV, In: *Cometary exploration* III, Budapest, Central Res. Inst. Phys. Press., 333–350

Gringauz, K.I., Gombosi, T.I., Remizov, A.P., Apáthy, I., Szemerey, I., Verigin, M.I., Denchikova, L.I., Dyachkov, A.V., Keppler, E., Klimenko, I.N., Richter, A.K., Somogyi, A.J., Szegö, K., Szendrö, S., Tátrallyay M., Varga, A., & Vladimirova, G.A., (1985) First Results of plasma and neutral gas measurements from Vega 1/2 near Comet Halley, *Adv. Space Res* **5** 12, 165–174

Gringauz, K., Klimenko, I., Remizov, A., Verigin, M., Vladimirova, G., Apáthy, I., Szegö, K., Szemerey, I., Szendrö, S., Tátrallyay, M., Keppler, E., & Richter, A. (1986a) The Vega Plasmag-1 experiment: description and first results, In: *Field, particle and wave experiments on cometary missions*, Graz, Austrian Academy of Sciences Publication, 203–216

Gringauz, K.I., Gombosi, T.I., Remizov, A.P., Apáthy, I., Szemerey, I., Verigin, M.I., Denchikova, L.I., Dyachkov, A.V., Keppler, E., Klimenko, I.N., Richter, A.K., Somogyi, A.J., Szegö, K., Szendro, S., Tátrallyay, M., Varga, A., & Vladimirova, G.A., (1986b) First *in situ* plasma and neutral gas measurements at Comet Halley, *Nature* **321** 282–285

Gringauz, K.I., Gombosi, T.I., Tátrallyay, M., Verigin, M.I., Remizov, A.P., Richter, A.K., Apáthy, I., Szemerey, I., Dyachkov, A.V., Balakina, O.V., & Nagy, A.F. (1986c) Detection of a new 'chemical' boundary at Comet Halley, *Geophys. Res. Lett,* **13** 613–616

Gringauz, K.I., Verigin, M.I., Richter, A.K., Gombosi, T.I., Szegö, K., Tátrallyay, M., Remizov, A.P., & Apáthy, I. (1986d) Cometary plasma region in the coma of Comet Halley: Vega-2 measurements, In: *Exploration of Halley's Comet, ESA SP-250* **1** 93–97

Gringauz, K.I., Remizov, A.P., Verigin, M.I., Richter, A.K., Tátrallyay, M., Szegö, K., Klimenko, I.N., Apáthy, I., Gombosi, T.I., & Szemerey T. (1986e) Electron component of the plasma around Halley's Comet measured by the electrostatic electron analyzer of PLASMAG-1 on board Vega-2, In: *Exploration of Halley's Comet, ESA SP-250* **1** 195–198

Gringauz, K.I., Remizov, A.P., Verigin, M.I., Richter, A.K., Tátrallyay, M., Szegö, K., Klimenko, I.N., Apáthy, I., Gombosi, T.I., & Szemerey, T. (1987a) Analyses of electron measurements from the PLASMAG-1 experiment on board Vega-2 in the vicinity of Comet Halley, *Astron. Astrophys.* **187** 287–289

Gringauz, K.I., Verigin, M.I., Richter, A.K., Gombosi, T.I., Szegö, K., Tátrallyay, M., Remizov, A.P., & Apáthy, I. (1987b) Quasi-periodic features and the radial distribution of cometary ions in the cometary plasma region of Comet Halley, *Astron. Astrophys.* **187** 191–197

Ip W.-H. (1976) Currents in cometary atmosphere, *Planet. Space Sci.* **27** 121–125

Ip W.-H. & Axford W.I. (1982) Theories of physical processes in the cometary comae and in ion tails, In: *Comets,* ed. L.L. Wilkening, Univ. of Arizona Press, Tucson, Arizona, 588–634

Ip W.-H., & Axford W.I. (1986) Metallic ions in cometary comae and plasma tails, *Nature* **321** 682–684

Ip W.-H., & Mendis D.A. (1976) The generation of magnetic fields and electric currents in the cometary plasma tails, *Icarus* **29** 147–151

Johnstone, A., Coates, A., Kellock, S., Wilken, B., Jockers, K., Rosenbauer, H., Studeman, W., Weiss, W., Formisano, V., Amata, E., Cerulli-Irelli, R., Dobrowolny, M., Terenzi, R., Egidi, A., Borg, H., Hultquist, B., Winningham, J., Gurgido, C., Bryant, D., Edwards, T., Feldman, W., Thomsen, M., Wallis, M.K., Biermann, L., Schmidt, H., Lust, R., & Haerendel, G., Paschmann, G. (1986) Ion flow at Comet Halley, *Nature* **321** 344–347

Kaminsky, M., (1965) *Atomic ionic impact phenomena on metal surface,* Springer-Verlag, Berlin

Kissel, J., Brownlee, D.E., Bühler, K., Clark, B.C., Fehtig, H., Grün, E., Hornung, K., Igenbergs, E.B., Jessberger, E.K., Krueger, F.R., Kuczera, H., McDonnell, J.A.M., Monfil, G.M., Rahe, T., Schweur, G.H., Sekanina, Z., Utterback, N.G., Völk, H.J., & Zook, H.A. (1986) Composition of Comet Halley dust particles from Giotto observations, *Nature* **321** 336–337

Korth, A., Richter, A.K., Anderson, K.A., Carlson, C.W., Cartis, D.W., Lin, R.P., Rème, H., Sauvaud, J.A., d'Uston, C., Cotin,

F., Cos, A., & Mendis, D.A. (1985) Cometary ion observations at and within cometopause — region of Comet Halley, *Adv. Space Res* **5** No. 12, 221–225

Korth, A., Richter, A.K., Leidl, A., Anderson, K.A., Carlson, C.W., Curtis, D.W., Lin, R.P., Rème, H., Sauvaud, J.A., d'Uston, C., Cotin, F., Cros, A., & Mendis, D.A. (1986) Mass spectra of heavy ions near Comet Halley, *Nature* **321** 335–336

Korth, A., Richter A.K., Mendis D.H., Anderson K.A., Carlson C.W., Curtis, D.W., Lin R.P., Mitchell D.L., Rème, H., Sauvaud, J.A., d'Uston, C., (1987) The composition and radial dependence of cometary ions in the coma of Comet P/Halley, *Astron. Astrophys* **187** 149–152

Krankowsky, D., Lämmerzahl, P., Herrwerth, I., Woweries, J., Eberhardt, P., Dolder, U., Herrmann, U., Schulte, W., Berthelier, J.J., Illiano, J.M., Hodges, R.R., & Hoffman, J.H. (1986) *In situ* gas and ion measurements at Comet Halley, *Nature* **321** 326–329

Krasnopolsky, V.A., Gogoshev, M., Moreels, G., Moroz, V.I., Krysko, A., Gogosheva, Ts., Palazov, K., Sargoichev, S., Clairemidi, J., Vincent, M., Bertaux, J.L., Blamont, J.E., Troshin, V.S., & Valnicek, B. (1986) Spectroscopic study of Comet Halley by the Vega-2 three-channel spectrometer, *Nature* **321** 269–271

Mendis, D.A., Houpis, H.L.F., & Marconi, M.L. (1985) The physics of comets, *Fundam. Cosm. Phys* **10** 1–380

Moreels, G., Gogoshev, M., Krasnopolsky, V.A., Clairemidi, J., Vincent, M., Parisot, J.P., Bertaux, J.L., Blamont, J.E., Festou, M.C., Gogosheva, Ts., Sargoichev, S., Palasov, K., Moroz, V.I., Krysko, A.A., & Vanysek, V. (1986) Near-ultraviolet and visible spectrometry of Comet Halley from Vega-2, *Nature* **321** 271–273

Olivero, J.J., Stagat, R.W., & Green, A.E.S. (1972) Electron deposition in water vapour, with atmospheric application, *J. Geophys. Res.* **77** 4797–4811

Rème, H., Sauvaud, J.A., d'Uston, C., Cotin, F., Cross, A., Anderson, K.A., Carlson, C.W., Curtis, D.W., Lin, R.P., Mendis, D.A., Korth, A., & Richter, A.K. (1986) Comet Halley — solar wind interaction from electron measurements aboard Giotto, *Nature* **321** 349–352

Remizov, A.P., Verigin, M.I., Gringauz, K.I., Apáthy, I., Szemerey, I., Gombosi, T.I., & Richter, A.K. (1986) Measurements of neutral particle density in the vicinity of Comet Halley by Plasmag-1 on board Vega-1 and Vega-2, In: *Exploration of Halley's Comet, ESA SP-250* **1** 387–390

Riedler, W., Schwingenschuh, K., Yeroshenko, Ye.G., Styashkin, V.A., & Russell, C.T. (1986) Magnetic field observations in Comet Halley's coma, *Nature* **321** 288–289

Sagdeev, R.Z., Shapiro, V.D., Shevchenko, V.I., & Szegö, K. (1986) MHD turbulence in the solar wind — comet interaction region, *Geophys. Res. Lett.* **13** 85–88

Savin, S., Avanesov, G., Balikhin, M., Klimov, S., Sokolov, A., Oberc, P., Orlowski, D., & Krawczyk, Z. (1986) ELF waves in the plasma regions near the comet, In: *Exploration of Halley's Comet, ESA SP-250* **3** 433–436

Schwingenschuh, K., Riedler, W., Schelh, G., Yeroshenko, E.G., Styashkin, V.A., Luhman, J.G., Russell, C.T., & Fedder, J.A. (1986) Cometary boundaries: Vega observations at Halley, *Adv. Space Res.* **6** No. 1, 217–268

Shmidt, R., & Azens, A. (1987) Measurements of integral yields of changed secondary particles using neutral beams simulating a cometary flyby, In: *The Giotto spacecraft impact-induced plasma environment, ESA SP-227*, 15–19

Shmidt, H.U., & Wegmann, R. (1982) Plasma flow and magnetic fields in the comets, In: *Comets*, Tucson, Univ. of Arizona Press, 538–634

Verigin, M.I., Gringauz, K.I., Richter, A.K., Gombosi, T.I., Remizov, A.P., Szegö, K., Apáthy, I., Szemerey, I., Tátrallyay, M., & Lezhen, L.A. (1986) Characteristic features of the cometosheath of Comet Halley: Vega-1 and Vega-2 observations In: *Exploration of Halley's Comet, ESA SP-250* **1** 169–173

Verigin, M.I., Gringauz, K.I., Richter, A.K., Gombosi, T.I., Remizov, A.P., Szegö, K., Apáthy, I., Szemerey, I., Tátrallyay, M., & Lezhen, L.A. (1987a) Plasma properties from the upstream region to the cometopause of Comet Halley: Vega observations, *Astron. Astrophys.* **187** 89–93

Verigin, M.I., Axford, W.I., Gringauz, K.I., & Richter, A.K. (1987b) Acceleration of cometary plasma in the vicinity of Comet Halley associated with an interplanetary magnetic field polarity change, *Geophys. Res. Lett.* **14** 987–990

Woods, T.N., Feldman, P.D., Dymond, K.F., Sahnow, D.J. (1986) Rocket ultraviolet spectroscopy of Comet Halley and abundance of carbon monoxide and carbon, *Nature* **324** 436–438

Woods, T.N., Feldman, P.D., & Dymond, D.F. (1986) The atomic carbon distribution in the coma of Comet Halley, In: *Exploration of Halley's Comet, ESA SP-250* **1** 431–435

Detection of water group ions from comet Halley by means of Sakigake

Koh-ichiro Oyama and Takumi Abe

1 OVERVIEW OF THE EXPERIMENT

Sakigake, Japan's first interplanetary spacecraft, was launched on 9 January 1985. Although Sakigake is dedicated to verification of several engineering topics such as deep space communication, spacecraft manoeuvering, and orbit determination, three scientific experiments were carried on-board to measure weak interplanetary magnetic fields, plasma waves, and solar wind. As Fig. 1 shows, the magnetic field was measured by means of ring core type sensors (IMF). Plasma waves were measured by means of a dipole antenna 8 m from the tip to tip (PWP). Solar wind measurements were made by means of a 4-mesh Faraday cup (SOW), whose function was substantially the same as that used by the MIT group (Oyama 1984). Sakigake made measurements before the

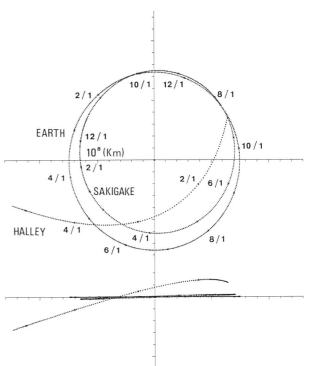

Fig. 2 — Trajectories of Sakigake, comet Halley and Earth in 1986 which are projected on the X-Y plane of the ecliptic (top) and Z-XY plane. (bottom).

encounter with comet Halley, and approached the comet with a closest distance of 7 million kilometres on 11 March 1986. After the closest encounter, Sakigake continued observation in 1987, and will continue until it returns to Earth on 9 January 1992, exactly seven years after its launch. Sakigake will then join the International Solar-Terrestrial Phenomena (ISTP) programme together with the Japanese

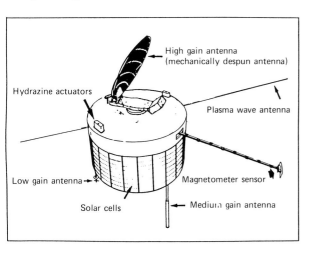

Fig. 1 — The Sakigake spacecraft.

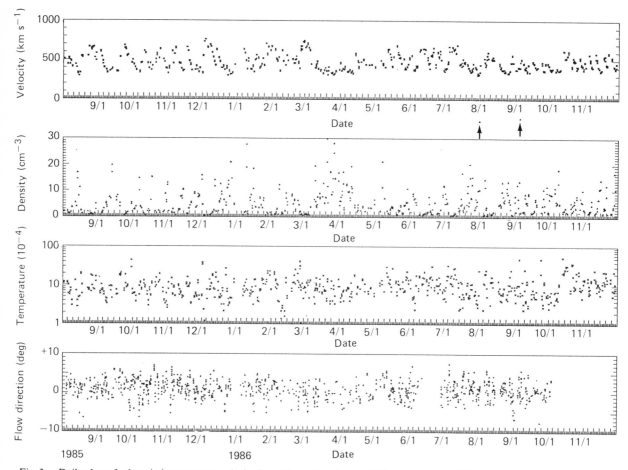

Fig. 3 — Daily plots of solar wind parameters; velocity (km s^{-1}), ion density (cm^{-3}), ion temperature (K) and flow direction (degrees) from the top to the bottom between the period of August 1985 and November 1986. '9/1' means 1 September, and so on.

'Geotail'.

The trajectories of Sakigake, comet Halley, and Earth are shown in Fig. 2.

In this paper we describe solar wind parameters which were measured on 10–12 March 1986 when the closest approach to comet Halley occurred. The parameters which were measured by means of the Faraday cup were ion density, ion temperature, solar wind velocity, and direction of plasma flow. The data were calculated by fitting a theoretical curve to a characteristic curve obtained during one spacecraft spin period (10 seconds), by assuming that the plasma consists only of protons and has one ion temperature. However, transmission of one data set to Earth takes 64 s because of the very slow data rate of 64 bits/s.

2 HELIOMAGNETOSPHERE AT CLOSEST ENCOUNTER

Fig. 3 shows the solar wind parameters which were obtained between 1 August 1985 and 1 November 1986. The velocity is the component which is projected onto the orbital plane of Sakigake. The flow direction is also only in the orbital plane of Sakigake, and no information can be obtained for the elevation angle from the orbital plane. The average speed of the solar wind is ~ 460 km s^{-1} during the period (see Fig. 6). The ion density sometimes rose to 40 ions cm^{-3}, as on the 4 August and 5 September 1986 (Fig. 3). The temperature is in the range 10^4–10^5 K. The positive flow direction shown here indicates plasma flows

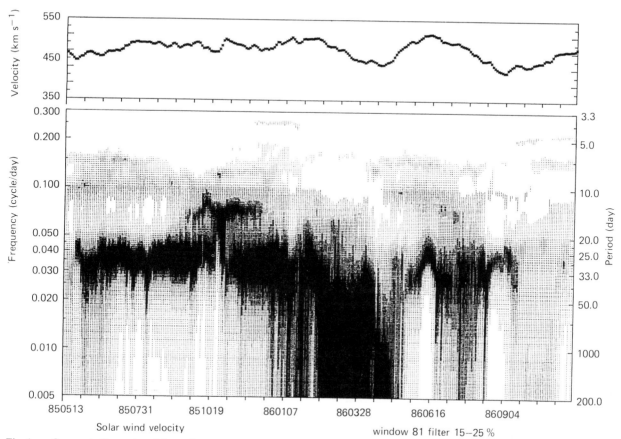

Fig. 4a — Grey scale illustration of dynamic spectrum of solar wind velocity which was calculated by using the data from 3 April 1985 to 4 November 1986. The top panel shows the solar wind velocity averaged over 81 days. '850513' means 13 May 1985, and so on.

from the lefthand side from Sakigake when looking toward the Sun, and negative flow from the righthand side, zero flow meaning that the solar wind flows radially from the Sun along the Sun–Sakigake line.

The data plotted in Fig. 6 were taken randomly from several hours' observation during a day, therefore the data point shown is not necessarily an average value for that day. However, each point can be considered to be representative, since solar wind values do not usually change much within a single day, solar activity being low.

Fig. 3 shows that solar wind velocity is about 300 km s^{-1} around the time of comet Halley encounter in the middle of March 1986. The velocity is the lowest obtained in 1985 and 1986, which suggests that Sakigake was in a low latitude of the heliomagnetosphere. To see this in more detail, 81 days' observation (3 solar rotations) were picked up as one data block,

and the maximum entropy method was applied in order to find frequency components hidden in one data block.

Figure 4a shows, in grey scale form, the dynamic spectrum of the solar wind velocity. Our analysis was based on 40-day observation periods, centred on chosen dates. Fig. 4a shows a 27-day's recurrence between 13 May 1985 (850513) and 7 January 1986 (860107). However, this periodicity starts to disappear from February 1986. The periodicity reappeared in June for a short time, but again became unclear in July. The dynamic spectrum of ion density behaved almost in the same way as that of the solar wind velocity. However, the dynamic spectrum of ion temperature measurement (Fig. 4b) does not show a 27-day periodicity so clearly between May 1985 and February 1986. However, ion temperature periodicity suddenly appears clearly from the beginning of 1986,

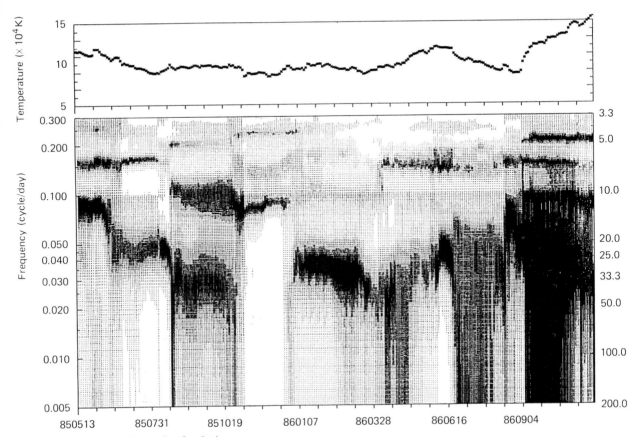

Fig. 4b — Same as for Fig. 4a, but for the ion temperature.

after the periodicity of the solar wind velocity had disappeared during the comet Halley encounter. This interesting and important behaviour of ion temperature could be attributed to the physics of the solar surface, and is still being studied.

The disappearance of the periodicity in solar wind velocity shows that Sakigake is located near the neutral sheet of the heliomagnetosphere, which is flat along the ecliptic plane, and the neutral sheet does not cross the ecliptic plane. This could mean that the effective dipole axis of the heliomagnetosphere is almost vertical to the solar ecliptic plane. Fig. 5a shows the occurrence rate of different solar wind velocities. Between 1 January 1986 (860101) and 30 April 1986 (860430) the solar wind has two main maxima at 400 km s^{-1} and 650 km s^{-1}. The 650 km s^{-1} peak starts to merge slowly with the lower velocity peak, as demonstrated by Fig. 5b, and finally disappears between 1 May (860501) and 31 August

1986 (860831) — as shown in Fig. 5c. Finally, only low-velocity components remain: the average velocity between 860501 and 860831 is 450 km s^{-1} (Fig. 5d).

Fig. 5 shows that even after the comet Halley encounter, the heliomagnetosphere is flatter with respect to the solar ecliptic plane, assuming that solar wind speed is lowest at the equinox of the Sun. This is consistent with the fact that Sakigake measured the lowest solar wind velocity in March 1986. The structure of the heliomagnetosphere was also examined by using interplanetary magnetic field data, and it is concluded that the neutral sheet of the heliomagnetosphere was quite flat at the time of closest encounter with comet Halley (Saito *et al.* 1986).

3 SOLAR WIND MEASUREMENTS AROUND CLOSEST APPROACH

Sakigake made its closest approach to comet Halley at around 0400 hours on 11 March 1986 with a distance

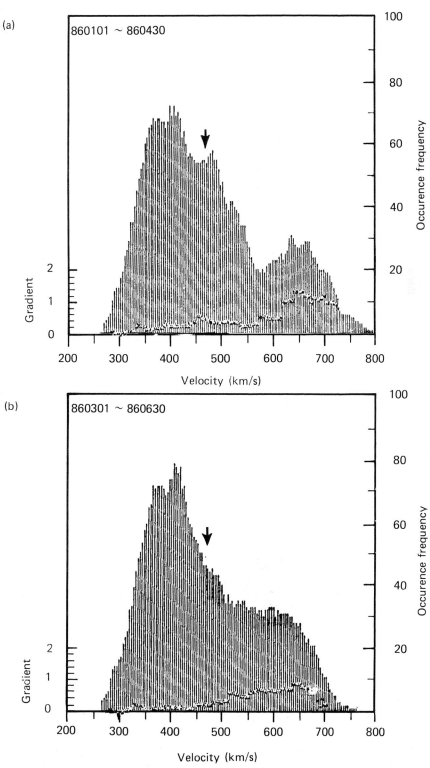

Figs. 5a and 5b — Occurrence frequency of the solar wind velocity. Occurrence frequency at 400 km s⁻¹ is the number of days when the solar wind velocity is, for example, between 350 km s⁻¹ and 450 km s⁻¹. From top to bottom, the periods covered are from 1 January to 30 April 1986 (a) and from 1 March to 30 June 1986 (b). Meaning of gradient at left of the figure follows Lopez (1986). Arrow shows average value of the solar wind velocity during the period noted at the top of each figure.

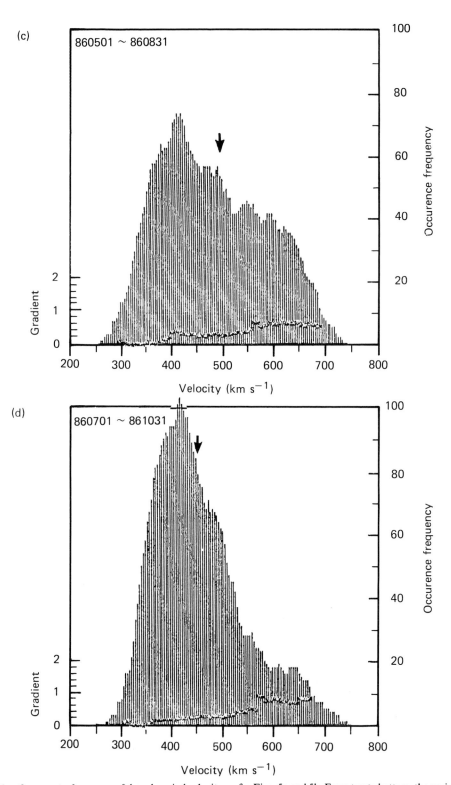

Figs. 5c and 5d — Occurrence frequency of the solar wind velocity, as for Figs. 5a and 5b. From top to bottom, the periods covered are from 1 May to 31 August 1986 (c) and from 1 July to 31 October 1986 (d).

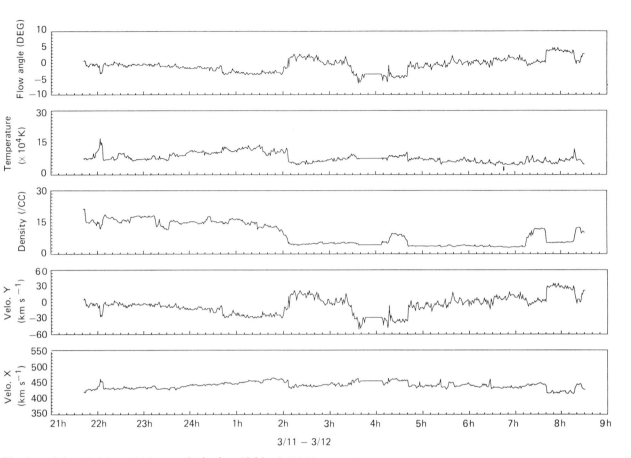

Fig. 6a — Solar wind data which were obtained on 10 March 1986 between 2100 and 0900.

of 7 million kilometres. Figs. 6a and 6b show the solar wind parameters which were obtained every 64 seconds from 10 to 12 March 1986. From the top to the bottom are shown flow angle, ion temperature, ion density (cm^{-3}), and the solar wind velocity divided into X and Y components. The X component is along the Sun–Sakigake line, and the Y component is vertical to it. At 0200 hours on 11 March the ion density suddenly changed from 15 ions cm^{-3} to 3 ions cm^{-3}. The total force of the interplanetary magnetic field increased at this time. At 0415 and 0715 ion density increased, while the magnetic field decreased. It seems that, in these three instances, the interplanetary magnetic field is controlled by the plasma cloud. The most remarkable change is the Y component of the solar wind velocity. The velocity changed in antiphase with the ion density at these three times. Although we do not know that the several changes of

ion density and magnetic field which occurred on 10 to 11 March are related to comet Halley, it should be noted that an intense plasma wave was detected at closest encounter by means of the dipole antenna, and the antenna had not detected such waves before (Oya, 1986).

Between 2230 on 11 March and 0300 on 12 March, the Y component of the solar wind velocity fluctuated with a maximum amplitude of 60 km s^{-1}, which is much greater than at closest encounter, but there is no significant fluctuation in the ion density.

To answer the question, 'is the fluctuation related to comet Halley?', periodic components were searched for in the fluctuating solar wind velocity. First, 50 data points were taken from the continuous observations to form the first data set, starting from 0000:42 on 11 March; and second, 50 data points were successively taken by shifting two data points. As one measure-

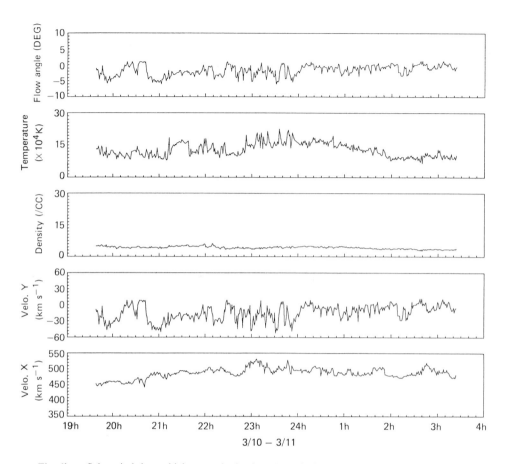

Fig. 6b — Solar wind data which were obtained on 11 and 12 March 1986 between 1900 and 0400.

ment takes 1 minute and 4 seconds, sampling for 50 data points takes 53 minutes and 20 seconds. The maximum entropy method was applied to the first data set to find various frequency components, and then to the successive data sets.

Figs. 7a and 7b show the spectra of Y component of solar wind velocity which was calculated around the comet Halley encounter. The horizontal axis shows the frequency (and period). The vertical axis is the spectral amplitude of the solar wind velocity. The data are plotted for two periods, 0027:22 to 0528:04 on 11 March (Fig. 7a) and from 2004:04 on 11 March to 0103:24 on the 12 March. Periodic components which are lower than 600 sec period have no physical meaning.

The five lines in the figure are the gyrofrequencies (and periods) which correspond to the mass numbers 13 (CH^+), 16 (O^+), 18 (H_2O^+), 28 (CO^+), and 44

(CO_2^+). These lines were calculated by using interplanetary magnetic field values which were simultaneously measured by IMF. The magnetic field used here is an average over 53 minutes and 20 seconds. Between 0319 and 0445 on 11 March, which corresponds to Sakigake's closest approach to comet Halley, the dynamic spectrum in the solar wind fluctuation shows a peak at the frequencies between 16^+ and 18^+. Between 0319 and 0402 a rather strong peak appears at the gyrofrequency of mass number 44. Between 0506 and 0528, the peak appears to correspond to the gyrofrequency of mass number 28 (CO^+), but the peak is rather sporadic. Ion components which produced a clear peak near the mass number of 24 between 0258 and 0402 might correspond to Mg^+ or $^{12}C^+_2$ (Krankowsky *et al.* 1986). The amplitude of the dynamic spectrum becomes higher on the following day, as is shown in Fig. 7b. In particular, the Y

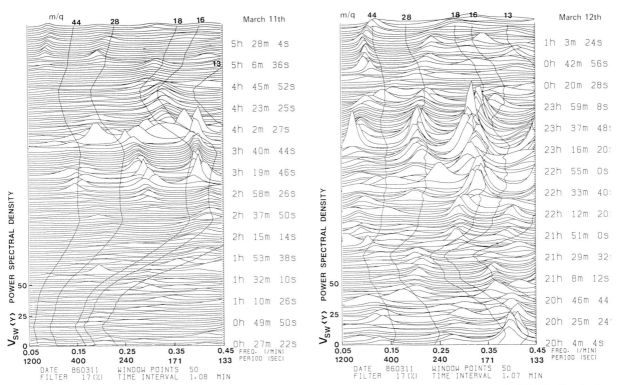

Fig. 7a (left) — Dynamic spectrum of Y component of solar wind velocity (V_{sw}) between 0027 and 0528 on 11 March 1986. Five lines in the figure are gyrofrequencies which correspond to the mass/charge of 13, 16, 18, 28, and 44.
Fig. 7b (right) — Same as for Fig. 7a, but between 2004 on 11 March and 0103 on 12 March 1986.

component of the solar wind velocity fluctuates most remarkably between 2212 and 2359 on the 11 March near the frequency which corresponds to gyrofrequency of mass number 18.

Fig. 8 will be useful to elucidate the close relationship between the dominant spectral frequencies of the solar wind velocity (the Y component) and *in situ* gyro-frequencies of several ion species; *viz*, whole spectral powers from each window segment of the dynamical spectra during the period from 2123:40 on 11 March to 0136:56 UT on 12 March, have been totalled and plotted in the figure. The best three of the distinguished spectral peaks are observed at 0.444, 0.321, and 0.266 (1/minute). These roughly agree with the gyro-frequencies for the mass/charge ratios of 13(CH^+), 18(H_2O^+), and 28(CO^+), respectively. Slight deviation of the peaks from the mass/charge of 13, 18 and 28 is being studied.

For the peaks which appeared in the dynamic spectra shown in Figs. 7a and 7b, a cross power spectrum (Figs 9a,. b) between interplanetary magnet-

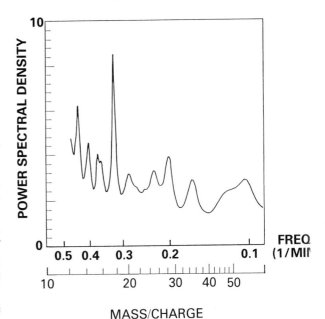

Fig. 8 — Contribution of various substances to the fluctuation of Y component of solar wind velocity.

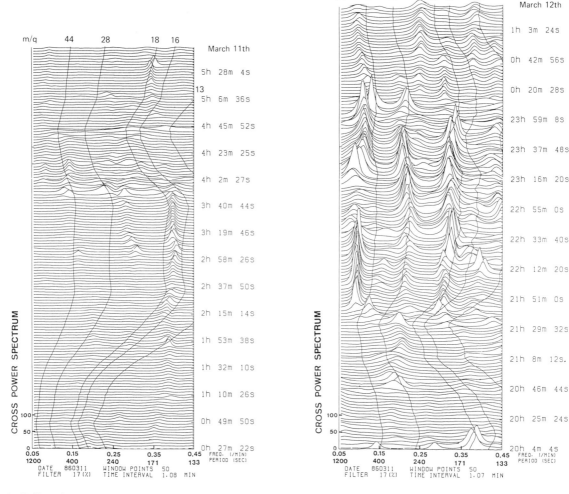

Fig. 9a (left) — Cross-power spectrum on 11 March 1986 between 0027 and 0545.
Fig. 9b (right) — Same as for Fig. 9a, but on 11 and 12 March, 1986 between 2004 and 0120.

ic field and solar wind velocity was calculated for the same Y components. Between 0205 and 0352 on 11 March, which is near the closest approach to comet Halley, a weak correlation can be seen between 16^+ and 18^+. There is a stronger correlation on the following day (Fig. 9b) at the gyrofrequencies of 18^+ and 28^+ between 2151 and 2359. The peak near the mass number of 44 is calculated to be the mass number 56 (Fe^+).

The significantly good correlation between the fluctuations of solar wind velocity and interplanetary magnetic field, suggests that the fluctuation which appeared in the Y component of solar wind is an Alfvén wave. This conclusion is confirmed when we recall that the intensity of solar wind fluctuation is nearly equal to the intensity which is calculated from IMF fluctuation by assuming that the fluctuation is Alfvénic.

4 COMETARY MOLECULES ALONG THE TRAJECTORY OF COMET HALLEY

The next question is, 'why is the fluctuation stronger between 21h and 24h on 11 March than at the time of the closest approach which occurred at 4h on 11 March?'.

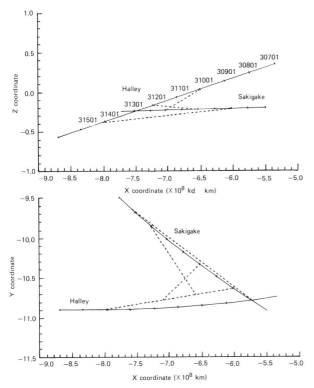

Fig. 10 — Trajectory of Sakigake around the closest approach for Y-Z plane and for X-Y plane. Z axis is vertical to the solar ecliptic plane. Minimum division is 1×10^7 km.

The trajectory of Sakigake around the time of closest approach is shown in Fig. 10 for the X-Y and X-Z planes.

The distance at 2100 on 11 March between Sakigake and the trajectory along which comet Halley passed is 5.11 million kilometres, and this instance corresponds to 41 hours after comet Halley passed (Fig. 11). The closest approach of Sakigake to the trajectory of comet Halley occurred on 14 March with 4 million kilometres, and comet Halley had already passed through this point 128 hours earlier. Although we do not have a firm answer to the above question, there is a possibility that comet Halley left neutral cometary materials such as H_2O and CO along its trajectory, and therefore components which we detected by means of Sakigake seem to have been generated as a result of charge exchange of neutral molecules with solar wind protons.

ACKNOWLEDGEMENTS

We are very grateful to the launching staff of the Institute of Space and Astronautical Science. Interplanetary magnetic field data during the closest encounter, which are necessary to compute gyrofrequencies of several cometary matters, are taken from the special issue of *Nature* (**321**, No. 6067, 15–21, May 1986), and the details were supplied by Prof T. Saito and Dr K. Yumoto. Typing of the article was done by Mrs K. Peschke while one of the authors (K.I. Oyama) was in the Max-Planck-Institut für Aeronomie in Lindau.

REFERENCES

Saito, T., Yumoto, K., Hirao, K., Nakagawa, T., & Saito K., (1986). Interaction between Comet Halley and the interplanetary magnetic field observed by Sakigake, *Nature* **321** 303–307.

Oyama, K., Akai, K., Nakazawa, K., Hirao, K. & Tei, S., (1984) *Development of Faraday cup onboard MS-T5 for solar wind measurement*, ISAS, Report No. 616

Oya, H., Morioka, A., Miyake, W., Smith, E.J., & Tsurutani, B.T., (1986). Discovery of cometary kilometric radiations and plasma waves at Comet Halley, *Nature* **321**, 307–310.

Krankowsky, D., Lämmerzahl, P., Herrwerth, I., Woweries, J., Eberhardt, P., Dolder, U., Herrmann, U., Schulte, W., Berthelier, J.J., Illiano, J.M., Hodges, R.R., & Hoffmann, J.H., (1986.) *In situ* gas and ion measurements at Comet Halley, *Nature* **321** 326–329.

Lopez, R.E., (1986). The relationship between proton temperature and momentum flux density in the solar wind, *Geophys. Res. Letters*, **13** 640–643.

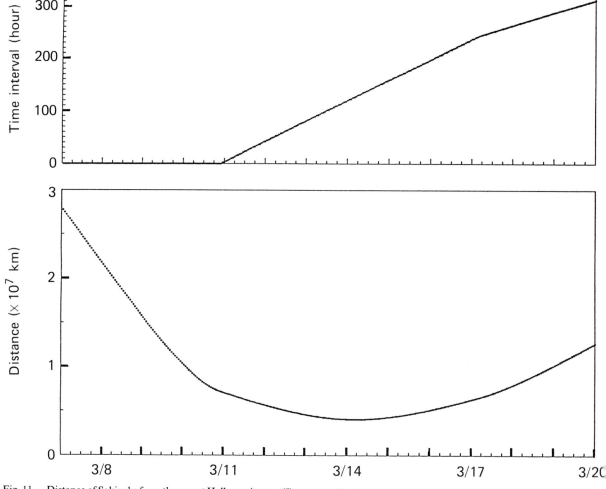

Fig. 11 — Distance of Sakigake from the comet Halley trajectory ("lower panel'). Upper panel show the time after Sakigake approaches the comet Halley trajectory. For example, at 21h on 11 March, the distance between Sakigake and the comet Halley trajectory is 5.11 million kilometres which corresponds to 41 h after comet Halley passed. '3/8' means 8 March, and so on.

11

Energetic pick-up ions in the environment of comet Halley

Susan M.P. McKenna-Lawlor

1 SPACE MISSIONS TO HALLEY'S COMET

The 1986 apparition of Halley's comet provided a stimulus to three space agencies, namely ESA (the European Space Agency), Intercosmos (IKI — the USSR Academy of Science) and ISAS (the Japanese Institute of Space and Astronautical Science) to send spacecraft to encounter this object at flyby distances that varied between 600 and 7 million km. ESA sent one spacecraft Giotto (closest approach 600 km on 13 March). Intercosmos sent Vega-2 and Vega-1 (closest approach 8030 km on 9 March and 8890 km on 6 March respectively). ISAS also sent two spacecraft, Suisei and Sakigake (closest approach 151 000 km on 8 March and 6.99 million km on 11 March respectively. In addition, NASA (National Aeronautics and Space Administration) used their existing spacecraft ICE (the International Cometary Explorer) to make observations of Halley from an upstream distance of 30 million km.

A certain overlap between the experiments carried aboard these different spacecraft now provides a stimulating basis for the intercomparison of data and allows the rich variety of phenomena occurring in the cometary environment to be appreciated over a time scale greatly exceeding that pertaining to any particular set of measurements performed aboard a particular probe.

In the present review, an account is presented of what was generally known about comet/solar wind interactions in advance of the making of detailed *in situ* measurements at comets Giacobini–Zinner and Halley using spacecraft-borne instrumentation. More recent theoretical studies complementary to the new observations from space are also cited. Plasma phenomena recorded at Halley's comet aboard various spacecraft during the 1986 encounter are then interpreted in the light of these considerations and new, but as yet unanswered, questions raised by these *in situ* observations are discussed.

The reader is referred, to the monograph by Mendis, Houpis and Marconi (1985), to the book *Comets*, edited by Wilkening (1982), to *Exploration of Halley's Comet*, edited by Grewing, Praderie and Reinhard (1988) and to *Physics of Comets in the Space Age*, edited by Huebner (1988) for background reading. Also, to the comprehensive review of the plasma measurements made at comet Giacobini-Zinner by the ICE spacecraft contained in McComas *et al.* (1987).

2 OVERVIEW OF THE EXPECTED CHARACTERISTICS OF COMETARY–SOLAR WIND INTERACTIONS

As a cometary body approaches the Sun, its nucleus begins to lose mass owing to sublimation processes and, in the condition of extremely low gravity pertaining, an expanding atmosphere composed of dust, molecules, radicals and molecular ions is formed which escapes from the collision-dominated inner region with a velocity of the order of ~ 1 km s^{-1}. The extent of the neutral atmosphere produced, is determined by the size of the nucleus and by the production rate of neutrals from the constituent ices, (Schmidt & Wegmann 1982). The gas production rate itself is a function of the position of the comet along its orbit and can vary significantly even between medium-bright and bright comets.

The radially expanding molecules are subjected to complicated chemical reactions in the inner reaches of the cometary coma, while, further out, ions are created from cometary neutrals through photoionization by solar EUV radiation, charge exchange processes with solar wind ions, and electron impact ionization, (Wallis 1973, Huebner & Giguere 1980, Ip 1985, Cravens *et al.* 1986). Bombardment by energetic electrons created in a cometary aurora has also been suggested by Smith *et al.* (1985) to comprise a plausible ionization mechanism.

In addition, anomalous ionization, proposed by Alfvén, (1954, 1960) to be triggered if a neutral gas and a plasma are in relative motion with a velocity exceeding a critical value which depends on the relevant plasma and gas parameters, has recently been re-studied as a possible contributor to the ionization mechanisms occurring at Halley's comet, (Formisano *et al.* 1982, Klimov *et al.* 1986).

2.1 Pick-up ions

Freshly created cometary ions, which are practically at rest in the comet frame of reference, are accelerated in the local interplanetary electric field (E) along a trochoidal trajectory perpendicular to the local magnetic field (B) direction. In the frame of the solar wind, the motion of the particle is a gyration with speed $E/B = V \sin \theta$, where θ is the angle between the magnetic field and solar wind velocity vectors. There is also a motion of the gyrocentre along the field line towards the Sun at a speed $V \cos \theta$. The total energy in this frame is $\frac{1}{2} Am V^2$ (where A is the atomic number of the species concerned and m is the mass of a proton). For a water ion, if $V = 400$ km s^{-1} and $\theta = 45°$, the total resulting energy would be 15 keV.

In the frame of the comet, the particle motion is perpendicular to the magnetic field and away from the Sun and the energy acquired varies between zero and a maximum value of $2 Am V^2 \sin^2 \theta$ (that is 30 keV for

a water group ion under the particular conditions described above), (Ip & Axford 1982, Galeev *et al.* 1985, Cravens 1986).

In velocity space, the distribution of the ions forms a ring in the plane perpendicular to the magnetic field with a radius equal to the component of the solar wind velocity perpendicular to the magnetic field and centred on the solar wind velocity parallel to the magnetic field. The maximum velocity in the comet frame is twice the solar wind velocity, while the minimum value is the original outward streaming velocity.

Since there is excess free energy in the ring distribution, it is highly unstable and it generates Alfvén waves via the ion-cyclotron instability, thereby making the magnetic field turbulent. Scattering in this turbulent field then acts to isotropize the ion distribution through the production, within a time scale of several gyration periods, of a shell like distribution of particles (Sagdeev *et al.* 1986, Gary *et al.* 1986).

In the frame of the solar wind (to the extent that second order Fermi acceleration may be neglected), the pick-up ions become isotropic without change of energy. In the comet frame, however, the energies observed range from zero to a maximum value of $2 Am V^2$, corresponding to a relative speed of $2V$, so that, again, a water group ion, under the conditions previously described, would attain a maximum energy of 60 keV (independent of the magnetic field direction).

2.2 Shock formation

The energy and momentum imparted to cometary ions in the 'pick-up' process described above, comes from the solar wind, which is therefore decelerated by increasing amounts as the comet's nucleus is approached. As estimated by Axford (1964), the solar wind may accrete no more than a few percent (by number) of heavy ions before a shock transition is formed which may be relatively strong or weak depending on the solar distance and on the gas production rate concerned. The existence of such a shock was calculated by Biermann *et al.* (1967), using a hydrodynamic model, and its basic nature was further elucidated by Wallis (1973). Several authors, including Schmidt & Wegmann (1982) constructed

MHD models of the comet–solar wind interaction, and all of them predicted the shock to be weak (M = 2, where M is the Mach no.). Also, Alfvén (1957) predicted that the interplanetary magnetic field would drape around the cometary ionosphere to form an extended magnetotail.

2.3 Contact surface

The solar wind plasma, contaminated (mass loaded) by cometary ions, flows through the cometary bow shock and fills the transition region down to the contact surface (also called the ionopause), which is a tangential discontinuity comprising a sharp transition between regimes characterized by strong magnetic fields on the one hand (the magnetic pile-up region) and zero field on the other (the magnetic cavity). The location of this contact surface, which was first predicted theoretically by Biermann *et al.* (1967), later interpreted in detail by Ip & Axford (1982) and physically observed in association with the 'artificial comets' of the AMPTE mission, Valenzuela *et al.* (1986), and associated papers, is determined by the establishment of an equilibrium between the magnetic pressure of the solar wind, ram pressure existing in the pile-up region (supplemented by frictional force between the stagnating cometary ions and the outflowing neutrals), and the thermal pressure of the cometary ionosphere existing within the cavity.

Flammer *et al.* (in press) further predicted the existence, upstream of the contact surface, of a 'collisionopause' (also referred to as a cometopause), characterized by a rapid increase in the collision rate between contaminated solar wind ions and outflowing cometary neutrals. Such a region has indeed been detected within comet Giacobini-Zinner, Mendis *et al.* (1986), Mendis (1988), and within comet Halley, Réme *et al.* (1986); Gringauz *et al.* (1986a, b), in the latter case at a location generally corresponding with the beginning of that enhancement in field strength defining the magnetic barrier/pile-up identified by Neubauer *et al.* (1986). There is presently no general agreement as to the mechanism for the formation of the cometopause, see however Sagdeev *et al.* (1987).

2.4 Acceleration mechanisms additional to the pick-up process in cometary environments

In addition to the acceleration of ions in the vicinity of comets by the pick-up process, the traversal of the bow shock by ions created upstream may lead, as proposed by Axford (1981) to further acceleration and energization of the pick-up particles by some combination of gradient B drift, Fermi acceleration, and adiabatic compression. Downstream of the bow shock, yet further acceleration could be caused by first and second order Fermi processes; by adiabatic compression, in particular near the magnetic pile-up region; and by magnetic reconnection upstream and downstream of the nucleus (Cravens 1986, Sagdeev *et al.* 1986, Ip & Axford 1986). It has been estimated by Ip & Axford that, as a result of such processes, water group ions could be accelerated to energies of several hundreds of keV and even up to MeV energies. A more recent study by the same authors, Ip & Axford (1987), indicates that, near the bow shock front, transit time damping by compressive magnetosonic waves and statistical acceleration by lower hybrid waves may play a significant rôle in producing ion energization. Their paper also provides an estimation of the efficiency of stochastic acceleration in contributing to the energization of cometary pick-up ions.

Downstream of the bow shock, the influence of processes such as stochastic acceleration tends at first to maintain energetic ion fluxes at an elevated level. However, continued mass loading of the solar wind produces an ever waxing opposite influence. In consequence, gradual deceleration of the plasma flow, with an associated decrease in the energy capable of being imparted to individual cometary ions, leads to a progressive decline in the number of particles present at the high-energy end of the spectrum. Charge exchange and collisional processes in the gradually stagnating solar wind contribute further to the reduction in energetic particle fluxes, and some additional depletion is consequent on the escape of energetic particles along open magnetic field lines (inbound) (McKenna–Lawlor *et al.* 1985).

3 SPACECRAFT OBSERVATIONS OF ENERGETIC PARTICLES UPSTREAM OF THE COMET HALLEY BOW SHOCK

Cometary ions picked up by the solar wind were detected by the TÜNDE-M (Energetic Particle Analyser) and PLASMAG Cometary Plasma Spectrometer) experiments on the Vega mission as well as by the ESP (Energy Spectrum of Particles) instrument aboard Suisei, at a distance of about 10 million km from the comet (see Somogyi *et al.* 1986, Gringauz *et al.* 1986b, and Terasawa *et al.* 1986 respectively).

On Giotto, the first hydrogen ions of cometary origin were detected by the JPA (Johnstone Plasma Analyser) at a distance of 7.8×10^6 km from the nucleus (Johnstone *et al.* 1986). Also, aboard Giotto, the EPA (Energetic Particle Analyzer) detected the first clear cometary signature at a distance of 7.5×10^6 km from the nucleus. Corresponding sector data show that the particles recorded were, as would be expected of pick-up ions, strongly beamed in the antisolar direction, (McKenna–Lawlor *et al.* 1986). It has been suggested by Sanderson *et al.* (1986) and Wenzel *et al.* (1986) that bursts of heavy ions detected on 25 March 1986 aboard the ICE spacecraft, which was then located at 28×10^6 km from the nucleus of Halley, may also have originated at the comet.

An analysis of JPA (Johnstone plasma analyzer) data from Giotto by Coates *et al.* (1989a) indicates that, far from Halley, the particle distribution comprised a relatively narrow ring. This was observed to develop into a shell via pitch angle diffusion, with some acceleration as a result of energy diffusion, as the spacecraft flew closer. A filled shell was not observed until the spacecraft reached a distance of 2×10^6 km from the comet.

The HERS detector of the Ion Mass Spectrometer on Giotto also provided evidence of the pick-up of cometary ions by the solar wind through the onboard detection of cometary protons, non-uniformly distributed over a spherical shell in velocity space, 6 million km upstream from the Halley bow shock. (Neugebauer *et al.* 1986, 1987 and 1989). Plasma observations from Suisei further clearly demonstrate the assimilation process of cometary ions in the solar wind flow, since, within 2.3×10^5 km from the nucleus, ion pick-up shells of several species were identified in velocity space (Mukai *et al.* 1986).

A beam of pick-up ions in the solar wind would excite low-frequency Alfvén waves owing to the ion-cyclotron instability, and such waves were indeed detected by magnetometers aboard Vega and Giotto at several million km from the Halley nucleus (see Riedler *et al.* 1986 and Neubauer *et al.* 1986). Yet higher frequency waves due to plasma instabilities associated with the pick-up process were also ob-

served out to 10 million km from Halley, using the plasma wave experiment on Sakigake (see Oya *et al.* 1986). It is possible that kilometric signatures recorded aboard the ICE spacecraft at a distance of 30 million km from Halley were also comet related (see Scarf *et al.* 1986).

Four separate spacecraft recorded bow shock crossings, namely Vega-1 and -2, Giotto, and Suisei. See Galeev *et al.* (1986), Gringauz *et al.* (1986a), Mukai *et al.* (1986), Johnstone *et al.* (1986), Coates *et al.* (1987) and the review by Ip and Axford, (1988). It has been demonstrated by Galeev (1986) that the bow shock microstructure thereby observed, as well as the presence of upstream waves and strong plasma turbulence, appears to agree in general with existing, although as yet incomplete, theoretical models of solar wind–cometary plasma interaction. These latter numerical simulations, based on quasilinear theory, indicate that a large fraction of the kinetic energy of cometary ions picked up by the solar wind is transferred into Alfvén wave turbulence. See also the review of this topic by Galeev, (1987) and references therein.

4 FROM THE BOW-SHOCK TO THE MAGNETIC PILE-UP REGION (INBOUND)

Behind the bow shock (inbound), continuing solar wind loading by heavy ions resulted, as would be expected, in a gradual deceleration of the plasma flow, complemented by magnetic field build-up. Both Vega-1 and Vega-2 penetrated to the pile-up region, and the onboard magnetometers individually registered 75 nT and 80 nT at their closest approach. (Riedler *et al.* 1986). The Giotto magnetometer observed peak magnetic fields of 57 nT inbound and 65 nT outbound (Neubauer *et al.* 1986).

Data obtained by the SDA (Solar Direction Analyser) of PLASMAG 1 on Vega-1, indicate that, at a distance of about 8×10^5 km from the nucleus, a single broad peak dominated the particle spectra and corresponded to a shocked solar wind proton flow with a velocity of ~ 350–400 km s^{-1}. Closer to the nucleus, a second peak in the SDA spectra appeared which corresponded to a much higher energy than that of the solar wind protons. This second peak was interpreted by the PLASMAG experimenters, Gringauz *et al.* (1986a) to represent cometary (probably) water group ions.

At a distance of 3×10^5 km from the nucleus the solar wind proton population became comparable with that of the cometary implanted ions, and it was concluded, on the basis of the Vega-1 and Vega-2 data, that, at 1.5×10^5 km from the nucleus, the original solar wind population effectively disappeared from the solar direction and was replaced by a broad distribution of slow cometary ions. This region was called the 'heavy ion mantle.' Shortly afterwards, at $\sim 1 \times 10^5$ km inbound, no further particles were detected by SDA. On the outbound pass, cometary ions reappeared at 7×10^4 km from the nucleus, after which shocked solar wind protons were again observed.

At a distance of a few tens of thousands of kilometres from the nucleus, PLASMAG data indicate that 70–80% of the ion content belonged to the water group, 15–20% stemmed from the CO/CO_2 group, and 2–5% had a mass of ~ 44 (CO^+_2) (Gringauz et al. 1986b). These ratios are in reasonable agreement with complementary observations made aboard Giotto, using the heavy ion analyser RPA2-PICCA and the ion mass spectrometer (IMS) (see Korth et al. 1986, Balsiger et al. 1986). An interesting feature of the PLASMAG observations was a quasi-periodic modulation in the density of the heavy ions with a period of 10 seconds, and the occurrence of a well-defined peak around $m/q = 56$ in the CRA, Cometary Ram Analyzer, spectra (Gringauz et al. 1986b).

Electron energy spectra measured at a distance of 1.5×10^4 km from the nucleus revealed, in addition, the presence of high fluxes of suprathermal electrons (Gringauz et al. 1986a). Simultaneous observations of low-frequency 15 Hz plasma waves provide some support for the view that the critical velocity ionization phenomenon of Alfvén occurs in the Halley interaction region (see Formisano et al. 1982).

Reviews by Galeev (1986, 1987) of the spacecraft cometary plasma and field obvservations, highlight the identification of various boundaries defining regions with different flow characteristics between the bow shock and the cometary ionopause and relate them to existing theory; see also Balsiger et al. (1986), Johnstone et al. (1986), Mukai et al. (1986), Neubauer et al. (1986). Réme et al. (1986) and Savin et al. (1986).

Two sharp surfaces of discontinuity whose nature is not presently understood, which are referred to as 'mysterious boundaries' were also identified by the experimenters; see Réme, (1990) chapter 6 in this volume, for his discussion of the 'mystery boundary' where keV electron fluxes disappear once there is an increase in the ion fluxes in the ram direction.

5 CLOSEST APPROACH

Fig. 1 provides an overview of ion intensities recorded (60 s averages) by the energetic particles experiment EPONA on Giotto, viewing in four directions, over an interval spanning 6 h before and 6 h after closest approach. (A detailed description of this instrument is provided by McKenna–Lawlor et al. 1987a). Even before crossing the inbound bow shock (BS), EPONA registered high intensities coming from the east as well as from the Sun (see Daly al. 1986). At the bow shock itself, these latter intensities sharply increased, as did also (slightly later on) the intensities measured in other directions. The extreme particle anisotropy in the solar wind was then seen to decline inside the interaction region as the solar wind decelerated until, by 2300 UT on 13 March, it showed a further substantial decrease and almost vanished.

At closest approach to Halley, the ion intensities increased by three orders of magnitude, see Fig.2. At count rates $> 10^3$ s^{-1}, saturation occurred, creating a 1-minute gap. The spatial size of the increase, which began to build up in the magnetic pile-up region, was 10 000 km. The contact surface was crossed at 4 700 km from the nucleus inbound and at 3 800 km from the nucleus outbound. While EPONA was not designed to distinguish the mass of an incident ion at a given energy, it can easily be calculated that water ions, the predominant pick-up species, could acquire sufficient energy (~ 70 keV) from the pick-up process acting alone to be detected by the instrument, although the energies attainable by the mechanism (see section 6) are not sufficient to account for the obtained flux of yet higher energy particles. Since the gyroradius of a 100 keV H_2O^+ ion in a 50 nT field is 4 000 km, it may be suggested that the relatively high magnetic fields in the pile-up region may bind certain ions to within a gyro-diameter of the cavity (see Daly et al. 1986).

An examination of the particle data contained in all 16 energy channels of EPONA at closest approach,

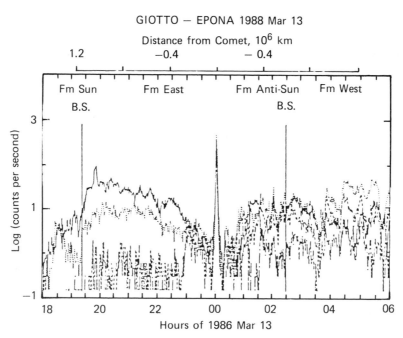

Fig. 1 — Ion intensities from EPONA on Giotto from 18 UT 13 March to 06 UT 14 March 1986, 60 s averages. Energy range 97–145 keV if interpreted to be ions of the water group.

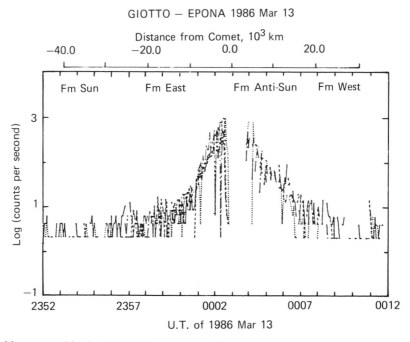

Fig. 2 — Ion intensities measured by the EPONA instrument on Giotto for 20 min about closest approach. 1 s resolution.

shows that energetic protons with E > 3.5 MeV, electrons and protons with E > 300 keV, heavier ions, or a mixture of these particles, were recorded within the magnetic cavity. The only complementary detection of energetic particles was made by the TÜNDE-M experiment at 8 700 km from Halley in association with the occurrence of maximum ambient magnetic field intensity, further characterized by rapid changes in field direction. (Somogyi *et al*. 1986).

6 PROBLEM AREAS

A preliminary study by Kirsch *et al*. (1987) reviews the possibility that the particle enhancements recorded by EPONA at closest approach might have been produced by extraneous effects, such as a microphonic response to dust impact on the detectors. The preliminary conclusion of these authors is, however, that an acceleration mechanism rather than an environmental effect must be invoked in order to explain the observed ion energies. If the enhancements recorded are indeed truly 'cometary', then the identification of the nature of the accelerating process at work presently poses one of the important unanswered questions of the Giotto mission. A report that energetic electrons (up to at least 300 keV) were detected for about one hour before closest approach and for several hours thereafter, McKenna–Lawlor *et al*. (1985) raises a further question as to how, if the identification of these particles as electrons is correct, they could have become so energized in the cometary environment. A recent paper by Kirsch *et al*. (1989), suggests that a possible mechanism for accelerating ions and electrons at Halley up to 300 keV may be provided by magnetic re-connection.

Again, estimates of the energies that ions of the water group and other typical ions could acquire from the solar wind pertaining at the times of encounter of the Giotto probe and of Vega-1 with comet Halley, indicate that the pick-up process, acting alone, is insufficient to account for the observed fluxes of high-energy particles. (note that, exclusive of the special enhancement near closest approach, particles of energies >250 keV if interpreted to be protons, and > 144 keV if interpreted to be water group ions, were recorded by telescope 3 of the EPONA instrument in the Halley environment). Thus, an additional acceleration mechanism or mechanisms, or the presence of

heavy ion species, must be postulated to account for the observations, (McKenna–Lawlor *et al*. 1986, Somogyi *et al*. 1986). This result is in line with similar observations made earlier at comet Giacobini–Zinner by instruments on the ICE spacecraft (Hynds *et al*. 1986, Ipavich *et al*. 1986).

Particle data secured aboard Giotto by the EPONA experiment at energies >60 keV and by the JPA Implanted Ion Sensor at energies >2 keV and <90 keV have recently been combined and composite energy spectra plotted for locations upstream of the

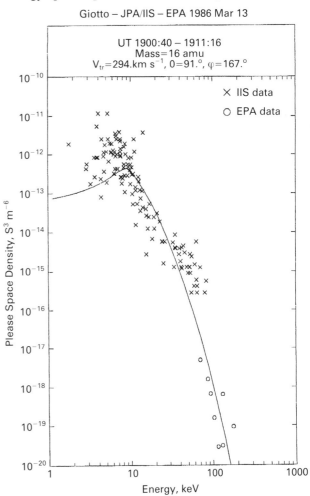

Fig. 3 — Ion spectrum in the solar wind frame from EPONA and IIS on Giotto at a distance (inbound) of 1.5×10^6 km from the nucleus, compared with a theoretical distribution calculated (see Ip and Axford, 1987) on the assumption that the mean free path corresponding to random scattering by Alfvén waves is of the order of 30 gyroradii.

bow shock, at the bow shock crossing and in the down-stream region (McKenna–Lawlor *et al.* 1987b and 1989a). These spectra are shown replotted in Figs. 3 and 4 to take into account an underestimate by Coates (1989b) of the values of the phase space densities derived from the JPA data which were used in the two earlier papers. This revision only requires the original values to be multiplied by seven. A comparison of the revised data with theoretical energy spectra for water group ions derived by Ip and Axford (1987), which take into account the effect of adiabatic

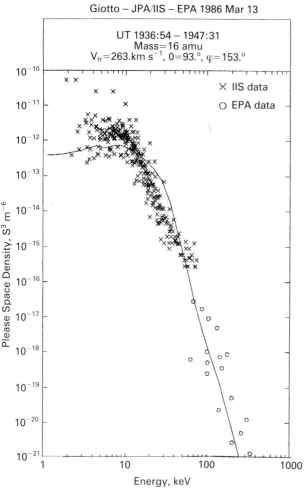

Fig. 4 — Ion spectrum in the solar wind frame from EPONA and IIS on Giotto obtained over an interval spanning the bow shock crossing inbound at 1.1×10^6 km from the nucleus compared with a theoretical distribution calculated (see Ip and Axford, 1987) on the assumption that the mean free path corresponding to random scattering by Alfvén waves is of the order of 5 gyroradii.

compression in the cometary accretion flow, and assume varying degrees of scattering by hydrodynamic waves, indicate a reasonable (and even improved) qualitative fit to the energy spectrum obtained upstream of the shock, if it is assumed that the mean free path corresponding to random scattering by Alfvén waves is of the order of 30 gyroradii. An enhanced level of stochastic acceleration must however be assumed in the turbulent region immediately upstream of the shock in order to adequately fit the observations. It is noted that, if particles with $A = 1$ are assumed, the EPONA observations will not match the water group ion measurements of IIS.

Returning to Vega data, an analysis by Gribov *et al.* (1986) in which the observed ion temperatures from TÜNDE-M using magnetic field (MISCHA) and solar wind (PLASMAG-1) records were compared with calculated ion temperatures based on stochastic Fermi acceleration, indicate that, while the second order Fermi mechanism is capable of explaining the temperature increase observed, the value predicted greatly exceeds that actually measured. It is thus unclear as to whether experimental uncertainties can account for the discrepancy, or if an as yet unidentified cooling mechanism may be present.

It is noteworthy that the ion fluxes detected by TÜNDE-M display peaks that appear to be periodic in time with a period of roughly four hours (Kecskemety *et al.* 1986). A recent analysis by Richter *et al.* (1988) shows that a similar periodicity was detected by the energetic particles experiment EPONA on Giotto during the inbound Halley pass. Evidence from EPONA data for complementary quasi-periodic variations of cometary ion fluxes during the outbound pass has recently been provided by McKenna–Lawlor *et al.* (1989b).

It has been suggested by Richter *et al.* (1988) that the periodic energetic ion flux enhancements are the result of periodic neutral gas production from a rotating cometary nucleus. (Possible candidates for the neutrals include hydrogen atoms created by the photodissociation of OH and oxygen from the dissociation of CO_2, Daly (1988). This relatively simple kinematic model, although it is able to reproduce the positions of the ion maxima recorded aboard Vega-1 and Giotto, runs into several severe difficulties, and more sophisticated models that incorporate Keplerian orbits and radiation pressure effects are being developed by Daly *et al.* (1988)

REFERENCES

Alfvén, H. (1954) *On the origin of the solar system* Oxford Univ. Press, Oxford

Alfvén, H. (1957) On the theory of comet tails, *Tellus* **9** 92

Alfvén, H. (1960) Collision between a non-ionized gas and magnetised plasma *Rev., Mod. Phys.* **32** 710

Axford, W.I., (1964) The interaction of the solar wind with comets *Planet. Space Sci.* **12** 719

Axford, W.I. (1981) Acceleration of cosmic rays by shock waves. *Proc. of 17th Int. Cosmic Ray Conf., Paris* **12** 155, Service de Docum. du CEN, Saclay

Balsiger, H., Altwegg, K., Buhler, F., Geiss, J. Ghielmetti, A.G., Goldstein, B.E., Goldstein, R., Huntress, W.T., Ip, W-H, Lazarus, A.J., Meier, A., Neugebauer, M., Rettenmund,. U., Rosenbauer, H., Schwenn, R., Sharp, R.D., Shelley, E.G, Ungstrup, E., & Young, D.T. (1986) Ion composition and dynamics at Comet Halley. *Nature* **321** 330

Biermann, L., Brosowski, B., & Schmidt, H.U. (1967) The interaction of the solar wind with a comet, *Solar Physics* **1** 254

Coates, A.J., Johnstone, A.D., Thomsen, M.F., Formisano, V., Amata, E., Wilken, B., Jockers, K., Winningham, J.D., Borg, H., & Bryant, D.A. (1987) Solar wind flow through the comet P/Halley bow shock, *Astron. and Astrophys.* **187**, 55.

Coates, A.J. (1989b) Private communication.

Glassmeier, K.-H. (1989a) Velocity space diffusion of pickup ions from the water group at comet Halley, *J. Geophys. Res.*, **94**, No. A8, 9983.

Coates, A.J. (1989b) private communication.

Cravens, T.E., (1986) Ion distribution function in the vicinity of Comet Giacobini–Zinner, *Geophys. Res. Lett.* **13** 275

Cravens, T.E., Kozyra, J.U., Nagy, A.F., Gombosi, T.I., & Kurtz, M. (1986) Electron impact ionization in the vicinity of comets, preprint of a paper submitted to JGR

Daly, P.W., (1988) Private communication

Daly, P.W., (1989) The use of Kepler trajectories to calculate ion fluxes at multigigameter distances from comet Halley, *Astron. Astrophys.* (in press).

calculate ion fluxes at multigigameter distances from comet Halley, *Astron. Astrophys.* (in press).

Daly, P.W., McKenna–Lawlor, S., O'Sullivan, Sanderson, T.P., Thompson, A., & Wenzel, K-P., (1986) Comparison of energetic ion measurements at Comets Giacobini–Zinner and Halley *Proc. 20th ESLAB Symposium on the Exploration of Comet Halley*, ESA SP-250, 179

Daly, P., Richter, A.K., Verigin M.I., Gringauz K.I., Erdos, G., Kecskemety, K., Somogyi, A.J., Szegö, K., Varga, A., & McKenna–Lawlor, S. (1989) The use of Kepler trajectories ;to calculate ion fluxes at multigigameter distances from comet Halley, *Astron. Astrophys.* (in press).

Flammer, K.R., Mendis, D.A., & Houpis, H.L.F. Variable nature of the solar wind interaction with Comet Halley *Proc. STIP Symposium on Retrospective Analysis and Future Co-ordinated intervals*, D. Reidel Publishing Co. (in press)

Formisano, V., Galeev, A.A., & Sagdeev, R.Z. (1982) The role of the critical ionization velocity phenomena in the production of inner coma cometary plasma *Planet. Space Sci.* **30** 491

Galeev, A.A., Griov, B.E., Gombosi, T., Gringauz, K.I., Klimov, S.I., Oberz, P., Remizov, A.P., Riedler, E., Sagdeev, R.Z., Savin, S.P., Sokolov, A., Yu, Shapiro, V.D., Shevchenko, V.I., Szego, K., Verigin, M.I. & Yeroshenko, Ye. G., (1986) The position and structure of comet Halley bowshock: VEGA-1 and VEGA-2 measurements *Geophys. Res. Lett.* **13**, 841

Galeev, A.A. (1986) Theory and observations of solar wind/cometary plasma interaction processes Proc. 20th ESLAB Symposium on the Exploration of Halley's Comet, *ESA SP-250* **1** 3

Galeev, A.A., Cravens, T.E., & Gombosi, T.I. (1985) Solar wind stagnation near comets *Astrophys. J.* **289** 807

Galeev, A.A. (1987) Encounters with comets; discoveries and puzzles in cometary plasma physics *Astron. and Astrophys.* **187**, 12.

Gary, S.P., Hinate, S., Madland, C.D., & Winske, D. (1986) The development of shell-like distributions from newborn cometary ions *Geophys. Res. Lett.* **13** 1364

Grewing, M., Pradierie, F. & Reinhard, R. (Eds) Exploration of Halley's comet (1988) Springer-Verlag Berlin, Heidelberg, New York, London, Paris, Tokyo.

Gribov, B.E., Kecskemety, K., Sagdeev, R.Z., Shapiro, V.D., Shevchenko, V.I., Somogyi, A.J., Szegö, K., Erdos, G., Eroschenko, E.G., Gringauz, K.I., Keppler, E., Marsden, R., Remizov, A.P., Richter, A.K., Riedler, W., Schwingenschuh, K., & Wenzel, K-P. (1986) Stochastic Fermi acceleration of ions in the pre-shock region of Comet Halley *Proc. 20th ESLAB Symposium on the Exploration of Halley's Comet*, ESA SP-250 **1** 271

Gringauz, K.I., Gombosi, T.I., Remizov, A.P., Apáthy, L., Szemerey, I., Verigin, M.L., Denschikova, L.L., Dyachkov, A.V., Keppler, E., Kimenko, I.N., RIchter, A.K., Somogyi A.J., Szego, K., Szendro, S., Iatrallyay, M., Varga, A. and Vladimirova, G.A. (1986a) First in situ plasma and neutral gas measurements at Comet Halley' *Nature* **321** No 6067, 282

Gringauz, K.I., Verigin, M.I., Richter, A.K., Gombosi, T.I., Szegö, K., Tátrallyay, M., Remizov, A.P., & Apáthy, I. (1986b) Cometary plasma region in the coma of Comet Halley: VeGa-2 measurements *Proc. 20th ESSAB Symposium on the exploration of Halley's Comet*, ESA SP-250 **1** 93

Huebner, W.F., & Giguere, P.T. (1980) A model of comet comae II. Effects of solar photodissociative ionization *Astrophys. J.* **238** 753

Hynds, R.J., Cowley, S.W.H., Sanderson, T.R, Wenzel, K-P, & van Rooijen, J.J. (1986) Observations of energetic ions from comet Giacobini–Zinner *Science* **232** 361

Ip, W-H, (1985) A preliminary consideration of the electron impact ionization effect in cometary comas *Adv. Space Sci.* **3** No. 12, 47

Ip W-H., & Axford, W.I. (1982) Theories of physical processes in the cometary comae and ion tails, In: *Comets*, Ed. Laurel L. Wilkening, University of Arizona Press, 588

Ip, W-H & Axford, W.I. (1986) The acceleration of particles in the vicinity of comets *Planet. Space Sci.* **34** 1061, also MPAE-W-100–86–07,

Ip, W-H., & Axford, W.I., (1987) A numerical simulation of charged particle acceleration and pitch angle scattering in the turbulent plasma environment of cometary comas *Proc. 20th International Cosmic Ray Conference, Nauka Moscow* **3** 233

Ip, W.-H., & Axford, W.I., (1988) Cometary Plasma Physics, a chapter in the book; Physics of Comets in the Space Age, Ed. W.F. Huebner, pre-print of the Max Planck Institut fur Aeronomie, Lindau, Germany.

Ipavich, F.M., Galvin, A.B., Gloeckler, G., Hovestadt, D., Blecker, B., & Scholer, M., (1986) Comet Giacobini–Zinner: *In situ* observations of energetic heavy ions *Science* **232** 366

Johnstone, A., Coates, A., Kellock, S., Wilken, B., Jockers, K., Rosenbauer, H., Studemann, W., Weiss, W., Formisano, V., Amata, E., Cerulli-Irelli, R., Dobrowolay, M., Terenzi, R., Egidi, A., Borg, H., Hultquist, B., Winningham, J., Gurgiolo, C., Bryant, D., Edwards, T., Feldman, W., Thomsen, M., Wallis, M.K., Biermann, L., Schmidt, H., Lust, R., Haerendel, G., & Paschmann, G., (1986) Ion flow at Comet Halley *Nature* **321** 344

Kecskemety, K., Cravens, T.E., Afonin, V., Erdos, G., Eroshenko, E.G., Gan, L., Gombosi, T.I., Gringauz, K.I., Keppler, E., Klimenko, I.N., Marsden, R.G., Nagy, A.F., Remizov, A.P., Richter, A.K., Reidler, W., Schwingenschuh, K., Somogyi, A.J., Szego, K., Tátrallyay, M., Varga, A., Verigin, M.I., & Wenzel, K-P. (1986) Energetic pick-up ions outside the Comet Halley bow shock *Proc. 20th ESLAB Symposium on the Exploration of Halley's Comet*, ESA SP-250 **1** 109

Kirsch, E., McKenna–Lawlor, S., O'Sullivan, D., Thompson, A., & Daly, P.W. (1987) Observation of energetic particles ($E > 30$ keV) by the Giotto Experiment EPA in the magnetic cavity of Comet Halley *Proc. of the Brussels Symposium on the Diversity and Similarity of Comets*, ESA SP-278, 145

Kirsch, E., McKenna–Lawlor, S., Daly, P., Korth, A., Neubauer, F.M., O'Sullivan, D., Thompson, A. and Wenzel, K.-P. (1989) Evidence for the field line reconnection process in the particle and magnetic field measurements obtained during the Giotto-Halley encounter *Ann. Geophys.* **7** (2) 107.

Klimov, S., Savin, S., Aleksevich, Ys, Avanesova, G., Balebanov, V., Balikhin, M., Galeev, A., Gribov, B., Nozdrachev, M., Smirnov, V., Sokolov, A., Vaisberg, O., Oberc, P., Krawczyk, Z., Grzedzielski, S., Juchniewicz, J., Nowak, K., Orlowski, D., Pafinovich, B., Wozniak, D., Zbyszynski, Z., Voita, Ya, & Triska, P., (1986) Extremely low frequency plasma waves in the environment of Comet Halley *Nature* **321** 292

Korth, A., Richter, A.K., Loidl, A., Anderson, K.A., Carlson, C.W., Curtis, D.W., Lin R.P., Rème, H., Sauvaud, J.A., d'Uston, C., Cotin, F., Cros A., & Mendis, D.A. (1986) Mass spectrum of heavy ions near Comet Halley, *Nature* **321** 335

McComas, D.J., Gosling, J.T., Bame, S.J., Slavin, J.A., Smith, E.J. & Steinberg, J.T. (1987) The Giacobini–Zinner magnetotail configuration and current sheet, *J. Geophys. Res.* **92**, 1139.

McKenna–Lawlor, S., Kirsch, E., Thompson, A., O'Sullivan, D., & Wenzel, K-P (1985) Energetic particles in the Comet Halley environment *Adv. Space Res.* **5** 211

McKenna–Lawlor, S., Kirsch, E., O'Sullivan, D., Thompson, A., & Wenzel, K.-P., (1986) Energetic ions in the environment of Comet Halley *Nature* **321** 347

McKenna–Lawlor, S., Kirsch, E., Thompson, A., O'Sullivan, D., & Wenzel K-P. (1987a) The lightweight energetic particle detector EPONA and its performance on Giotto *J. Phys. E: Sci. Instrum.* **20** 732

McKenna–Lawlor, S., Wilken, B., Daly, P., Ip, W-P, Kirsch, E., Coates, A., Johnstone, A., Thompson, A., O'Sullivan, D., & Wenzel, K-P. (1987b) Energy spectra of pick-up ions recorded during the encounter of Giotto with Comet Halley *Proc. of the Brussels Symposium on the Diversity and Similarity of Comets*, ESA SP-278 133

McKenna–Lawlor, S., Daly, P., Kirsch, E., Wilkin, B., O'Sullivan, D., Thompson, A., Kecskemety, K., Somogyi, A. & Coates, A. (1989a) *In situ* energetic particle observations at comet Halley

recorded by instrumentation aboard the Giotto and Vega 1 missions *Ann. Geophys.* **7**, (2), 121

McKenna–Lawlor, S., Kirsch, E., Daly, P., O'Sullivan, D., Thompson, A., Somogyi, A. & Kecskemety, K. (1989b) A comparison of quasi-periodicity in the ion flux enhancements recorded in-bound and out-bound at Halley's comet by the EPONA instrument aboard Giotto and by the Tunde-M instrument aboard Vega-1 *Adv. Sp. Res.* **9**, No. 3, (3) 325.

Mendis, D.A. (1988) Symposium Summary, Exploration of Halley's comet, Eds. M. Grewing, F. Praderie, R. Reinhard, Supplement Science Summary, 939. Publ. Springer-Verlag.

Mendis, D.A., Houpis, H.L.F. & Marconi, M.L. (1985) The Physics of comets, *Fund. Cosmic Phys.*, **10**, 1.

Mendis, D.A., Smith, E.J., Tsurutani, B.T., Slavin, J.A., Jones, D.E., Siscoe, G.L. (1986) Comet-solar wind interaction: dynamical length scales and models. *Geophys. Res. Lett.* **13**, 239.

Mukai, T., Miyake, W., Teresawa, T., Kitayama, M., & Kirao, K., (1986) Plasma observation by Suisei of solar-wind interaction with Comet Halley *Nature* **321** 299

Neubauer, F.M., Glassmeier, K.H., Pohl, M., Raeder, J., Acuña, M.H., Burlaga, L.F., Ness, N.F., Musmann, G., Mariani, F., Wallis, M.K., Ungstrup, E., & Schmidt (1986) First results from the Giotto magnetometer experiment at Comet Halley *Nature* **321** 352

Neugebauer, M., Lazarus, A.J., Altwegg, K., Balsiger, H., Goldstein, B.E., Goldstein, R., Neubauer, F.M., Rosenbauer, H., Schwenn, R., Shelley, E.G., & Ungstrup, E. (1986) The pick-up of cometary protons by the solar wind *Proc. 20th ESLAB Symposium on the Exploration of Halley's Comet*, ESA SP-250 **1** 19

Neugebauer, M., Lazarus, A.J., Altwegg, K., Balsiger, H., Goldstein, B.E., Goldstein, R., Neubauer, F.M., Rosenbauer, H., Schwenn, R., Shelley, E.G., & Ungstrup, E. (1987) The pick-up of cometary protons by the solar wind *Astron. and Astrophys.* **187**, 21.

Neugebauer, M., Lazarus, A.J., Balsiger, H., Fuselier, S.A, Neubauer, F.M., & Rosenbauer, H. (1989) The velocity distributions of cometary protons picked up by the solar wind. *J. Geophys. Res.*, **94**, No. A5, 5227.

Oya, H., Morioka, A., Miyake, W., Smith, E.J., & Tsurutani, B.T. (1986) Discovery of cometary kilometric radiations and plasma waves at Comet Halley *Nature* **321** 307

Réme, H. (1990) Regions of interaction between the solar wind plasma and the plasma environment of comets (this volume).

Réme, H., Sauvaud, J.A., d'Uston, C., Cotin, F., Cros, A., Anderson, K.A., Carlson, C.W., Curtis, D.W., Lin, R.P., Mendis, D.A., Korth, A. & Richter, A.K. (1986) Comet Halley-solar wind interaction from electron measurements aboard Giotto *Nature*, **321**, 349.

Richter, A.K., Daly P.W., Verigin, M.I., Gringauz, K.I., Erdos G., Kecskemety, K., Somogyi, A.J., Szegö, K., Varga A., & McKenna–Lawlor, S. (1989) Quasi-periodic variations of cometary ion fluxes at large distance from Comet Halley *Ann. Geophys.* **7**, 115

Riedler, W., Schwingenschuh, K., Yeroschenko, Ye G., Styashkin, V.A., & Russell, C.T. (1986) Magnetic field observations in Comet Halley's coma *Nature* **321** 288

Sanderson, T.R., Wenzel, K-P., Daly, P.W., Cowley, S.W.H., Hynds, R.J., Richardson, I.G., Smith, E.J., Bame, S.J., & Zwickl, R.D. (1986) Observations of heavy energetic ions far upstream from Comet Halley *Proc. 20th ESLAB Symposium on the Exploration of Halley's Comet*, ESA SP-250 **1** 105

Sagdeev, R.Z., Shapiro, V.D., Shevchenko, V.I., & Szego, K., (1986) MHD turbulence in the solar wind–comet interaction region *Geophys. Res. Lett.* **13** 85

Sagdeev, R.Z., Shapiro, V.D., Shevchenko, V.I., & Szegö, K. (1987) Plasma phenomena around comets: interaction with the solar wind *KFKI-1987–50/C* Hungarian Academy of Sciences, Central Research Institute for Physics, Budapest

Savin, S.P., Avanesova, G., Balichin, M., Klimov, S., Sokolov, A., Oberc, P., Orlowskii, D., Krawczyk, Z., (1986) ELF waves in the plasma regions near the comet, *20th. ESLAB Symp. ESA SP-250*, Vol. 3, 433.

Scarf, F.L., Coroniti, F.V., Kennel, C.F., Ip, W-H, Gurnett, D.A., & Smith, E.J. (1986) Observations of cometary plasma phenomena *Proc. 20th ESLAB Symposium on the Exploration of Halley's Comet,* ESA SP-250 **1** 163

Schmidt, H.U., & Wegmann, R. (1982) *Plasma flow and magnetic fields in comets* Ed. Wilkening, L.L., University of Arizona Press, 538

Smith, E.J., Hanson, W.B., Ip, W-H, Neugebauer, M., & Scarf, F.L., (1985) *Plasma science report of the comet rendezvous science working group*, JPL, PD 699–10 <cf3>II

Somogyi, A.J., Gringauz, K.I., Szego, K., Szabö, L., Kozma, Gy., Remizov, A.P., Ero, J. Jr., Klimenko, I.N., Szucs, I.T., Verigin, M.I., Windberg, J., Cravens, T.E., Dyachkov, A., Erdos, G., Farago, M., Gombosi, T.I., Kecskemety, K., Keppler, E., Kovacs, T. Jr., Kondor, A., Logachev, Yu I., Lohonyai, L., Marsden, R., Redl, R., Richter, A.K., Stolpovskii, V.G., Szabö, J., Szenpetery, I., Szepesvary, A., Tátrallyay, M., Varga, A., Vladimirova, G.A. Wenzel, K-P, & Zarandy, A. (1986) First observations of energetic particles near Comet Halley *Nature* **321** 285

Terasawa, T., Mukai, T., Miyake, W., Kitayama, M., & Hirao, K. (1986) *Detection of cometary pickup ions up to 10^7* km from Comet Halley: Suisei observation ISAS Research note 344 (also submitted to *Geophys. Res. Lett.*)

Valenzuela, A., Haerendel, G., Foppl, H., Melzner, F., Nuess, H., Rieger, E., Stocker, J., Bauer, O., Hofner, H., & Loidl, J. (1986) The AMPTE artificial comet experiments *Nature* **320** 700

Wallis, M.K. (1973) Weakly shocked flows of the solar wind plasma through atmospheres of comets and planets *Planet. Space Sci.* **21** 1647

Wenzel, K.-P., Sanderson, T.R., Richardson, I.G., Cowley, S.W.H., Hynds, R.J., Bame, S.J., Zwickl, R.D., Smith, E.J., & Tsurutani, B.T. (1986) *In situ* observations of cometary pick-up ions ⩾ 0.2 AU upstream of comet Halley; *ICE observations, G.R.L.* **13**, 861.

Wilkening, L.L. Editor (1982) 'Comets', Publ. University of Arizona Press.

12

Plasma flow and pick-up ions around comet Halley: Suisei observations

W. Miyake, T. Mukai, T. Terasawa, M. Kitayama and K. Hirao

1 INTRODUCTION

The interaction of the solar wind with cometary gases is one of the most important topics not only in cometary physics but also in Solar System plasma physics. Because of the weak gravitational force of comets, neutral atmosphere sublimated from the nucleus can escape freely from the collision dominated inner coma. Biermann *et al.* (1967) pointed out that the solar wind interaction with freely expanding cometary atmosphere commences at distances of millions of kilometres from the tiny cometary nucleus in contrast to the interaction with planets. Cometary ions created by various ionization processes (see Mendis & Ip 1977) of neutral molecules begin to rotate around the solar wind magnetic field and are eventually assimilated into the solar wind. As cometary ions are added, the solar wind flow is slowed down by the mass loading effect, which may influence conditions for bow shock formation.

Suisei, one of Japan's two Halley probes (Hirao & Ito 1986), made a closest approach of 151 000 km to comet Halley at 1306 UT on 8 March 1986. Plasma observations by Suisei have demonstrated the existence of the assimilation process of cometary ions inside the cometosheath of comet Halley as well as in the upstream region. In this paper we give a summary of the observed results, which have already been published elsewhere (Mukai *et al.* 1986a, b, 1987a, Terasawa *et al.* 1986). In section 2, a brief description of the plasma instrument onboard Suisei is given for better understanding of the results. The observed results of the plasma flow around the closest approach with comet Halley are summarized in section 3. Suisei

is the only probe which observed the plasma flow in the sunward region of the cometosheath. Its results complement those of Vega (Gringauz *et al.* 1986) and of Giotto (Johnstone *et al.* 1986). In section 4, we present the identification of the shell structure in the phase space distribution of cometary protons and water group ions as evidence of the pick-up process. From the Giotto observation, Neugebauer *et al.* (1986) also obtained the three-dimensional shell structure of cometary protons over a distance of 6×10^6 km upstream of the comet. The Suisei observation demonstrates the existence of the shell structure for both protons and water group ions inside the cometosheath as well as in the upstream region. Spatial distribution of these cometary ions up to 10^7 km from the nucleus is summarized in section 4. A brief summary of the observation results and the remaining problems is presented in the final section.

2 INSTRUMENTATION

The plasma instrument onboard Suisei (ESP instrument: energy spectrum of particles, see Mukai *et al.* 1986a, 1987b) was operated in the two-dimensional mode during the encounter with comet Halley. The geometrical factor ($<S\Omega\Delta E>/E$) is 2.3×10^{-4} cm^2 sr, and the field of view is $5° \times 60°$, with the longer dimension perpendicular to the ecliptic plane. The energy range from 30 eV/q to 15.8 keV/q (where q is the charge state) is divided into 96 steps, equally spaced on a logarithmic scale. During one spin period (9.18 s), ion fluxes in four successive energy steps are measured at 22 azimuthal points. The separation between neighboring points is $5°.625$ within $\pm 22°.5$ from the solar direction, and is $22°.5$ outside. For the encounter mode of operation, the energy scan was made alternately every 16-spin periods (147 sec) for the lower 64 steps (30 eV/q–1.92 keV/q; scan mode L) or for the higher 64 steps (248 eV/q 15.8 keV/q; scan mode H). For the solar wind mode, the energy was scanned every 16-spin periods only for the scan mode H. Since the downlink telemetry requires 512 s to send one sample of the phase-space distribution function (in 64 steps $\times$ 22 sectors), there was a 365-sec gap between samples.

3 PLASMA FLOW AROUND COMET HALLEY

Figs 1 and 2 show the results of plasma flow observation during the Suisei encounter with comet Halley. In Fig. 1, the observed flow pattern is shown by arrows starting from the Suisei position at each epoch. We use the cometocentric solar ecliptic (CSE) coordinate system, in which the X-axis is taken in the ecliptic plane toward the Sun, and the Z-axis toward the north. The Y-axis is taken so as to make a right-hand coordinate system. Fig. 2 shows variations of flow velocity, proton density, and flow direction. The measured density includes contributions from the protons of the solar wind as well as of cometary origin. From 1232 UT, when the ESP instrument was switched on, until 1443 UT, Suisei was in the cometosheath, as is evident from the characteristic

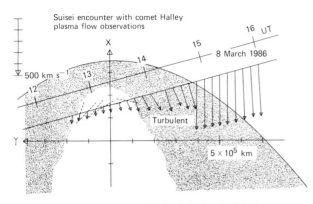

Fig. 1 — Plasma flow vectors obtained during the Suisei encounter with comet Halley. The flow vectors and angles are represented in the CSE coordinate system. Dashed lines represent the estimated directions of magnetic field.

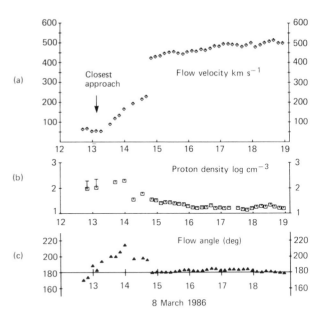

Fig. 2 — Time variation of (a) bulk flow velocity, (b) proton density, and (c) flow direction of the solar wind.

pattern of the flow around an obstacle. The minimum flow velocity, 56 ± 10 km s^{-1} was observed around 1316 UT, when Suisei was in the subsolar direction from the cometary nucleus. The flow pattern inside the cometosheath is roughly symmetric with respect to the direction 4–6° from the solar direction. Since the velocity of the comet with respect to the Sun was $(-24, 37, -12)$ km s^{-1}, then the flow direction of the solar wind velocity of 400–500 km s^{-1} is expected to be inclined by 4–6° from the solar direction in the comet frame. The estimation of proton density around the closest approach suffers a large error (a factor of 2) because of the low flow speed.

During the observation gap between 1443 and 1449 UT, when Suisei was around 4.5×10^5 km away from the nucleus, the plasma parameters changed remarkably. The flow velocity increased from 240 km s^{-1} to 440 km s^{-1}, and the flow direction changed by $\sim 16°$. This change is interpreted to be due to an outbound crossing of the cometary bow shock. The upper limit of the shock thickness set by this observation is 2.6×10^4 km. A possible shape of the cometary bow shock is drawn as a parabola in Fig. 1, where the nose of the bow shock is assumed to be at 3.5×10^5 km in the direction of the symmetry axis. Beyond the shock crossing, the solar wind continued to change gradually. The velocity and proton density, 440 ± 10 km s^{-1} and 35 ± 5 cm^{-3} just after the bow shock crossing, were 510 ± 10 km s^{-1} and 15 ± 3 cm^{-3} around 1710 UT, when the spacecraft was 1×10^6 km away from the nucleus. This change is probably due to the mass loading effect in the upstream solar wind. This conclusion is further supported by the fact that until ~1700 UT we could identify a specific feature (i.e., shell structure; see next section) in the phase space distribution of ions expected from the assimilation process of the cometary ions.

A quantitative problem, however, arises if we calculate the momentum flux carried by the solar wind. The lower limit of the ratio between the momentum flux near and far from the comet is calculated as $(35 \times 440^2)/(15 \times 510^2) = 150\%$. This ratio gives the lower limit, since we do not consider the change of the solar wind. Thus the momentum flux near the comet is at least 50% larger than that far away from the comet. Note that the mass loading process in the upstream solar wind is expected to conserve the momentum flux approximately. (This is because the momentum changing terms, namely the pressure term and Lorentz force, in the equation of plasma motion are not effective in the supersonic and super-Alfvénic solar wind. These terms can become effective only inside the cometosheath where the flow becomes subsonic and sub-Alfvénic.) Therefore, the above difference in the momentum flux may be probably due to intrinsic change of solar wind properties. Further

study is necessary to resolve this momentum flux problem.

It takes 147 s (16 spin periods) to get a full ion distribution in 64 energy steps. However, since angular distributions in four successive energy steps are obtained in each spin period (9.18 s), short-term fluctuations, if they exist, can be detected as fluctuations of the anisotropy direction. In Fig. 3, each of the bars represents the anisotropy direction observed in one spin period (averaged over four energy steps). Bars are drawn only if the count rate averages are higher than 200 counts s^{-1}. This is to avoid errors associated with statistical fluctuations. The times below the figure indicate when the energy scans (from lower to higher energies) started. Around intervals 2–5, good alignment among bars is seen, suggesting that the flow is laminar. From intervals 6 to 16 (all within the cometosheath), especially from 11 to 16, the anisotropy directions are fluctuating, which suggests that the flow is turbulent. The above observation is summarized in Fig. 1, where the inferred 'turbulent' region is shown by hatching.

4 PICK-UP PROCESS OF COMETARY IONS

Fig. 4 illustrates the structure in ion velocity space expected for the cometary ion pick-up process. The rest frame of the spacecraft is used here. Since the outflow velocity of the neutral gas from the cometary nucleus (at most several km s^{-1}) can be neglected in the present analysis, newly produced cometary ions

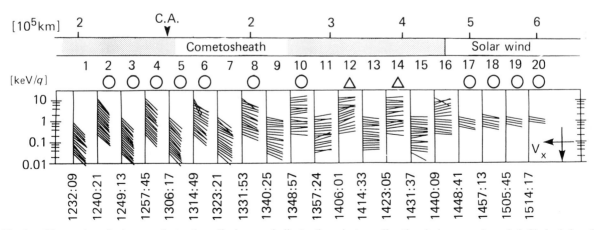

Fig. 3 — Observation of anisotropy fluctuations. Each arrow indicates the anisotropy direction during one spin period. Circles below the column numbers indicate intervals when the pick-up shell is observed (see text). Distance from the nucleus is given at the top.

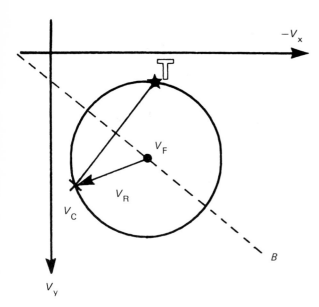

Fig. 4 — A schematic illustration of the theoretical model of the assimilation process of the cometary ions.

start to rotate around the magnetic field, forming a torus in the velocity space. V_c is the projection of this initial velocity onto the ecliptic plane. When the magnetic field line projected to the ecliptic plane is along a dashed line B within the ecliptic plane, the other side of the torus is to be found at the point shown by a star symbol marked with $\mathbb{T}$. These ions are then scattered in pitch angles to form a velocity–space shell (ion pick-up shell), where the scattering process is caused either by waves excited by themselves through various instability processes (e.g. Wu & Davidson 1972) or by existing turbulent magnetic waves. The circle in Fig. 4 shows the intersection of the pick-up shell with the ecliptic plane. The radius, V_R, of the shell in velocity space is given by the difference $|V_c\text{-}V_F|$, where V_F is the bulk flow velocity. While the torus geometry depends on the magnetic field direction, the shell geometry does not.

Fig. 5 shows the contour plot of phase space density $f(v)$ observed at $3.4\text{–}2.7 \times 10^6$ km away from the nucleus on the inbound orbit. In this plot, $f(v)$ is plotted as if all the ions are protons. Ions of different mass-to-charge ratio (m/q) appear at the velocity $(m/q)^1$ times larger than their actual values. The contour curves are drawn at logarithmic steps, e.g. at 10^{-15}, $10^{-14.5}$, 10^{-14}, $10^{-13.5}$,.....s^3/m^6. These contour levels should be rescaled by multiplying a factor of $(m/q)^2$ for

the ions other than protons. This is a typical example which shows the existence of the pick-up shell of both protons and water group ions (likely to be mainly O^+). The regions where $f(v) > 10^{-13.1}$ s^3/m^6 for O^+ $(>10^{-15.5}$ s^3/m^6 for protons) are emphasized by light hatching. The regions where $f(v) > 10^{-13.5}$ s^3/m^6 for protons are darkly shaded. The highest peak around $(-450, 50)$ km s^{-1} represents the solar wind protons. A dashed circle around this peak shows the theoretical positions of the pick-up shell of cometary protons. A corresponding structure is seen in the dark shaded region, where $\sim 50\%$ of the shell seems to be filled. It is seen in Fig. 5 that the symmetry axis of this partial shell is roughly parallel to the radial direction. Though we have no magnetometer onboard Suisei, this symmetry direction agrees with the magnetic field direction projected to the ecliptic plane determined from the pressure anisotropy of solar wind protons $(P_\parallel/P_\perp \simeq 1.5\text{–}2)$. A large circle in Fig. 5 shows the theoretical position of the pick-up shell for O^+. Enhancement of the phase space density is seen along this circle. Close examination of the phase space distributions reveals that the water group ions became identifiable from ~ 2200 UT on 7 March at $\sim 4.0 \times 10^6$ km from the nucleus, while the pick-up shell of cometary protons was first identified around 10^7 km from the nucleus.

Fig. 6a* is a colour presentation of the phase space density of ions ($0.2\text{–}2.6$ keV/q) obtained in the intervals $1257{:}45\text{–}1300{:}12$ UT on 8 March, 1986, around 1.53×10^5 km from the nucleus, 9 to 7 min before the closest approach. The colour code is made by assuming that all ions are O^+. For the other ion species, this code should be rescaled by multiplying by a factor $((m/q)/16)^2$. In Fig. 6a*, a ringlike structure is seen on the lower right. We interpret this as representing the ecliptic cross-section of the pick-up shell for water group ions. The expected position for the initial velocity of O^+ (the cross) is found to fall on the observed ringlike structure. The velocity scales in the figure are given for H_2O^+, OH^+, and O^+ from the top. The result of the fitting procedure (the dashed circle) yields a flow velocity of $(-75 \pm 10, 65 \pm 10)$ km s^{-1}. The part of the shell below ~ 0.8 keV/q (~ 100 km s^{-1} for O^+) is masked by the proton contribution.

* to be found in the colour illustration section in the centre of the book.

7 MARCH 2351–8 MARCH 0251

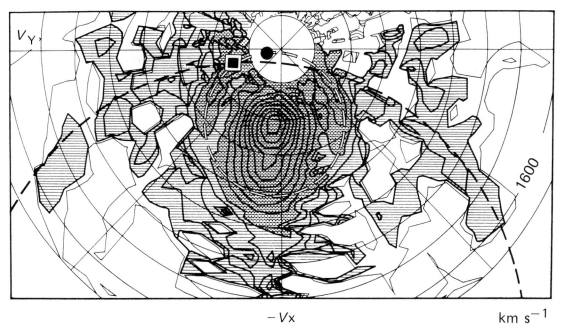

Fig. 5 — Phase space density distribution of ions in the range of $200 > V_y > -1000$ km s^{-1} and $600 > V_x > -600$ km s^{-1}. The CSE frame is used except that the origin of the phase space is shifted to the spacecraft frame. The smaller circle shows the position of pick-up shell for cometary protons, and larger circle (partial) that for O$^+$ ions of cometary origin. A solid circle and a square show the initial velocities of cometary protons and water group ions, respectively.

In Fig. 6a*, the shell seems to be incompletely filled: The phase space density is not uniform along the circle, but is highest around the torus. This partial shell structure suggests two possibilities: the rate of the pitch angle scattering is not enough to obscure all the initial features of the torus, or the fast energy diffusion takes place simultaneously with the pitch angle scattering. From the symmetry axis of this partial shell (shown by a dashed line), which is 140° ± 10° counterclockwise from the $+V_X$ direction (the left-ward arrow), we can estimate the magnetic field direction. (There is, of course, a 180° ambiguity in this estimation.)

Fig. 6b* shows the phase space density of ions (0.03–0.33 keV/q) obtained in the interval 1306:17–1308:44 UT around the closest approach (1.51 × 10^5 km from the nucleus). Again, a ringlike structure is seen on the lower right, which represents the partly filled pick-up shell of cometary protons. The contrast between the shell and the region inside is much weaker for protons than for the water group ions shown in Fig. 6a* (note a

compressed colour scale for protons (heated at the bow shock) inside the shell). The estimated flow velocity vector and symmetry axis are (−74 ± 10, 63 ± 10) km s^{-1} and at 120° ± 10°, respectively.

Fig. 6c* shows the phase space density of ions (0.33–7.5 keV/q) obtained in the interval 1331:53–1334:20 UT around 1.92 × 10^5 km from the nucleus (outbound). A clear enhancement of the density along a dashed circle represents a pick-up shell of water group ions. The flow velocity vector is estimated to be (105 ± 10, 30 ± 10) km s^{-1} from the fitting procedure. The shell seems to be only partly filled, and the symmetry axis is at 80° ± 30°. A large error in this estimation of the symmetry direction is due to masking proton contribution below ~0.8 keV/q (~100 km s^{-1} for O$^+$).

The magnetic field direction estimated from the symmetry axes of the pick-up shells are shown by dashed arrows in Fig. 1. As seen in the figure, the field direction seems to rotate continuously around the nucleus. This rotation is likely to represent the effect

* to be found in the colour illustration section in the centre of the book.

of the magnetic field draping around the comet
(Alfvén 1957). These directions, however, can be
estimated only within $\sim 2 \times 10^5$ km from the nucleus.
Outside this region in the cometosheath, estimation is
not possible, because the fluctuations in the flow
direction (see Fig. 3) become too large.

Within the cometosheath, the pick-up shell of water
group ions is clearly identified in the even intervals
(0.248–15.8 keV/q) 2, 4, 6, 8, and 10 (shown by circles).
In the midst of the 'turbulent region' (see Figs. 1 and
3), the pick-up shell for water group ions is still seen,
but is largely obscured by the anisotropy fluctuation
(shown by triangles). The pick-up shell for protons is
identified only in the intervals (0.03–1.92 keV/q) 3 and
5, nearest to the closest approach. The higher
temperature of heated solar wind protons outside the
above region as well as the larger anisotropy fluctuations, prevent us from identifying the pick-up shell of
protons.

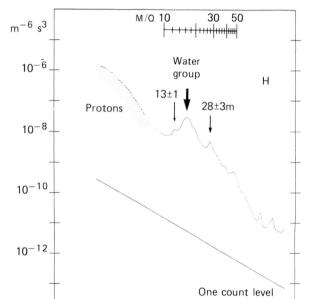

Fig. 7 — A cross-section of the distribution function taken in the
direction of the flow velocity. The scale for m/q is shown at the top,
where the peak of water group ions is adjusted to $m/q = 16$–18. A
small bar shows a possible error in the identification of m/q
corresponding to a 10% fluctuation in the flow velocity.

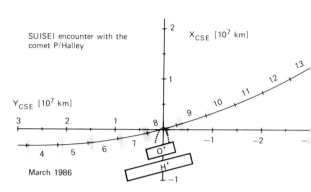

Fig. 8 — Suisei's orbit around comet Halley. The hatched regions
show when the plasma observations were performed. Cometary
protons and water group ions expected to be mainly O^+ were
detected during the period shown by bars with H^+ and O^+,
respectively.

Fig. 7 shows a cross-sectional plot of distribution
function in Fig. 6a*, taken in the direction of the flow
velocity. In Fig. 7, the abscissa is in units of
energy/charge, and the ordinate is scaled as if all the
ions are H_2O^+. For different ion species, the ordinate
should be rescaled by multiplying by a factor
$(m/q)/18)^2$. Ions around 1.7 keV/q have been identified as water group ions. Ions below 1 keV/q are the
thermal-tail protons. (There is also a possibility that
the solar wind He^{++} ions may contribute to this
range.) Pick-up shells of other ions (of mass-to-charge
ratio m/q) are to be found at the energy/charge of
$(\frac{1}{2})m_p(|V_F| + V_R)^2 \times (m/q)$ (m_p is the proton mass). A
scale for m/q is drawn at the top of Fig. 7, where the
peak of water group ions is adjusted to $m/q = 16$–18.
Since an interval of 147 s is needed to complete one
energy scan, velocity fluctuations during this interval
may cause misidentifications of ion species. A small
bar in Fig. 7 shows a possible error in m/q
determination corresponding to 10% fluctuation of
the flow velocity. The peak at $m/q = 28 \pm 3$, whose
likely candidates are CO^+ and N_2^+, is statistically
significant. Other less significant peaks are also seen,
among them a peak around 13 ± 1 which may
represent a contribution of C^+ or CH^+.

5 SPATIAL DISTRIBUTION OF COMETARY IONS

Fig. 8 shows the orbit of Suisei of from 4 to 13 March
1986, drawn in the cometocentric solar ecliptic
coordinate system of comet Halley (the same as used
in Fig. 1). A dotted parabola shows the approximate

shape of the cometary bow shock. Shaded areas show the intervals when the plasma observations were made. Bars with H+ and O+ respectively show the intervals when the cometary protons and water group ions are clearly identified from the existence of the pick-up shell in the velocity space.

When the shell structure is identified, we can estimate the density of cometary ions which form the shell. To determine the number density of the cometary ions, n_c, we need to extrapolate the observed ion distribution by taking into account the geometrical relation between the shell and the field of view, since the instrument field of view did not cover the whole area of the pick-up shell. We calculate n_c as

$$n_c = N_{obs}\frac{V_{shell}}{V_{obs}}, \qquad (1)$$

where V_{shell} is the total volume of the pick-up shell in the velocity space, and n_{obs} the number density of ions distributed inside the subvolume of the shell (V_{obs}) which is within the instrument coverage. The shell thickness needed for the calculation of V_{shell} and V_{obs} is defined as the average thickness of the ringlike

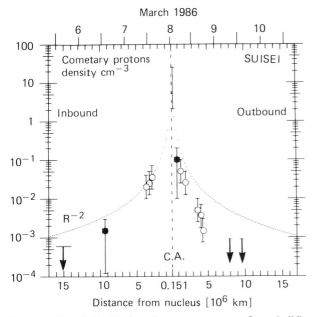

Fig. 9 — Spatial distribution of cometary protons. Open (solid) circles correspond to the intervals when the direction of the magnetic field can (cannot) be determined. Note that only protons forming phase space shell are counted here. Contributions from the protons that have diffused out of the shell are not known.

structure where the phase space density is above 10% of the maximum value inside the ring.

Fig. 9 shows the spatial density profile of cometary protons, where we estimate the density by assuming that the pick-up shell is 50% filled. For the intervals when the field direction is known from the pressure anisotropy observation of the protons in the upstream solar wind (open circles), the uncertainty in the density estimation mainly comes from the uncertainty of the determination of the shell thickness (error of −50% ~ +100%). For the intervals when the field direction is not known (solid circles), the field might be perpendicular to the ecliptic plane so that the observed 'shell' would be a torus lying in the ecliptic plane. This possibility gives the lowest limit for the density estimation for these data. For the data obtained inside the cometosheath, the filling factor ~ 50% seems inadequate (see Fig. 6b*), and only upper (100% shell) and lower (ecliptic torus) limits are given in the diagram. Observation around 1.5×10^7 km (inbound) and around 8×10^6 and 9.5×10^6 km (outbound) from the nucleus only give the upper limit. As a scale, R^{-2} dependence (where R is the radial distance) is shown by dotted curves, which are normalized to be 3×10^{-2} cm^{-3} at 3×10^6 km.

Fig. 10 summarizes the estimated density of water group ions. Within the cometosheath, we estimate the upper limit by assuming that the shell is filled uniformly. The lower limit corresponds to a case where the cometary ions form a velocity space torus lying within the ecliptic plane. The dashed bar corresponds to interval 14 in Fig. 3, where large anisotropy fluctuations make the density estimation less reliable. In the upstream solar wind, the shell-filling factor for cometary protons is found to be 50%, which is adopted for the density estimation of water group ions in Fig. 10 (open/solid circles). Note that in the upstream solar wind the density estimation for water group ions is less reliable than for protons, since only part of the shell is within the energy range of the instrument, as shown in Fig. 5.

We discuss the observed spatial distribution of water group ions in terms of a simple pick-up model. We assume that the parent neutral molecules escape isotropically from the cometary nucleus, since we are concerned with the neutral density far away from the collision dominated inner coma. Then the neutral density n cm^{-3} at the radial distance R is given by,

$$n(R) = n_0(R_0/R)^2(V_0/V_n(R))\exp\left(-\int_{R_0}^{R} \lambda(R)/V_n(R)\,dR\right),$$

$$(2)$$

where n_0 is the neutral density at the radial distance R_0, $\lambda(R)$ is the loss rate of neutral molecules, and V_0 and $V_n(R)$ are the escape velocities of neutral molecules at the radial distances R_0 and R, respectively. The neutral density is related to the gas production rate Q s^{-1} as follows,

$$n_0 = (Q/(4\pi R_0^2 V_0))\exp\left(-\int_{0}^{R_0} \lambda(r)/V_n(R)\,dR\right). \quad (3)$$

The above formulation is essentially the same as that used in Ipavich *et al.* (1986), but includes dependence of the loss rate and the escape velocity on the cometocentric distance. During radial expansion, the neutral water molecules are lost by dissociation into H

and OH and ionization including dissociative ionization. In the present model, however, we take only the ionization process into account; i.e., the loss rate due to dissociation is not taken here, since the dissociated products OH and O also belong to the parent molecules of to observed water group ions.

The number density n_i (P_{obs}) of pick-up ions at the observation point P_{obs} is calculated from the integral along the solar wind stream line,

$$N_i(P_{obs}) = (1/V_c) \int_{P_{obs}}^{P_{limit}} n(R)\lambda(R)\,ds, \quad (4)$$

where P_{limit} is the upstream limit and V_c is the convection velocity. Since we are concerned only with the ion density in the phase space shell, the upstream limit should be set such that the convection velocity does not change significantly over the path of integration. In the present calculation, the upstream limit is determined, based on the result of the plasma flow observation. For the density calculation inside the bow shock, we take the upstream as $X_{limit} = 3 \times 10^5$ km (a little inside the cometary bow shock). For the density calculation corresponding to observation points up to the cometocentric distance of $\sim 10^6$ km beyond the bow shock crossing, we take the integration path length of 1.2×10^6 km. The upstream limit is selected as $X_{limit} = 3 \times 10^6$ km for the 3×10^6 km points. We neglect the loss of pick-up ions due to recombination with electrons, since the transportation time from P_{limit} to P_{obs} (order of 10^3 s) is much shorter than the recombination time.

The ionization rate would vary according to variation in the plasma parameters and the neutral composition, depending on the cometocentric distance. We divide the whole region of interest into several sub-regions with the cometocentric distance ranges of (a) <1.5, (b) 1.5–2.0, (c) 2.0–4.0, and (d) 4.0 $\times 10^5$ km, according to our plasma observation. We have calculated the ion densities corresponding to the observation points for various combinations of ionization rates in these sub-regions. First we tried to see if a good agreement can be reached between observation and calculation for $Q = 1 \times 10^{30}$ s^{-1}, but failed to find global coincidence. An enhanced ionization rate in the sub-regions (b) and (c) can give a result in good agreement with the observed values only around the

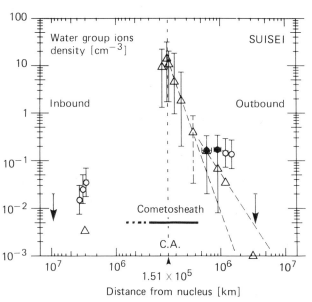

Fig. 10 — Similar to Fig. 9, except that it is for water group ions of cometary origin. Note that the abscissa (distance from the nucleus) is scaled logarithmically. A horizontal bar indicates the intervals when Suisei was in the cometosheath. Triangles show the calculated densities under the condition of gas production rate $Q = 2 \times 10^{30}$ s^{-1}. Two dashed lines show trends of the calculated density profiles inside the cometosheath region and just outside the bow shock region, respectively.

closest approach ($< 2 \times 10^5$ km), but yields much lower ion densities than the observed values at greater distances. In Fig. 10, one of the best fitted results (marked by triangles) is shown in comparison with the observed values. This result is obtained under the condition of $Q = 2 \times 10^{30}$ s^{-1} and the ionization rates of 1.5, 2.5, 1.5, and 1.0 $\times$ 10^{-6} s^{-1} corresponding to the sub-regions (a)–(d), respectively. The escape velocity of neutral molecules is assumed to be 1 km s^{-1}. These values for the ionization rates are reasonably interpreted when the effective ionization mechanisms are mainly charge exchange with solar wind protons and photoionization due to solar EUV radiation. It is noted that a combination of Q and the ionization rate in the sub-region (a), which gives the identical neutral density at $R_0 = 1.5 \times 10^5$ km from equation (2), can give the same result as the present one, since we are concerned with the ion density in the regions beyond the cometocentric distance of R_0; the observation closest to the Halley nucleus was made at the distance of 151 000 km. The escape velocity of 1 km s^{-1} is reasonable for the water group neutrals, but some part of the parent molecules for O$^+$ may be atomic oxygen dissociated from CO, especially at greater distances; in this case the escape velocity is higher than 1 km s^{-1}.

The water production rate of $Q = 2 \times 10^{30}$ s^{-1} is higher when compared to the values measured by Vega-1 ($Q = 1.3 \times 10^{30}$ s^{-1}; Gringauz et al. 1986) and by Giotto ($Q = 6.9 \times 10^{29}$ s^{-1}; Krankowsky et al. 1986). However, the parent molecules of cometary ions observed at greater cometocentric distances would be ejected from the nucleus much earlier than the observation period; e.g., about 10 days earlier (or, around 26 February 1986) for the 10^6 km observation if the escape velocity of the neutral molecules is 1 km s^{-1}. The water production rate of Halley at that time might be higher than the values measured by Vega-1 and Giotto. Weaver et al. (1986) also reported the water production rate of 2.3×10^{30} s^{-1} on 24 March 1986, when comet Halley was located at the heliocentric distance of ~ 1 AU (post-perihelion), even further away from the Sun than for our observation period.

It is noted that the calculated density profile has a more gradual slope just outside the bow shock region when compared to the slope in the cometosheath region. This arises from the longer path length of the integration (Eq. 3) outside the bow shock region than that adopted inside the cometosheath. However, the observation shows a still more gradual slope which cannot be explained by the present calculation. The breathing of comet Halley (Kaneda et al. 1986) cannot account for this, either, since each observed point around 10^6 km distance represents an average density in a 2×10^5 km width of cometocentric distance. This average width roughly corresponds to the period of breathing (~ 53 h), if the escape velocity of neutral molecules is 1 km s^{-1}.

Finally, the observation shows that the water group ions were detected as far as 3–4 $\times 10^6$ km in the inbound orbit, whereas the outbound observation gives only the upper limit. Similar asymmetry is also seen in the density profile of cometary protons (see Fig. 10). It is interesting to point out that the comparison of the calculated densities at 3×10^6 km between the inbound and outbound orbits also shows an inbound–outbound asymmetry by a factor of ~ 3. This asymmetry is caused mainly by the orbital difference in reference to the CSE frame. The inbound observations were made in the downstream ($X < 0$), where the cometocentric distance along the integral path in equation (3) decreases during negative values of X. Therefore, the integrand of equation (3) first increases (during ($X < 0$) and then begins to decrease. On the other hand, the integrand corresponding to the outbound observation decreases monotonically along the integration path. Another cause for the asymmetry arises from the flow velocity difference between the inbound and outbound orbits: The flow velocity was 430 km s^{-1} on the inbound and 570 km s^{-1} on the outbound. Thus the observed asymmetry in the density distribution of cometary ions between the inbound and outbound orbits can be reasonably interpreted.

6 SUMMARY

The ion pick-up process plays an essential rôle in the interaction of the solar wind with cometary atmosphere. The shapes of the ion distribution functions in the phase space give us insight on the governing physics of the assimilation process itself. One of the major results of plasma observation by Suisei is the identification of the characteristic shell structure of both protons and water group ions of cometary origin as an evidence of the pick-up process. The analysis of the ringlike distribution enables us to estimate the plasma flow velocity and the magnetic field direction inside the cometosheath, and the density of cometary ions.

There remain a few problems which have not yet been solved by the analysis described in the paper. In section 3 we discussed the problem that the momentum flux was not conserved along the outbound orbit of Suisei. We attributed the difference in the momentum–flux conservation between theory and observation to an intrinsic change of the solar wind property. Vega-1, however, did not observe such a change in the solar wind parameters at the Suisei encounter with comet Halley (Apáthy *et al.* 1986). Another problem is that the pick-up shell is incompletely filled. Quantitative discussion of theoretical work is needed to clarify the partly filled structure.

ACKNOWLEDGEMENTS

We thank all the members of the Planet-A (Suisei) project team for their contributions to the success of the Suisei plasma experiment, especially T. Itoh (Project Manager) and M. Oda (Director General of ISAS). We are also grateful to T. Abe, A. Nishida, O. Ashihara, M. Shimizu, and T. Yamamoto for valuable discussions and comments. We appreciate the assistance of JPL (Jet Propulsion Laboratory)/NASA in acquiring the data presented in this paper.

REFERENCES

Alfvén, H. (1957) On the theory of comet tail *Tellus* **9** 92

Apáthy, I. *et al.* (1986) PLASMAG-1 experiment solar wind measurements during the closest approach to Comet Giacobini–Zinner by the ICE probe and to Comet Halley by the Giotto and Suisei spacecraft, *Proc. 20th ESLAB symposium on the exploration of Halley's comet*, 65

Biermann, L. *et al.* (1967) The interaction of the solar wind with a comet, *Solar Phys.* **1** 254

Gringauz, K.I. *et al.* (1986) First *in situ* plasma and neutral gas measurements at Comet Halley *Nature* **321** 282

Hirao, K., & Ito, T. (1986) The Planet-A Halley encounter *Nature* **321** 294

Ipavich, F.M. *et al.* (1986) Comet Giacobini–Zinner: *In situ* observation of energetic heavy ions, *Science* **232** 366

Johnstone, A. *et al.* (1986) Ion flow at Comet Halley, *Nature* **321** 344

Kaneda, E. *et al.* (1986) Strong breathing of the hydrogen coma of Comet Halley, *Nature* **320** 140

Krankowsky *et al.* (1986) *In situ* gas and ion measurements at Comet Halley, *Nature* **321** 326

Mendis, D.A, & Ip, W-H (1977) The ionospheres and plasma tails of comets, *Space Sci. Reviews* **20** 145

Mukai, T. *et al.* (1986a) Plasma observation by Suisei of solar wind interaction with Comet Halley *Nature* **321** 299

Mukai, T. *et al.* (1986b) Ion dynamics and distribution around Comet Halley: Suisei observation, *Geophys. Res. Lett.* **13** 829

Mukai, T. *et al.* (1987a) Spatial distribution of water group ions near Comet Halley observed by Suisei, *Astron. Astrophys.* **187** 129

Mukai, T. *et al.* (1987b) Observation of solar wind ions by the interplanetary spacecraft Suisei (PLANET-A) *J. Geomag. Geoelectr* **39** 377

Neugebauer, M. *et al.* (1986) The pick-up of cometary protons by the solar wind, *Proc. 20th ESLAB Symposium on the exploration of Halley's Comet* 19

Terasawa, T. *et al.* (1986) Detection of cometary ions up to 10^7 km from Comet Halley: Suisei observation, *Geophys. Res. Lett.* **13** 837

Weaver, H.A. (1986) Post-perihelion observations of water in Comet Halley *Nature* **324** 441

Wu, C.S., & Davidson, R.C. (1972) Electromagnetic instabilities produced by neutral-particle ionization in interplanetary space *J. Geophys. Res.* **77** 5399

13

Particle acceleration in the plasma fields near Comet Halley

A.J. Somogyi, W.I. Axford, G. Erdös,
W.-H. Ip, V.D. Shapiro, V.I. Shevchenko

Notes

- All event times quoted throughout this review are spacecraft event times (never ground-station-received-time)
- SI units are used in all equations in this review (unless stated otherwise)

1 INTRODUCTION

During their journey through the inner Solar System cometary nuclei emit a large amount of gas. Ions of cometary origin are created from the expanding neutral gas coma by photoionization and by charge exchange with solar wind particles; at smaller distances from the cometary nucleus also by direct electron impact and by critical velocity ionization (Mendis *et al.* 1985, Galeev & Khabibrakhmanov 1986, Cravens 1986). Once the particles are ionized, they interact with the electric and magnetic field of the interplanetary plasma carried by the solar wind. As a consequence, the ions are accelerated and diverted tailward whilst the mass density of the solar wind increases and the solar wind itself is slowed down. These processes involve generation of various kinds of plasma instabilities together with turbulent magnetic and electric field fluctuations.

As well as the first observation of a cometary nucleus by the two Vega and the Giotto missions and the establishing of its basic features (shape, rotation, surface properties, etc.) the most surprising result of the 1985 (ICE) and 1986 (Vega, Giotto, Suisei, Sakigake) missions was to prove that comets Giacobini–Zinner and Halley were surrounded by a hot plasma cloud to distances as great as 10^7 km and 10^6

km from the nucleus, consisting of ions with energies extending to the MeV region (Halley) and 0.5 MeV region (Giacobini-Zinner) (Hynds *et al.* 1986, Ipavich *et al.* 1986 for Giacobini-Zinner, and Somogyi *et al.* 1986, McKenna–Lawlor *et al.* 1986b, for Halley).

In this review the high-energy (> 50 keV for protons, > 100 keV for O^+ ions) plasma processes (production, propagation, interaction) will be dealt with in the context of low-energy plasma, magnetic field, and plasma wave phenomena, as observed by the 1986 spacecraft missions to comet Halley. Emphasis will be given to observations made by the Vega mission. For more detailed treatment of the Giotto observations see McKenna–Lawlor (1986a, b), McKenna–Lawlor (1989). This review reflects the state of affairs at the time of writing. The data observed are not yet fully analyzed. Many questions are as yet unanswered, and some answers are still preliminary. They will be pointed out in the text.

2 A SHORT REVIEW OF ACCELERATION PROCESSES NEAR COMETS

The first reports from the 1985/86 cometary missions have been most stimulating in directing theoretical effort in the investigation of acceleration mechanisms operating in the vicinity of comets. In reviewing these processes we follow the main lines set up in the papers Ip & Axford (1986), Sagdeev et al. (1987), and Galeev et al. (1987).

2.1 Ion pick-up by the solar wind

Freshly created cometary ions have very low velocities ($v_i \approx 1$ km s^{-1}) as compared to that of the solar wind ($u \gtrsim 300$ km s^{-1}), thus they experience an electric field $E = -u \times B$ due to the magnetic field, B, frozen into the solar wind and are accelerated to move on trochoidal orbits with guiding centre velocities $v_{gc} = E \times B/B^2$, that is $v_{gc} = u \sin \phi$, where ϕ is the angle between u and B (solar wind pick-up process). The velocity distribution of the pick-up ions forms a ring in velocity space with the radius $v_{gc} = u \sin \phi$ in a plane perpendicular to B and centred around the end point of the vector v_{gc} (Fig. 1).

In a frame of reference moving together with the solar wind, the guiding centre has the velocity $u \cos \phi$ directed parallel to the magnetic field lines (Fig. 1). Such a formation is, however, unstable, and gives rise to resonant ion-cyclotron waves on which the accelerated ions are pitch angle scattered (Sagdeev & Shafranov 1960, Wu & Davidson 1972, Winske et al.

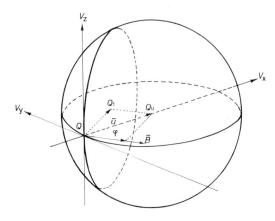

Fig. 1 — Velocity distribution of pick-up ions shown in velocity space. Solar-ecliptic system of coordinates centred at the point Q
v_x: anti-sunward direction
v_z: ecliptic North
Q_u is the end point of the solar wind velocity vector $\mathbf{u} = \overrightarrow{QQ_u}$
Q_1 is the centre of the parallel circle perpendicular to the IMF direction, **B**

$$\overrightarrow{QQ_1} = -(\mathbf{u} \times \mathbf{B}) \times \mathbf{B}/B^2$$

1985, Galeev et al. 1986b, Sagdeev et al. 1986b). As a consequence, the velocity distribution of the pick-up ions is isotropized in the solar wind frame of reference: the new velocity distribution has the form of a spherical shell with the radius u centered around Q_u, the end point of the vector u (Fig. 1). The characteristic time of isotropization is of the order of $\omega_c^{-1} (\Delta B/B)^{-2} \approx 30$ to 300 s, where ω_c is the gyrofrequency of the pick-up ions and ΔB is the rms amplitude of magnetic field fluctuations (Ip & Axford 1987a, Sagdeev et al. 1986a, Gary et al. 1986). This means that, unless affected by mechanisms other than those mentioned above, all pick-up ions move in directions which form angles less than 90° with the solar wind direction. The solar wind thus 'takes along' the pick-up ions. This 'mass loading' of the solar wind is especially significant if the pick-up ions are heavy, which may be the case in cometary environments. Pick-up ions have maximum speed (equal to $2u$) in the solar wind direction (Fig. 1). For example, an O$^+$ ion picked up by a solar wind with $u = 550$ km s^{-1} and moving in the solar wind direction, has the energy of $T_{max} = 100$ keV; a CO$_2^+$ ion picked up by the solar wind of the same velocity would have $T_{max} = 276$ keV.

It should, however, be noted that the isotropization is not complete; the creation of new ions maintains a certain anisotropy of the velocity distribution, and it supports continuous excitation of Alfvén waves. According to measurements, the anisotropy degree does not exceed 20–30% (Neugebauer et al. 1986) in the upstream region.

2.2 Adiabatic heating

As a consequence of the mass loading by cometary ions, there is a gradual slow-down of the solar wind plasma flow as it streams through the cometary coma. Applying ideal gas approximation in this adiabatic process (Wallis 1973), the average energy (as measured in the solar wind frame of reference) of the pick-up ions will be increased by a factor of $F^{2/3}$, where F is the ratio of plasma speed at the point where the ion is produced and the local plasma flow speed. Suppose an O$^+$ ion is created in the high-speed ($u = 550$ km s^{-1}) flow region upstream of the bow shock; then at the time it reaches a low-speed flow region with $u = 300$ km s^{-1}, say, its energy of gyro-motion in the solar wind frame would be increased from 25 keV to 37 keV by adiabatic compression.

2.3 Shock-associated accelerations

2.3.1 Shock compression

Shock compression is similar to the adiabatic heating effect resulting from the gradual slow-down of the plasma. The sudden jump in the flow velocity at the bow shock also leads to an increase of ion thermal energies characterized by the factor F (shock) $= u_1/u_2$ where u_1 and u_2 are the pre-shock and post-shock solar wind flow velocities, respectively.

2.3.2 Shock drift acceleration

In the stationary frame of a bow shock, there is an electric field given by $\mathbf{u}_1 \times \mathbf{B}_1$ which could facilitate particle acceleration at the shock front, since the grad B drift motion of particles is parallel to that electric field (Hudson 1965, Sonnerup 1969).

2.3.3 Diffusive shock acceleration

Diffusive shock acceleration is essentially a first-order Fermi type process. As a source of high-energy ions near comets, it was first suggested by Amata & Formisano (1985). The process is more effective if the shock is quasi-parallel, in which case random scattering of the particles on both sides of the bow shock leads to the gain of energy (see Axford 1981), the amount of which depends essentially on the strength of the shock.

Multiple shock front crossing may occur also in quasi-perpendicular shocks if particles are trapped near the shock front: one turning point may be created by the shock potential, another by the magnetic field. At each crossing the particle gains energy and is thus accelerated along the shock front (Sagdeev 1964).

2.4 Second-order Fermi acceleration (SOFA)

The essential feature of this process is diffusion-like velocity change (diffusion in velocity space) via resonant scattering on magnetic irregularities (Fisk 1976). To be operational, the resonant waves must have different phase velocities. The process can be described by a Fokker–Planck type equation in the six-dimensional phase space. The result of it is that the original energy distribution spreads toward both higher and lower energies, i.e. both high-energy and low-energy tails are formed (Ip & Axford 1987a, Isenberg 1987).

In applying this method to explain ion acceleration near comets, two different approaches were made.

2.4.1 SOFA-1

One of them, let us call it SOFA-1, (Ip & Axford 1986, 1987a, b) takes the velocity–diffusion coefficient, D, in the form suggested by Fisk (1976), then introduces a parameter $f = \lambda/R$ (with λ denoting scattering mean free path and R the Larmor radius). It is found that the efficiency of the acceleration (i.e. the time needed to increase the ionic energy by a prefixed factor) depends essentially on f, and explicit analytical expression is given for the number density of ions as a function of kinetic energy and time, t. The expression, in principle, may be adequate to describe the energy spectrum right from the lowest energies up to the highest observed ones. The expression contains f and t but in the combination t/f.

2.4.2 SOFA-2

The other approach (Gribov et al. 1987), is more general in the sense that it uses a more general expression for D, derived by quasi-linear theory, which takes into account the actual power of the observed magnetic fluctuations taken at the cyclotron resonance frequency. If the power spectrum of the magnetic fluctuations has the form of a power function, the general solution of the Fokker-Planck equation may be given in an explicit analytical form. The solution takes a particularly simple form if the power index is -1. In this case the time dependence of the average kinetic energy (temperature, T) of the accelerated ions may be written as (Gribov et al. 1987), namely:

$$T(x) = T(x_0) + \frac{1}{\mu(x)} \int_{x_0}^{x} dx \frac{1}{3} m_i v_A^2 \omega_c \hat{B}^2/B^2 \quad (1)$$

where $x = u(x)t$ is measured along the solar wind velocity, $\hat{B}^2$ is the constant in the fluctuation power (B_k^2) spectrum of B with the index -1, that is:

$$B_k^2 = \hat{B}^2/k$$

and v_A is the Alfvén speed.

A closer consideration of the resonance condition, that is:

$$k = \omega_c/(v|\cos\theta|)$$

where k is the resonant wave number of the magnetic fluctuations and θ is the pitch angle of the particle velocity, v, reveals that for short wavelength ($k > \omega_c/u$) fluctuations, equation (2) can be satisfied but only for the part of the pitch angle range where $|\cos\theta| < u/v$, whereas in the case of long wavelengths ($k < \omega_c/u$) a much larger pitch angle range may be effective in the acceleration process. This very much restricts the efficiency of short wavelength fluctuations in regions where $v \gg u$.

Short wavelength magnetic fluctuations near comets may be produced by Alfvén waves or by oblique magnetosonic waves. The excitation of Alfvén waves by ion-cyclotron instability was observed experimentally near comets (Galeev et al. 1986a, Tsurutani et al. 1986b, Johnstone et al. 1986, Oberc et al. 1987). However, since they propagate along magnetic field lines, waves propagating in directions both parallel and antiparallel to B are needed for SOFA. The conditions for excitation of such waves are not clear, and as yet no such wave configuration has been clearly detected.

The contribution of oblique magnetosonic waves to SOFA is possible; it is, however, limited by the fact that such waves suffer strong Landau damping (due to Čerenkov resonance with solar wind protons) in regions where the solar wind β is near to 1 (Kennel 1986). They may be more effective closer to the comet, where the level of oscillations is large enough to form a plateau on the distribution function, and, in this way, switch off the Landau damping (Galeev et al. 1987).

Long-wavelength magnetic fluctuations are in fact observed near comets. Their weaker power density may be compensated by the larger pitch angle region available for resonant acceleration to high ($v \gg u$) energies. They may probably be excited by fire-hose instability, the conditions of which seem to be fulfilled upstream of the bow shock in its close vicinity ($2.5 \geq R \geq 1 \times 10^6$ km on the inbound leg). Further theoretical investigations into the production and power density of such waves are needed for estimation of their contribution to SOFA.

2.5 Field reconnection

Reconnection of magnetic field lines can occur on the upstream side of the comet as a result of change in the direction of the ambient interplanetary magnetic field (Niedner & Brandt 1978), or on the downstream side as a result of 'substorm'-like activity (Ip & Mendis 1976, Axford 1979, Ip & Axford 1982, Niedner & Schwingenschuh 1986, Gringauz et al. 1986c). Magnetic field reversals were observed both by the Giotto (Neubauer et al. 1986, Raider et al. 1986) and the Vega (Riedler et al. 1986) magnetometer experiments. Under steady state conditions, the maximum energy that ions may acquire as a result of field reconnection is approximately

$$E_m = Ze\,B\,v_A\,L \qquad (3)$$

where Ze is the ion charge, L is the characteristic length of the reconnection region, and v_A is the Alfven speed. (Axford 1984, Ip & Axford 1986). This process may play a rôle in regions where magnetic field draping is present, i.e. near to the magnetic barrier in the coma and in its tailward extension.

3 EXPERIMENTAL

Of the six spacecraft (Vega-1 and -2, Giotto, Suisei, Sakigake, ICE) which approached comet Halley in March 1986, four (Vega-1 and -2, Giotto, ICE) carried instruments to study the dynamics of high-energy ($\geq$ 30 keV) ion populations. ICE approached the cometary nucleus on the solar side at a minimum distance of about 28×10^6 km (Wenzel et al. 1986) which was too large to decide unequivocally whether the high-energy ion fluxes observed there were of cometary or interplanetary origin (Tsurutani et al. 1986a). On Vega-2, the high-energy ion device was not operating during the flyby period, owing to an intermittent contact failure. There were thus altogether two devices sending back detailed data on the high-energy ions around Halley's comet: one was the TÜNDE-M device on Vega-1, the other the EPONA device on Giotto. Both used two fully depleted silicon wafers as detectors. A schematic diagram of the detector telescope of TÜNDE-M is shown in Fig. 2. For further details see Somogyi et al. (1982, 1985). For

VEGA –1 AND –2
PARTICLE DETECTOR TÜNDE –M

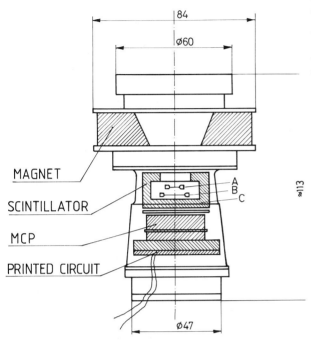

Fig. 2 — Cross-sectional schematic drawing of the particle detector TÜNDE-M flown on spacecraft Vega-1 and Vega-2.
A and B: fully depleted silicon detectors
C (also: SCINTILLATOR): plastic scintillator guard detector
MCP: multi-channel plate

details of the EPONA device, see McKenna–Lawlor et al. (1986a)

Encounter data (time and distance of closest approach, velocities and angles at closest approach) of Vega-1 with comet Halley are given in Table 1.

Table 1 — Encounter data of Vega-1 with comet Halley

Time of closest approach†	0720:06 s UT† on 6 March 1986
Distance of closest approach	8890 km

Spacecraft velocity at closest approach
  orbital
  relative to comet
	
	35.2 km s^{-1}
	79.2 km s^{-1}

Angle of spacecraft velocity
  with the Sun–comet line
  with the comet velocity
	
	112°
	150°

† Spacecraft event time

TÜNDE-M cannot distinguish between ions of different charges or masses. It measures the energy deposited in the front silicon wafer by ions stopping in that wafer, and counts the number of such ions arriving during a prefixed time Δt (= time resolution) in prefixed energy bins (channels). Information on the time resolution and energy channels is given in Table 2.

It is an intrinsic property of semiconductor detectors that, because of energy loss in the insensitive layer covering the detector, the incident energy of a particle coming to rest in the detector is higher than the energy deposited in the silicon wafer (pulse height defect), the difference being dependent on the identity and the energy of the particle. Unless stated otherwise, the ionic energies quoted throughout this review refer to incident oxygen ions. The front cover of the front detector was, in the case of TÜNDE-M on Vega-1, a 15 μg/cm^2 aluminum layer. The energy losses of O^+ ions in that layer were calculated on the basis of the papers by Ipavich et al. (1978) and Kecskeméty et al. (1989).

Vega-1 was three axis stabilized with one axis pointing towards the Sun and another to the star Canopus. The detector of TÜNDE-M pointed in the direction $r \times z$ where r is a vector pointing from the Sun toward the spacecraft and z points to the ecliptic north (the direction $r \times z$ is nearly opposite to that of the orbital velocity of the spacecraft). The half opening angle of the viewing cone was 25°, and the geometric factor of the telescope 0.25 ± 0.03 cm^2 sr.

4 OUTSIDE THE BOW SHOCK
4.1 Outlines

Fig. 3 presents a survey of the time history of high-energy ionic fluxes as observed by Tünde-M along an approximately 18 × 10^6 km (12.4 before, 5.3 after closest approach) track of Vega-1, passed between 1200 UT, 4 March 1986, and 0200 UT, 7 March 1986. Vega-1 crossed the cometary bow shock in the inbound direction at about 0350 UT on 6 March (Gringauz et al. 1986a, Riedler et al. 1986) at a distance of about 1.0 × 10^6 km from the cometary nucleus. From 0350 UT there was an approximately 30-minute gap in data transmission.

In this chapter we summarize the results concerning high-energy cometary ionic processes as observed by

Table 2 — Time resolutions and energy channels of Tünde-M in the period 4–7 March 1986

Period [UT]	Distance from nucleus [km]	Time resolution Δt, [s]	Number of energy bins	Total range of ion energy (for O^+) [keV]†	Notes
(1) 0900 March 4 to 0400 March 6	13.2×10^6 to 1×10^6	150	4	96–153	#1
(2) 0400 to 0715 March 6	1×10^6 to 25 000	4	52	96–766	
(3) 0715 to 0830 March 6	25 000 to 8 890, then increasing to 330 000	4	4 4 4 sum: 11	96–138 184–238 274–324	#2
(4) 0830 March 6 to 0640 March 7	330 000 to 6.7×10^6	150	2	96–125	#3

† Allowance is made for the pulse height defect. See text.
#1 The noise level in the lowest energy bin (96–106 keV) was higher than that of the signal until 2030 UT on 5 March.
#2 At 0715 UT, 41 bins (out of the 52) became unoperational owing to damage of the apparatus by a heavy cometary environment.
#3 The noise level in the 96–106 keV bin became higher than that of the signal at 2300 on 6 March.

TÜNDE-M on Vega-1 between 20h, 4 March and 0350, 6 March, i.e. in the outside-the-bow-shock region in the inbound pass.

The first clear sign of energetic ions of cometary origin was observed at about 2100 UT on 4 March, i.e. as far as $9.8 \pm 0.1 \times 10^6$ km from the nucleus of comet Halley. The ion fluxes generally increased as the spacecraft approached the comet, the increase becoming steeper at a distance of about 3×10^6 km from the nucleus, and reaching maximum values near the bow shock. An interesting feature of the counting rates detected by TÜNDE-M is that large flux enhancements, in some cases almost two orders of magnitude, are superimposed on the general trend. In the inbound pass, the enhancements show a quasi-periodic nature with a period of about 4 h.

4.2 Appearance of cometary ions at $R = 10^7$ km

Once a neutral particle is ionized, it is convected away with the solar wind (see section 2.1), therefore cometary particles detected as ions should have travelled as neutrals close to, or upstream from, the observer.

Since the expansion velocity of the neutral gas is estimated to be about 1 km s^{-1}, travelling to 10^7 km arouses serious concern about the value of 10^6 s characteristic time of ionization loss of neutrals (Ip & Axford 1987b). The minimum velocity of neutral particles required to reach a particular observer has been calculated by assuming that neutral gas particles move on Keplerian orbits around the Sun (Daly 1987, Erdös & Kecskeméty 1987). It has been shown that some of the observations of pick-up ions in distant regions (Somogyi et al. 1986, Yumoto et al. 1986, Wenzel et al. 1986) were inconsistent with 1 km s^{-1} gas outflow velocity, suggesting that a high-speed velocity component of neutral gas outflow should be considered. Photodissociation of CO_2 and/or CO molecules (or dissociative recombination of $CO_2{}^+$ and/or CO^+ ions) can provide fast C and O atoms (Ip & Axford 1987b) with velocities of several km s^{-1} and reasonable densities for us to understand, in a semi-quantitative way, the appearance of high-energy ions

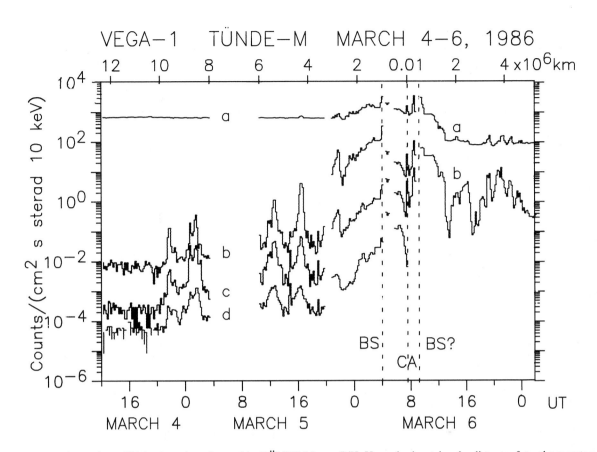

Fig. 3 — Intensity vs time of high-energy ions detected by TÜNDE-M near P/H. Upper horizontal scale: distances from the cometary nucleus.
BS: bow shock crossing (inbound)
CA: closest approach
BS?: bow shock crossing (outbound, uncertain)
Limits of energy channels (for O⁺ ions):
(a)   96–106 keV   (c)   125–138 keV
(b)   106–125 keV   (d)   138–153 keV
For more data inside the bow shock see Figs 6 and 7

at $R = 10^7$ km. Another source of fast neutrals may be the photodissociation of water molecules into neutral H and OH, and/or that of OH radicals into H and O. In these exoterm processes, the H atoms leave the parent molecules with velocities of 20 and 8 km s⁻¹ thus being fast enough for a considerable number of them to reach distances of the order of 10^7 km from the nucleus. Here, however, the difficulty is met with the post-acceleration process which, in this case, must be more effective by about an order of magnitude to be able to accelerate pick-up protons (instead of pick-up O⁺ ions) to the energies observed by TÜNDE-M.

4.3 Insufficiency of the solar wind pick-up process

Contribution of pick-up ions to the high-energy fluxes observed by TÜNDE-M may be ruled out, unless the pick-up ions undergo further acceleration by some mechanism different from the pick-up process. This follows from the distribution of the velocities an ion may gain by the pick-up process (see section 2.1) and the viewing direction of the TÜNDE-M telescope. The effect of the latter is explained in Fig. 4, which is essentially the projection of the velocity–space diagram of Fig. 1 onto the X, Y (i.e. the ecliptic) plane.

The cone of acceptance of TÜNDE-M is added, together with areas containing end points of velocity vectors of O^+ ions which may be detected in the lowest energy channels of TÜNDE-M (see section 3). In the spacecraft frame of reference, the maximum velocity that a pick-up ion flying within the cone of acceptance may have, is about 345 km s^{-1}, if $u = 500$ km s^{-1}. Actually, u varied between 580 km s^{-1} and 430 km s^{-1} in the interval considered (Verigin et al. 1986 and unpublished data, courtesy of Prof. K.I. Gringauz, measured by the PLASMAG-1 experiment on Vega-1). An O^+ ion with $v = 345$ km s^{-1} has a kinetic energy of 9.9 keV. Even a CO_2^+ ion (the number density of which is very small along the part of the spacecraft track under consideration) would have kinetic energy of only about 27 keV, much too low to be detected by TÜNDE-M. Unrealistically high ($u \geq 1400$ km s^{-1}) solar wind speeds would be needed for pick-up O^+ ions to be detected by TÜNDE-M, without additional acceleration by a different process.

The mere existence of high-energy ions recorded by TÜNDE-M proves the operation of one or more acceleration mechanisms in addition to that of solar wind pick-up.

4.4 Adiabatic and shock-associated acceleration

According to solar wind plasma measurements made by the instrument PLASMAG-1 onboard Vega-1 (Gringauz et al. 1986a, Verigin et al. 1986) the solar-wind bulk speed, u, decreased from about 580 km s^{-1} (at about $R = 107$ km) to about 520 km s^{-1} (at about $R = 3 \times 10^6$ km), and then to about 430 km s^{-1} at the outer boundary of the bow shock (about $R = 10^6$ km).

4.4.1 Compression

Compression of such orders of magnitude is insufficient to raise energies of pick-up cometary ions to a level detectable for TÜNDE-M or EPONA. A quantitative estimate must also take into account the slowing down of the solar wind (Ip & Axford 1986). A decrease of solar wind velocity from u_1 to u_2 ($<u_1$) would cause the velocity of the pick-up ions to be distributed on the surface of a sphere (in velocity space) with a radius $u_1(u_1/u_2)^{\frac{1}{3}}$ and centred at the end point of u_2. Taking into account the TÜNDE-M acceptance geometry shown in Fig. 4, the maximum energy, $E_m(O^+)$ in the spacecraft frame, of an O^+ ion

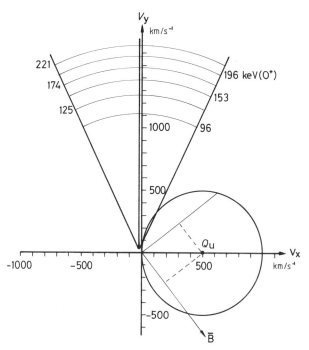

Fig. 4 — Projection of the velocity distribution sphere of pick up ions shown in Fig. 1 onto the (v_X, v_Y) plane, together with the projection of the cone of acceptance of the TÜNDE-M device. Q_u is the end point of the vector u.

arriving within the cone of acceptance of TÜNDE-M, assuming an initial solar wind velocity $u_1 = 580$ km s^{-1}, takes the following values:

at $F =$	1	580/520	580/430	2	5†
$E_m(O^+)$ in keV =	14.2	24.0	34.6	48.8	83.6

These values are lower than those observed in the lowest energy channel of TÜNDE-M (96–106 keV).

4.4.2 Diffuse shock

Diffuse shock acceleration also has limitations. First of all, the drop of u at the shock was too small. Its exact value cannot be estimated on the basis of the Vega-1 (PLASMAG-1) measurements because of the data gap mentioned in section 4.1. By inference from Giotto measurements, the drop might have been a factor of 1.4 (Coates et al. 1986), which, in itself, could cause only negligible acceleration.

Further serious limitation on the efficiency of diffusive shock acceleration is imposed by the fact

† unrealistic in cometary bow shocks.

that the convection of the particles by the solar wind is still strong enough to reduce the chances of diffusion back into the upstream region.

4.4.3 Gradient drift

The angle between the magnetic field vector and the shock normal determined from the coplanarity theorem is about 45° (Schwingenschuh *et al.* 1986), favourable for gradient-drift acceleration to operate. However, it should be noted that there is a low-energy cut-off in the reflection of particles by the shock wave (Decker 1983, Terasawa 1979), if particle velocity is less than about the velocity of the de Hoffman–Teller frame. For transmission of particles through the shock, the process is very similar to the adiabatic compression, therefore it seems to be ineffective.

4.5 Second-order Fermi acceleration

Experimental results at comets Giacobini-Zinner and Halley (Tsurutani *et al.* 1986b, Galeev *et al.* 1986a, Schwingenschuh *et al.* 1987) for the total energy of magnetic field fluctuations and for the spectral density of short wavelength ($k > \omega_c/u$) fluctuations are close to theoretical expectations. An extension of the fluctuation spectrum into the long wavelength ($k < \omega_c/u$) region has also been confirmed experimentally.

Comparison of energy spectra of ions as predicted by the method SOFA-1 (section 2.41) with observed spectra is dealt with by McKenna–Lawlor (1989). The agreement between theory and experiment is good if we take into account the large energy range (two orders of magnitude) of the relevant particles, although there are difficulties in fitting the experimental points simultaneously both at the low-energy and the high-energy ends of the range by one single value of the parameter.

The results obtained by the method SOFA-2 (section 2.42) are shown in Fig. 5 (Gribov *et al.* 1987). The upper panel of the figure shows time profiles of intensities of high-energy ions as observed by TÜNDE-M during the last five hours (1.4 × 10⁶ km) before crossing the bow shock (in the inbound direction); the lower panel shows temperatures of the high-energy ions (96–153 keV for O⁺ ions) as obtained on the basis of the high-energy ionic intensities

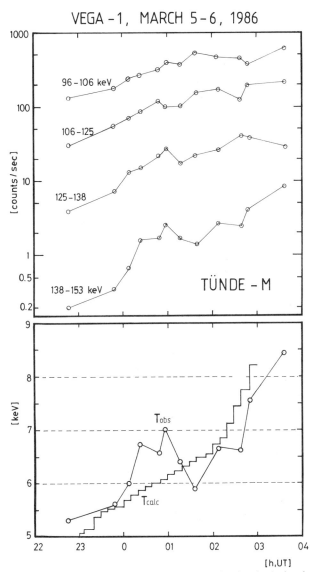

Fig. 5 — Lower panel: comparison of observed (T_{obs}) and calculated (T_{calc}) temperature values of high-energy ions in the pre-shock region (2.45 > R > 1.05 × 10⁶ km). Upper panel: Values of high-energy ionic intensities observed by TÜNDE-M and used in deriving the T_{obs} values.

observed (T_{obs}) and temperatures calculated on the basis of equation (1), using solar wind velocity data and magnetic field data measured onboard Vega-1 by PLASMAG-1 and MISCHA, respectively. The agreement is good if we take into account uncertainties of the measured quantities. For a more detailed discussion see Gribov *et al.* (1987).

It may thus be concluded that SOFA plays a

significant rôle in heating the ionic environment of comet Halley. This is the more interesting since this is most probably the first case where the operation of the SOFA process is proved by direct measurement. As for the level of significance, we refer to the theoretical difficulties dealt with in section 2.4, as well as the not quite satisfactory (nevertheless qualitatively good) agreement between theory and experiment outlined in this section. Some opinions (Richardson *et al.* 1986, McKenna–Lawlor *et al.* 1987, Ip & Axford 1987a) attribute less efficiency to SOFA than was previously thought.

4.6 Quasi-periodic intensity enhancements

No obvious correlation can be found between the quasi-periodic peaks in the counting rates and the measured plasma parameters (Kecskeméty *et al.* 1986, 1989). To explain the 4 h periodicity in the observed fluxes, an interesting possibility could be to relate it to the rotation of the nucleus itself. Suppose that the gas production rate is modulated by the rotation of the nucleus (due, for example, to active region(s) on the surface facing the Sun), then the neutral gas would be enhanced in concentric shell-like structures centred around the comet and expanding with the velocity of the gas, similarly to the hydrogen coma observed by Suisei (Kaneda *et al.* 1986). If gas-shells with larger neutral densities exist, a larger flux of pick-up ions is expected, when the spacecraft–Sun line is tangential to the inner surface of the shell, since it is the column density of neutrals along the solar wind flow direction which contributes to the flux of ions. This might explain the periodicity observed in the ion flux. However, a very effective post-acceleration is needed to explain the almost two orders of magnitude flux increases. Another limitation is that the 53 h rotation period of the nucleus (Sagdeev *et al.* 1986b) would require a large expansion velocity of neutrals ($\approx$ 7 km s^{-1}, Kecskeméty *et al.* 1989) to match the 4 h periodicity in the timing of the peaks as seen by the spacecraft.

5 INSIDE THE BOW SHOCK

5.1 Outlines

Fig. 6 ('TÜNDE-M' panel) shows a representative sample of the high-energy ionic intensities observed

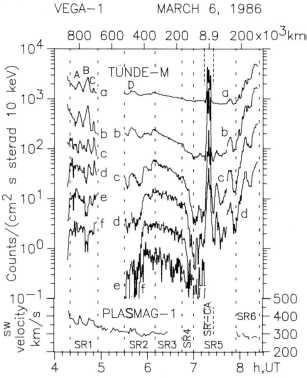

Fig. 6 — Upper curves: Sample of high-energy ionic intensity profiles observed by TÜNDE-M inside the bow shock. Limits of energy channels (for O$^+$ ions) in keV:
(a) 96–106   (c) 184–196   (e) 443–458
(b) 106–125   (d) 274–284   (f) 607–623
Lowest curve: solar wind velocity as measured by the SDA (solar direction analyzer) of PLASMAG-1 (courtesy of Prof. K.I. Gringauz).
Upper horizontal scale: distances from the cometary nucleus.

inside the cometary bow shock by the TÜNDE-M detector system.

The inside-the-bow-shock region may, from the point of view of high-energy ionic structures, be subdivided into several subregions (SR). A possible

Table 3 — Limits of subregions (SR) within the inside-the-bow-shock region (6 March 1986)

No. of SR	Time (UT) start	end	Distance† start	end
1	0420	0456	860	690
2	0530	0610	520	330
3	0610	0645	330	160
4	0645	0700	160	90
5	0700	0755	90 − 8.9 −	160
SR-CA	0714	0726	30 − 8.9 −	29
6	0755	0825	160	310

† In thousands of kilometres from the cometary nucleus

division is shown in Table 3 (compare with Fig. 6).

SR1 is characterized by large intensity fluctuations superimposed on a falling trend (at energies lower than $\approx$ 150 keV) and a nearly constant trend (over $\approx$ 150 keV), respectively.

There was no data transmission for TÜNDE-M between 0455 and 0530 UT. In SR2, intensities start at values $I_{2s}(E)$, considerably lower than $I_{1e}(E)$, that is, those observed at the end of SR1. The energy dependence of $I_{2s}(E)/I_{1e}(E)$ can, with good accuracy, be written in the form:

$$I_{2s}(E)/I_{1e}(E) = (53 \pm 9) \times 10^3 \, E^{-2.38 \pm 0.03}$$

with E measured in keV. After the start, the intensities show — apart from fluctuations — rapid rises to values, $I_{2e}(E)$, which with good approximation satisfy the relation

$$I_{2e}(E)/I_{2s}(E) = (63 \pm 10) \times 10^{-5} \, E^{+1.59 \pm 0.03}$$

(E in keV). For the time being there is no explanation for the enormous decreases of the ionic intensities and their partial recovery between 0455 and 0610 UT.

In SR3, intensities in all energy channels decrease more or less independently of energy, by a factor of 1.9 $\pm$ 0.35, except in the lowest energy channel, where the decrease is a factor of only 1.26.

The slow decrease observed in SR3 changes abruptly to a fast one in SR4. The energy dependence of the ratio I_{4e}/I_{4s} (that is, that of the intensities at the end and the beginning of the SR) may be expressed by the power function:

$$I_{4e}/I_{4s} = (290 \pm 60) \times E^{-1.31 \pm 0.04}$$

(with E in keV).

SR5 is characterized by a continuous increase of intensities with superimposed maxima. The largest is around the point of closest approach, and differs strongly from the others, not only in size. It will be dealt with separately under the name SR-CA.

SR6 is similar to SR5, but the increase of the intensities is much faster.

5.2 Discussion

5.2.1 Ion populations in SR1 and SR2

The PLASMAG-1 panel in Fig. 7 shows solar wind velocities as measured by the plasma detector PLAS-

MAG-1 onboard Vega-1 between 0420 and 0630 UT, and 0755 and 0825 UT, on 6 March 1986 (unpublished data, courtesy of Prof. K.I. Gringauz). Several maxima are superimposed on the general trend of high-energy ion intensities, the most conspicuous of which are denoted by A, B, C, and D in Fig. 6. Approximate times of the maxima are 0432, 0443, 0451, and 0540 UT, respectively. They show up — with varying amplitudes — in almost all energy channels, and are synchronous (within time differences less than 2 minutes) in the various energy channels and also with similar maxima in the solar wind velocity profile shown in the lower panel of Fig. 6 (the B and C maxima are not resolved in the solar wind velocity profile).

The correlation of the maxima A, B, C, and D of ionic intensities with those of the solar wind velocity may be understood qualitatively in the following way. Increased solar wind velocity means stronger convection and thus smaller loss of intensity of high-energy ions via escape along the field lines. Furthermore, increased solar wind velocity produces pick-up ions of higher energies which, if the post-acceleration is efficient enough, may also contribute to the ionic intensity increases observed.

5.2.2 Intensity decreases in SR3 and SR4: cometopause crossing

The boundary between SR2 and SR3 (0610 UT, $R \approx 330 \times 10^3$ km) coincides with the boundary M1 established on the basis of magnetic field measurements (Schwingenschuh et al. 1986). By inference from Giotto measurements (Balsiger et al. 1986), and with allowance made for the higher total gas output rate of comet Halley at the time of the Vega-1 encounter, $R \approx 330 \times 10^3$ km is also the region where low-energy heavy ions begin to be clearly distinguishable from the background.

PLASMAG-1 measurements onboard Vega-1 (Fig. 6) show that the solar wind velocity decreases fairly smoothly until the end of SR3, ($R \approx 160 \times 10^3$ km) which coincides with the region where the solar wind practically comes to rest and control of plasma processes is taken over by heavy cometary ions (cometopause: Gringauz et al. 1986a, b). The IMF strength increases slowly from about 15 nT at the

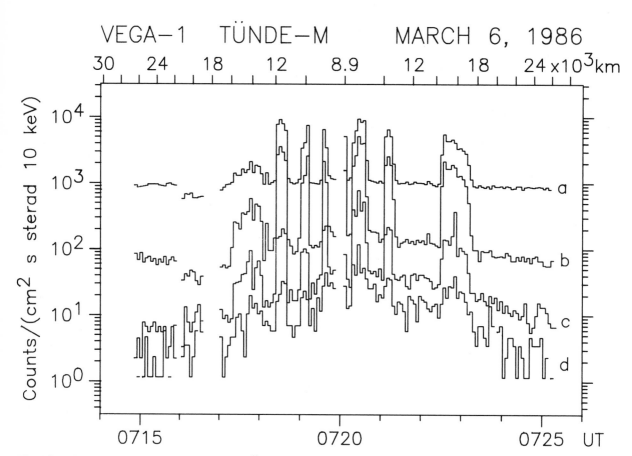

Fig. 7 Sample of intensity profiles as observed by TÜNDE-M near closest approach (0720:06 s UT).
Upper scale: Distances from the cometary nucleus. (a), (b), (c), and (d) denote the same energy channels as in Fig. 6.

beginning to about 40 nT at the end of SR3, where (that is, at the cometopause), an approximately 12 nT amplitude peak is superimposed on the increasing trend, and the IMF elevation angle changes from about $-15°$ to $+25°$. (Schwingenschuh *et al.* 1987).

No plasma data are available on Vega-1 for SR4, in which B increased fairly smoothly from 40 nT to about 50 nT without essential change in azimuth, but its elevation decreased to $0°$ by the end of the SR.

It is difficult to explain the ionic intensity decreases observed in SR3 and SR4 on the basis of the plasma and IMF behaviour just characterized. One of the causes may be gradual fading (in SR3) and then (in SR4) disappearance of pick-up ions and, in parallel, the increasing chance of (in SR4 almost free) escape of high-energy ions along the field lines. This effect may be enhanced by the **grad** B drift which causes a

displacement, $\Delta\mathbf{z}$, of the guiding centre in the direction of $\boldsymbol{B} \times \textbf{grad}\ B$ equal to,

$$\Delta z = \int \mathbf{v}_\mathrm{d}\, \mathrm{d}t,$$

where

$$\boldsymbol{v}_\mathrm{d} = v_\perp^2 (\boldsymbol{B} \times \textbf{grad}\ B)/(2\omega_\mathrm{c} B^2)$$

is the drift velocity, and $v_\perp$ is the particle velocity perpendicular to $\boldsymbol{B}$. Applying a very simple one-dimensional approximation, and assuming isotropy together with adiabatic compression, that is (since $\langle v_\perp^2 \rangle = \frac{2}{3} v^2$),

$$v_\perp^2 u^{2/3} = \mathrm{const},$$

and also assuming magnetic field freezing, that is,

$$\omega_\mathrm{c} u = \mathrm{const},$$

we arrive at

$$\Delta z = \frac{m_i}{2Ze} \int \frac{v_\perp^2}{B^2} \frac{dB}{dx} \frac{dx}{u} = v_2^2 [1 - (B_1/B_2)^{2/3}]/(2\omega_c u)$$

(4)

where the indices 1 and 2 refer to values further from and closer to the nucleus respectively. Note that Δz is proportional to the kinetic energy, and a depletion proportional to kinetic energy was observed in SR4. With reasonable values ($B_1 = 20$ nT, $u_1 \leq 200$ km s^{-1}, $B_1/B_2 = 0.4$) one arrives at $\Delta z \leq 35 \times 10^3$ km for ions with $E_{kin} = 200$ keV. This may provide a strong contribution to the depletion. More thorough analysis would, however, be necessary to see whether it is sufficient to explain the huge amplitudes of the decreases observed in SR4.

The contribution of charge exchange of high-energy ions in collisions with neutral cometary molecules may be neglected when considering intensity decreases in SR3 and SR4: it is too small by an order of magnitude, and is decreasing with energy, whilst the opposite was observed.

5.2.3 The giant peak of intensity around closest approach (SR-CA)

The most striking feature in the downstream region is the enormous increase of high-energy ionic intensities near the point of closest approach. Although the measuring apparatus suffered some damage at 0715 UT, resulting in the loss of 41 ionic energy channels out of 52, the possibility that the increase would be an instrumental artifact can be practically excluded, both because of the internal consistency of the data observed and also because of the fact that none of the housekeeping data showed any sign of damage to the remaining channels, and also that the high-energy ionic device EPONA on Giotto has observed similar increases in the closest vicinity of comet Halley (McKenna–Lawlor 1989).

Profiles of intensities observed in four energy channels are shown in Fig. 7. Each profile shows a single, flat maximum and eight superimposed narrow flashes. The flat maxima peak near the point of closest approach with amplitudes increasing from a factor of 2 at ≈ 100 keV to a factor of ≈ 10 at 320 keV. The superimposed narrow flashes have widths between 12

and 50 s and relative amplitudes, f, decreasing from about $10 < f < 40$ around 120 keV to about $2 < f < 10$ around 320 keV.

When comparing ionic intensities with magnetic field data, it should be noted that the B_z channel of the device MISCHA onboard Vega-1 was saturated from 0715:40 UT to 0723:00 UT (practically during the whole SR-CA) at a level of about 60 nT (Schwingens-chuh et al. 1987). On the basis of the trend of the B_z values outside the SR-CA, a peaking of B at a level of about 90 nT around closest approach would be a conservative estimate; higher maximum values cannot be excluded. We may thus suppose an increase of the magnetic field strength by a factor of 2 within SR-CA.

Adiabatic compression by a factor of 2 would result in a temperature increase by a factor of $2^{2/3} \approx 1.6$. This may contribute to the measured flat maximum of intensities, provided that the conditions are favourable for adiabatic compression (nearly isotropic velocity distribution). This may be the case if scattering is strong enough. If not, particles can still increase their perpendicular momentum via betatron acceleration. The conservation of the first adiabatic invariant involves an energy gain by a factor of $\leq B_2/B_1$ which, in our case, is ≈ 2. The actual gain depends on the angular distribution of the particle velocity.

Greater difficulties are encountered when trying to explain the intensity 'flashes' around closest approach. Since the spatial separation of flashes is of the order of the gyroradii of ions, neither the lack of any energy dependence of the onset and offset times, nor the steepness of the onsets and offsets of the flashes can be explained, except if TÜNDE-M has detected electrons. In the given conditions, the only possible mechanism able to produce the flashes seems to be magnetic field reconnection. Assuming $B = 80$ nT, an oxygen ion density of 500×10^6 m^{-3}, and $L = 50 \times 10^6$ m, and using equation (3), we arrive at $E_m \approx 80$ keV which is not enough to explain the intensity increases of 100 to 320 keV particles. To get $E_m \approx 300$ keV, we should need, for example, $B \approx 120$ nT and $L \approx 10^8$ m, values which cannot be excluded but which seem to be unlikely. However, it should be noted that spikes have been observed in the terrestrial magnetosphere in which the maximum energies of spike particles were an order of magnitude larger than the values calculated on the basis of equation (3) (Sarris *et al.* 1976).

Data observed along the outbound leg (after closest approach) have not yet been analyzed in detail.

6 CONCLUSIONS

Unexpected were the energy ranges and spatial extension of high-energy ions observed onboard the spacecraft that approached comet Halley in 1986. The energy range extends to at least 0.8 MeV (detected by TÜNDE-M), and probably even further (measurements of EPONA onboard Giotto). The spatial extent of ions exceeds 10^7 km from the cometary nucleus.

The main results of the high-energy ion measurements carried out by TÜNDE-M onboard Vega-1 may be summarized as follows:

(1) TÜNDE-M has observed high-energy ions in energy regions much higher than, and directions far from being accessible to, pick-up ions. Existence of such ions demonstrates that substantial and very effective post-acceleration of pick-up ions is taking place.

(2) The high level of magnetic turbulence near to the bow shock, and also the spectral characteristics of ion populations observed there, suggest that second-order Fermi acceleration is responsible for, or at least, strongly contributes to, the energization of ions. The direct observation of such processes at comets is also interesting by itself, since, to our knowledge, this is the first case where the effectiveness of the second-order Fermi acceleration is demonstrated by direct, *in situ* measurements.

(3) The existence of ions at great distances from the nucleus on the subsolar side of the comet gives information about the spatial extent of the neutral gas coma as well. Neutral cometary molecules must reach these distances, since once they are ionized, they are picked up by the solar wind and drifted toward the cometary tail. High-energy ionic measurements are thus sensitive probes of the spatial distribution of the neutral cometary molecules around comets. The difficulty connected with travelling of slow neutrals to great distances may suggest the existence of a fast ($v = 5$ to 8 km s^{-1}) component of neutral gas outflow.

(4) In addition to the Sun, the planetary magnetospheres, and interplanetary shocks, cometary environments — at least in the case of comet Halley — are now known to be a fourth type of localized source of high-energy ions. When comparing this source with astrophysical accelerators, the pick-up process (a natural injection mechanism) may be compared to cosmic ray acceleration at diffusive shocks, with, however, the big difference that the mass of pick-up ions substantially contributes to the dynamics of the cometary shock, unlike cosmic rays at astrophysical shocks.

(5) Many of the observed features of high-energy ion processes around comet Halley are still unexplained, such as the post-acceleration of ions during the periodic intensity enhancements between $R = 10^7$ km and the bow shock, or the enormous high-energy ionic intensity flashes observed at the magnetic barrier, and others. Some of the unsolved problems may perhaps be answered by pushing forward the performance of theory. Some of them will need further measurements which, in the case of Halley's comet, will have to be wait until AD 2061.

ACKNOWLEDGEMENTS

The authors wish to express their thanks to Dr K. Kecskeméty for his help with data handling, preparing figures of intensity profiles, and for discussions. They are indebted to Prof. K.I. Gringauz for the permission to use, and Dr Ms. M. Tátrallyay for handling, PLASMAG-1 (Vega-1) data.

REFERENCES

Abbreviations used for ESA publications:

ESA SP-250: *Proceedings of the 20th ESLAB Symposium on the exploration of Halley's Comet*, Heidelberg (FRG), Vols 1–3 (1986)

ESA SP-251: Plasma astrophysics. *Proceedings of the joint Varenna-Abastumani school and workshop, Sukhumi (USSR)* (May 1986)

ESA SP-278: *Proceedings of the Symposium on the diversity and similarity of comets, Brussels (Belgium)* (April 1987). (in press)

ESA SP-1077: *The Giotto Mission — its scientific investigations* (1986)

* * *

Amata, E., & Formisano, V. (1985) Energization of positive ions in the cometary foreshock region, *Planet. Space Sci.* **33** 1243

Axford, W.I. (1979) Comets and cometary missions — An introductory review, In: *Cometary missions,* (eds. W.I. Axford, H. Fechtig, & J. Rahe), 1–12, Remeis-Sternwarte Bamberg, Astron. Institut der Universität Erlangen-Nürnberg

Axford, W.I. (1981) Acceleration of cosmic rays by shock waves, *Proc. 17th International Cosmic Ray Conf., Paris,* **12** 155

Axford, W.I. (1984) Magnetic field reconnection, *Geophysical Monograph* **30** 1

Balsiger, H., Altwegg, K., Bühler, F., Fuselier, S.A., Geiss, J., Goldstein, R., Lazarus, A.J., Meier, A., Neugebauer, M., Rettenmund, U., Rosenbauer, H., Schwenn, R., Shelley, E.G.,

Ungstrup, E., & Young, D.T. (1986) The composition and dynamics of cometary ions in the outer coma of Halley. *ESA SP-250/1*, 99

Coates, A.J., Johnstone, A.D., Thomsen, M.F., Formisano, V., Amata, E., Wilken, B., Jockers, K., Winningham, J.D., Borg, H., & Bryant, D.A. (1986) Solar wind flow through the Halley bow shock, *ESA SP-250/1*, 263

Cravens T.E. (1986) Ion distribution function in the vicinity of Comet Giacobini–Zinner *Geophys. Res. Lett.* **13** 275

Daly, P.W. (1987) Can neutral particles from Comet Halley have reached the ICE spacecraft?, *Geophys. Res. Lett.*, **14** 648

Decker, R.B. (1983) Formation of shock-spike events at quasi-perpendicular shocks, *J. Geophys. Res.* **88** 9959

Erdös, G., & Kecskeméty, K. (1987) Distribution of neutral gas molecules at large distances from Halley's Comet, *ESA SP-278*, 223

Fisk, L.A. (1976) On the acceleration of energetic particles in the interplanetary medium, *J. Geophys. Res.* **81** 4641

Galeev, A.A., & Khabibrakhmanov, I.Kh. (1986) The critical ionization velocity phenomenon in astrophysics, *ESA SP-251*, 129

Galeev, A.A., Gribov, B.E., Gombosi, T.I., Klimov, S.I., Oberc, P., Remizov, A.P., Riedler, W., Sagdeev, R.Z., Savin, S.P., Sokolov, A.Yu., Shapiro, V.D., Shevchenko, V.I., Szegö, K., Verigin, M.I., & Yeroshenko, Ye, G. (1986a) The position and structure of Comet Halley bow shock: Vega-1 and Vega-2 measurements, *Geophys. Res. Lett.* **13** 841

Galeev, A.A., Sagdeev, R.Z., Shapiro, V.D., Shevchenko, V.I., & Szegö, K. (1986b) Mass loading and MHD turbulence in the solar wind/comet interaction region, *ESA SP-251*, 307

Galeev, A.A., Sagdeev, R.Z., Shapiro, V.D., Shevchenko, V.I., & Szegö, K. (1987) MHD turbulences in a mass-loaded solar wind and the cometary bow-shock — similarity and the difference with the cosmic ray shocks, *Proc. of the International Symposium on collisionless shocks, Balatonfüred, Hungary, June 1987*, 237

Gary, S.P., Hinate, S., Madland, C.D., & Winske, D. (1986) The developement of shell-like distributions from newborn cometary ions, *Geophys. Res. Lett.* **13** 1364

Gribov, B.E., Kecskeméty, K., Sagdeev, R.Z., Shapiro, V.D., Shevchenko, V.I., Somogyi, A.J., Szegö, K., Erdös, G., Eroshenko, E.G., Gringauz, K.I., Keppler, E., Marsden, R.G., Remizov, A.P., Richter, A.K., Riedler, W.R., Schwingenschuh, K., & Wenzel, K.-P. (1987) Stochastic Fermi acceleration of ions in the pre-shock region of Comet Halley, *Astronomy & Astrophysics* **187** 293–296

Gringauz, K.I., Gombosi, T.I., Remizov, A.P., Apáthy, I., Szemerey, I., Verigin, M.I., Denchikova, L.I., Dyachkov, A.V., Keppler, E., Klimenko, I.N., Richter, A.K., Somogyi, A.J., Szegö, K., Szendrö, S., Tátrallyay, M., Varga, A., & Vladimirova, G.A. (1986a) First *in situ* plasma and neutral gas measurements at Comet Halley, *Nature* **321** 282

Gringauz, K.I., Verigin, M.I., Richter, A.K., Gombosi, T.I., Szegö, K., Tátrallyay, M., Remizov, A.P., & Apáthy, I. (1986b) Cometary plasma region in the coma of Comet Halley: Vega-2 measurements, *ESA SP-250/1*, 93

Gringauz, K.I., Remizov, A.P., Verigin, M.I., Richter, A.K., Tátrallyay, M., Szegö, K., Klimenko, I.N., Apáthy, I., Gombosi, T.I., & Szemerey, T. (1986c) Electron component of the plasma around Halley's comet measured by the electrostatic electron analyzer of PLASMAG-1 on board Vega-2, *ESA SP-250/1* 195

Hudson, P.D. (1965) Reflection of charged particles by plasma shocks, *Mon. Not. Roy. Astron. Soc.* **131** 23

Hynds, R.J., Cowley, S.W.H., Sanderson, T.R., van Rooijen, J.J., & Wenzel, K.-P. (1986) Observation of energetic ions from Comet Giacobini–Zinner, *Science* **232** 361

Ip, W.-H., & Axford, W.I. (1982) Theories of physical processes in the cometary comae and ion tails, In: *Comets* (ed. L.L. Wilkening), 588, Univ. of Arizona Press, Tucson

Ip, W.-H., & Axford, W.I. (1986) The acceleration of particles in the vicinity of comets, *Planet. Space Sci.* **34** 1061

Ip, W.-H., & Axford, W.I. (1987a) A numerical simulation of charged particle acceleration and pitch angle scattering in turbulent plasma environment of cometary comas, *Proc. of the 20th International Cosmic Ray Conference, Moscow* **3** 233

Ip, W.-H., & Axford, W.I. (1987b) Stochastic acceleration of cometary ions: the effects of neutral composition and source strength distributions, *ESA SP-278*, 139

Ip, W.-H., & Mendis, D.A. (1976) The generation of magnetic fields and electric currents in cometary plasma tails, *Icarus* **29** 147

Ipavich, F.M., Lundgren, R.A., Lambird, B.A., & Gloeckler, G. (1978) Measurements of pulse-height defect in Au-Si detectors for H, He, C, N, O, Ne, Ar, Kr from $\approx$ 2 to 400 keV/nucleon, *Nuclear Instr. and Methods* **154** 291

Ipavich, F.M., Galvin, A.B., Gloeckler, G., Hovestadt, D., Klecker, B., & Scholer, M. (1986) Comet Giacobini–Zinner: *in situ* observation of energetic heavy ions, *Science* **232** 366

Isenberg, P.A. (1987) Energy of pick-up ions upstream of comets *J. Geophys. Res.* **92** 8795

Johnstone, A., Glassmeier, K.H., Acuña, M., Borg, H., Bryant, D., Coates, A., Formisano, V., Heath, J.W., Mariani, S., Musmann, G., Neubauer, F.M., Thomsen, M., Wilken, B., & Winningham, J. (1986) Waves in the magnetic field and solar wind flow outside the bow shock at Comet Halley, *ESA SP-250/1*, 277

Kaneda, E., Hirao, K., Shimizu, M., & Ashibara, G. (1986) Activity of Comet Halley observed in the ultraviolet, *Geophys. Res. Lett.* **13** 833

Kecskeméty, K., Cravens, T.E., Afonin, V.V., Erdös, G., Eroshenko, E.G., Gan, L., Gombosi, T.I., Gringauz.K.I., Keppler, E., Klimenko, I.N., Marsden, R.G., Nagy, A.F., Remizov, A.P., Richter, A.K., Riedler, W., Schwingenschuh, K., Somogyi, A.J., Szegö, K., Tátrallyay, M., Varga, A., Verigin, M.I., & Wenzel, K.-P. (1986) Energetic pick-up ions outside the Comet Halley bow shock, *ESA SP-250/1*, 109

Kecskeméty, K., Cravens, T.E., Afonin, V.V., Erdös, G., Eroshenko, E.G., Gan, L., Gombosi, T.I., Gringauz., K.I., Keppler, E., Klimenko, I.N., Marsden, R.G., Nagy, A.F., Remizov, A.P., Richter, A.K., Riedler, W., Schwingenschuh, K., Somogyi, A.J., Szegö, K., Tátrallyay, M., Varga, A., Verigin, M.I., & Wenzel, K.-P. (1989) Pick-up ions in the unshocked solar wind at Comet Halley, *J. Geophys. Res.* **94** 185

Kennel, C.F. (1986) Quasi-parallel shock *Advances in Space Research* **6**, Nr 1, 5

McKenna–Lawlor, S. (1990) (in the present volume)

McKenna–Lawlor, S., Thompson, A., O'Sullivan, D., Kirsch, E., Melrose, D., & Wenzel, K.-P. (1986a) The Giotto energetic particle experiment, *ESA SP-1077*, 53

McKenna–Lawlor, S., Kirsch, E., O'Sullivan, D., Thompson, A., & Wenzel, K.-P. (1986b) Energetic ions in the environment of Comet Halley, *Nature* **321** 347

McKenna–Lawlor, S., Wilken, B., Daly, P.W., Ip, W.-H., Kirsch,

E., Coates, A., Johnstone, A., Thompson, A., O'Sullivan, D., & Wenzel, K.-P. (1987) Energy spectra of pick-up ions recorded during the encounter of Giotto with Comet Halley, ESA SP-278, 133

Mendis, D.A., Houpis, H.L.F., & Marconi, M.L. (1985) The physics of comets, *Fund. Cosmic Phys.* **10** 1

Neubauer, F.M., Glassmeier, K.H., Pohl, M., Raeder, J., Acuña, M.H., Burlaga, L.F., Ness, N.F., Musmann, G., Mariani, F., Wallis, M.K., Ungstrup, E., & Schmidt, H.U. (1986) First results from the Giotto magnetometer experiment at Comet Halley, *Nature* **321** 352

Neugebauer, M., Lazarus, A.J., Altwegg, K., Balsiger, H., Goldstein, B.E., Goldstein, R., Neubauer, F.M., Rosenbauer, H., Schwenn, R., Shelley, E.G., & Ungstrup, E. (1986) The pick-up of cometary protons by the solar wind. ESA SP-250/1, 19

Niedner, M.B., & Brandt, J.C. (1978) Interplanetary gas. XXIII. Plasma tail disconnection events in comets: Evidence for magnetic field line reconnection at interplanetary sector boundaries, *Astrophys. J.* **223** 655

Niedner, N.B., & Schwingenschuh, K. (1986) Plasma-tail activity at the time of the Vega encounters, ESA SP-250/3, 419

Oberc, P., Parzydlo, W., Koperski, P., Orlowski, D., & Klimov, S. (1987) Some new features of plasma wave phenomena at Halléy: APV-N observations, ESA SP-278, 89

Raeder, J., Neubauer, F.M., Ness, N., & Burlaga, L.F. (1986) Macroscopic perturbations of the IMF by P/Halley as seen by the Giotto magnetometer, ESA SP-250/3, 173

Richardson, I.G., Cowley, S.W.H., Moore, V., Staines, K., Hynds, R.J., Sanderson, T.R., Wenzel, K.-P., & Daly, P.W. (1986) Spectra and bulk parameters of energetic heavy ions in the vicinity of Comet P/Giacobini–Zinner, ESA SP-250/3, 441

Riedler, W., Schwingenschuh, K., Yeroshenko, Ye.G., Styashkin, V.A., & Russell, C.T. (1986) Magnetic field observations in comet Halley's coma, *Nature* **321** 288

Sagdeev, R.Z., & Shafranov, V.D. (1960) The instability of plasma with the anisotropic velocity distribution in magnetic field. *ZhETF* **39** 181

Sagdeev, R.Z. (1964) Kollektivnye protsessy i udarnye volny v razrezhennoe plazme, In: *Voprosy teorii plazmy* (ed) M.A. Leontovich, Atomizdat, Moskva **4** 3

Sagdeev, R.Z., Shapiro, V.D., Shevchenko, V.I., & Szegö, K. (1986a) MHD turbulence in the solar wind–comet interaction region *Geophys. Res. Lett.* **13** 85

Sagdeev, R.Z., Szabö, F., Avanesov, G.A., Cruvellier, P., Szabö, L., Szegö, K., Abergel, A., Balázs, A., Barinov, I.V., Bertaux, J.-L., Blamont, J., Detaille, M., Demarelis, E., Dul'nev, G.N., Endröczy, G., Gárdos, M., Kanyö, M., Kostenko, V.I., Krasikov, V.A., Nguyen-Trong, T., Nyitrai, Z., Rényi, I., Rusznyák, P., Shamis, V.A., Smith, B., Sukhanov, K.G., Szabö, F., Szalai, S., Tarnopolskii, V.I., Tóth, I., Tsukanova, G., Valniček, B.I., Várhalmi, L., Zaiko, Yu.K., Zatsepin, S.I., Ziman, Ya.L., Zsenei, M., & Zhukov, B.S. (1986b) Television observations of Comet Halley from Vega spacecraft, *Nature* **321** 262

Sagdeev, R.Z., Shapiro, V.D., Shevchenko, V.I., & Szegö, K. (1987) *Plasma phenomena around comets: interaction with the solar wind*, Report KFKI-1987–50/C

Sarris, E.T., Krimigis, S.M., & Armstrong, T.P. (1976) Observations of magnetospheric bursts of high-energy protons and electrons at 35 R_e with IMP 7, *J. Geophys. Res.* **81** 2341

Schwingenschuh, K., Riedler, W., Schelch, G., Yeroshenko,

Ye.G., Styashkin, V.A., Luhmann, J.G., Russell, C.T., & Fedder, J.A. (1986) Cometary boundaries: Vega observations at Halley, *Adv. Space Res.* **6** No. 1, 217

Schwingenschuh, K., Riedler, W., Lichtenegger, H.I.M., Phillips, J.L., Luhmann, J.G., Russell, C.T., Fedder, J.A., Somogyi, A., & Yeroshenko, Ye. (1987) Variability of Comet Halley's coma: Vega-1 and Vega-2 magnetic field observations, ESA SP-278, 63

Somogyi, A.J., Szabö, L., Afonin, V.V., Bánfalvi, A., Erö, J., Faragö, M., Gombosi, T.I., Gringauz, K.I., Kecskeméty, K., Keppler, E., Klimenko, L.I., Kovács, T., Kozma, G., Logachev, Yu.I., Lohonyai, L., Marsden, R., Redl, R., Remizov, A.P., Richter, A., Skuridin, G.A., Stolpovskii, V.G., Szepesváry, A., Szücs, T.-I., Varga, A., Vladimirova, G.A., Wenzel, K.-P., & Windberg, J. (1982) TÜNDE-M apparatus of the SPF unit of the Vega program, In: *Cometary exploration* (ed. T.I. Gombosi), *Proc. of the International Symposium, Budapest* **3** 351

Somogyi, A.J., Afonin, V.V., Erdös, G., Erö, J., Gombosi, T.I., Gringauz, K.I., Kecskeméty, K., Keppler, E., Klimenko, L.I., Kovacs, T., Kozma, G., Logachev, Yu.I., Lohonyai, L., Marsden, R., Remizov, A.P., Richter, A., Skuridin, G.A., Stolpovskii, V.G., Szabö, L., Szegö, K., Szentpétery, I., Szücs, T.-I., Szepesváry, A., Tátrallyay, M., Varga, A., Verigin, M.I., Wenzel, K.-P., Windberg, J., & Vladimirova, G.A. (1985) First results of high energy particle measurements with the TÜNDE-M telescopes on board of the s/c Vega-1 and -2, *Proc. of the International Workshop on Field, Particle, and Wave Experiments on Cometary Missions, Graz,* (eds. K. Schwingenschuh & W. Riedler), 237

Somogyi, A.J., Gringauz, K.I., Szegö, K., Szabö, L., Kozma, Gy., Remizov, A.P., Erö, Jr, J., Klimenko, I.N., T.-Szücs, I., Verigin, M.I., Windberg, J., Cravens, T.E., Dyachkov, A., Erdös, G., Faragö, M., Gombosi, T.I., Kecskeméty, K., Keppler, E., Kovács Jr, T., Kondor, A., Logachev, Y.I., Lohonyai, L., Marsden, R., Redl, R., Richter, A.K., Stolpovskii, V.G., Szabö, J., Szentpétery, I., Szepesváry, A., Tátrallyay, M., Varga, A., Vladimirova, G.A., Wenzel, K.P., & Zarándy, A. (1986) First observation of energetic particles near Comet Halley, *Nature* **321** 285

Sonnerup, B.U.O. (1969) Acceleration of particles reflected at a shock front, *J. Geophys. Res.* **74** 1301

Terasawa, T. (1979) Energy spectrum and pitch angle distribution of particles reflected by MHD shock waves of fast mode, *Planet. Space Sci.* **27** 193

Tsurutani, B.T., Brinca, A.L., Smith, E.J., Thorne, R.M., Scarf, F.L., Gosling, J.T., & Ipavich, F.M. (1986a) MHD waves detected by ICE at distances $> 28 \times 10^6$ km from Comet Halley: Cometary or solar wind origin?, ESA SP-250/3, 451

Tsurutani, B.T., Smith, E.J., Thorne, R.M., Gosling, J.T., & Matsumoto, H. (1986b) Steepened magnetosonic waves in the high β plasma surrounding Comet Giacobini–Zinner, ESA SP-250/3, 457

Verigin, M.I., Gringauz, K.I., Richter, A.K., Gombosi, T.I., Remizov, A.P., Szegö, K., Apáthy, I., Szemerey, T., Tátrallyay, M., & Lezhen, L.A. (1986) Characteristic features of the cometosheath of Comet Halley: Vega-1 and Vega-2 observations, ESA SP-250/1, 169

Wallis, M.K. (1973) Weakly-shocked flows of the solar wind plasma through atmospheres of comets and planets, *Planet. Space Sci.* **21** 1647

Wenzel, K.-P., Sanderson, T.R., Richardson, I.G., Cowley,

S.W.H., Hynds, R.J., Bame, S.J., Zwickl, R.D., Smith, E.J., & Tsurutani, B.T. (1986) *In situ* observation of cometary pick-up ions ≥ 0.2 AU upstream of Comet Halley: ICE observations, *Geophys. Res. Lett.* **13** 861

Winske, D., Wu, C.S., Li, Y.Y., Mou, Z.Z., & Guo, S.Y. (1985) Coupling of newborn ions to the solar wind by electromagnetic instabilities and their interaction with the bow shock, *J. Geophys. Res.* **90** 2713

Wu, C.S., & Davidson, R.C. (1972) Electromagnetic instabilities produced by neutral particle ionization in interplanetary space. *J. Geophys. Res.* **77** 5399

Yumoto, K., Saito, T., & Nakagawa, T. (1986) Long-period HM waves associated with cometary O^+ (or H_2O^+) ions: Sakigake observations, ESA SP-250/1, 249

Part III
Gas

14

Hydrogen coma of comet Halley

K. Hirao

1 INTRODUCTION

As is well known, a comet is composed of many volatile materials which form a long tail when the comet approaches the Sun. Water ice has been thought to be the main material. Ejected water is photodissociated by solar ultraviolet radiation and produces hydrogen atoms in two successive processes. Total production of water molecules of the order of magnitude of $10^{29 \sim 30}$ molecules s^{-1} has been measured. The hydrogen atoms dissociated mainly from the evaporated water form a huge egg-shaped hydrogen coma, its greatest diameter being almost 10^7 km.

This hydrogen coma, however, can be observed only by ultraviolet (hydrogen Lyman-α) radiation. The hydrogen coma was first observed by the OAO-2 spacecraft at the apparition of comet Tago-Sato-Kosaka in 1970. Later, several observations were conducted by satellites as well as by sounding rockets. These data were analysed by many investigators, establishing the general model of the hydrogen coma and determining the production rate of hydrogen atom and hydroxyl radicals. (Keller & Lillie 1974, 1978, Keller & Meier 1980, Meier *et al.* 1976).

In the spacecraft Suisei, which was launched by the Institute of Space and Astronautical Science, Japan, to encounter Halley's comet with a closest approach of 1.5×10^5 km to the nucleus, an ultraviolet camera-/photometer (Ultraviolet Imager: UVI) was installed to study the hydrogen coma of the Halley's comet. It took many ultraviolet pictures of the hydrogen coma

from outside, and also traversed the coma, measuring its hydrogen density distribution versus cometocentric distance. Although the hydrogen Lyman-imaging has been carried out for several comets, this was mainly at the time of the most active phase of the comets. However, a long observing period from the start of formation of a detectable hydrogen coma to its final decay was planned for the 1985/86 Halley apparition. The imaging was carried out from the middle of November 1985 to the middle of April 1986.

The first scientific objective was to monitor the temporal change of evaporation of water molecules from the nucleus, i.e. the variation of water evaporation with changing heliocentric distance of comet Halley. In this observation, the detection of abrupt changes of the luminosity of the coma was also hoped for, although it was realized that this might not be possible, because the rather long time constant for photodissociation of water might smooth over such changes. However, such abrupt changes were indeed observed and were used for the determination of the spin period of the nucleus in the early period of observation. (Kaneda *et al.* 1986a)

The second objective was to determine the density distribution of hydrogen atoms in the coma versus cometocentric distance. Suisei is the first spacecraft which traversed the hydrogen coma to study its inner structure. Although the theoretical study of hydrogen density distribution inside the coma had been carried out by Haser and others (Haser 1957, Keller & Meier 1976), such a measurement has not yet been made.

Suisei is equipped with two independent scientific experiments, the ultraviolet imager and the plasma energy analyzer. As these two experiments are independent of one another in the spin condition of the spacecraft, the ultraviolet imager was operated only in the inbound part of the trajectory before closest encounter.

In this paper, results and their interpretation about the hydrogen coma of comet Halley are reviewed. The instrument used is also briefly described.

2 ULTRAVIOLET IMAGING INSTRUMENT

The hydrogen coma can be detected effectively by the resonantly scattered solar Lyman-α line. To fit the payload capability of the spacecraft, the instrument must be lighter than 10 kg. To meet various constraints, the optical system of the imager was designed as shown in Fig. 1. The hydrogen Lyman-α image of the coma is guided into the camera system by the spinning motion and guiding mirror. The image is focused by the telescopic lens onto the front face of the three-stage multichannel plate after passing through a MgF_2 window. The front face is sputtered with KBr to produce photoelectric conversion of the UV image. The electronic image is then produced and intensified through the channel plate. The electron beam hits a photoluminescence plate and forms a visible picture. The stage between the MgF_2 window and the photoluminescence plate is evacuated, forming an ultraviolet image intensifier. The image detector is a two-dimensional charge coupled device (CCD) with 153×122 pixels, developed by the Nippon Electric Company. The CCD must be cooled at least to $-20°C$ to reduce the dark current to allow detection of the lowest UV signal of 1 kR. The cooling of the CCD was carried out by a fin attached to the case of the CCD which is radiation coupled with the cold bottom plate of the spacecraft. The CCD is illuminated through a fibre plate by the coma image which appears on the photoluminescence plate. As the sensitivity of the CCD is not high enough, and the spacecraft is spin-stabilized, although at a low rate of 0.2 rev min^{-1}, the special technique known as SSS (Spin Synchronous Shift) was adopted. As the spacecraft rotates around

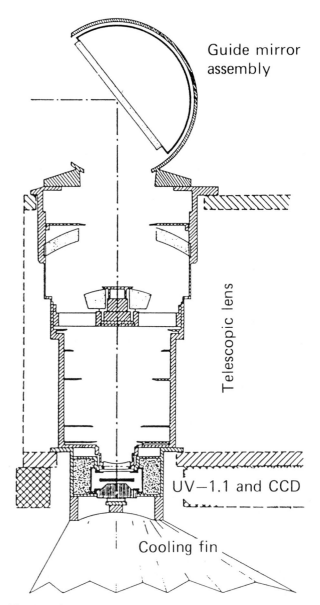

Fig. 1 — Cross-section of the ultraviolet imager (UVI) onboard Suisei.

the spin axis, the view direction of the camera also rotates with the same angular velocity. Therefore, the image on the CCD moves also at the same rate. The electric charge induced in each pixel corresponding to the image is shifted to the next pixel electronically synchronized with the movement of view direction. Therefore, the integration of charge, i.e. light intensity, is carried out automatically on the CCD.

The integration time is designed to be not greater than two seconds.

The imager can be operated in three modes: the first is the ordinary camera mode, the second the photometry mode, and the third the search mode. In camera mode, one can transmit the data of all pixels, or reduced data after data compression by an onboard microprocessor. In the latter case, one can take a snapshot every five minutes. In photometry mode, all pixel intensities are statistically processed by the microprocessor and telemetered as a histogram of brightness of the field of view. The search mode is used to find the object by dividing the whole field of view into 3 × 3, and the mean brightness of each of the nine sectors is transmitted.

The timing of imaging, the setting of the field of view, and selection of the operation mode can be telecommanded directly or by programmable command. The overall sensitivity and the blemishes of the camera system were calibrated before integration by means of the Jobin Yvon LHT 30 UV spectrograph. The inflight calibration was made once by imaging the ultraviolet geocorona after launch.

3 PERIODICITY OF THE BRIGHTNESS OF THE HYDROGEN COMA

The first detection of the hydrogen coma of Halley's comet was on 14 November 1985 and the following few days. However, it became impossible to find the coma during about 10 days after this first detection. From this result, it was expected that the brightness of the coma is quite variable. The first detection might correspond to the strong outburst of the nucleus. After 26 November 1985, the hydrogen coma was observed continuously until the middle of April 1986. Throughout this observation period, the hydrogen coma was always quite variable and sometimes showed outburst-like flare-up phenomena. Although the coma observations were rather intermittent owing to the time-sharing of operation between this imaging and plasma observation, and that of signal reception between Suisei and another spacecraft, Sakigake, the regularity of the brightness change was derived from the results of observation during the first 18 days from 26 November, as shown in Fig. 2. On each line, covering two days, the shaded rectangles indicate the observation periods. The superimposed triangles

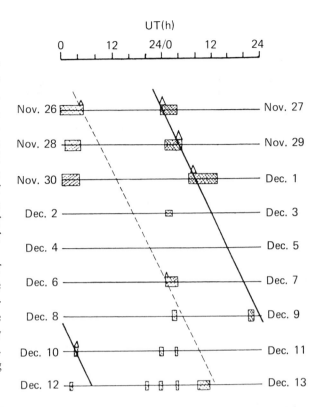

Fig. 2 — Observation history of Suisei's UVI from 26 November to 13 December 1985, showing the observation period (shaded rectangles) and flare-up time (triangles).

indicate the times when fairly bright pixels were observed at the position of the cometary nucleus in the image of the hydrogen coma. The appearance of such bright pixels, which represent outburst or flare-up of the coma, is so regularly distributed that triangles can be connected by the straight lines shown in the figure. It suggests that the flare-up occurs regularly with period of 2.2 days. From this fact, this period is concluded to correspond with the spin period of the nucleus of comet Halley. This period persisted to the end of observation by Suisei's UVI in the middle of April 1986. This is shown in Fig. 3, covering the whole period of UV observation during November 1985 and April 1986. This figure shows the time sequence of the activity peak corresponding to the strongest jet, S1, which will be explained later, in the cometary hydrogen coma. Time intervals between 1985 December 11.94 (UT) and 1986 February 22.85 are not shown, because there was no observation by the UVI because of severe interference from the solar radiation

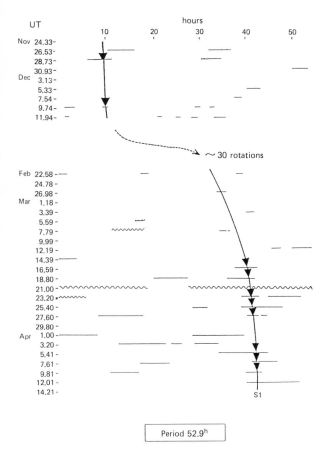

Fig. 3 — Activity diagram of comet Halley during November 1985 and April 1986. Triangles show peaks of activity.

7.37 days was found from the results of ground-based observations of the CN emission (Samarasinha *et al*. 1986). The discrepancy between these two periods for nucleus spin is at present unexplained. It is presumed that these two periods actually exist as a spinning motion of 2.2 days and nutation of 7.37 days. These two periods should be found in all observations. However, it is probable that the longer period is detected only in CN observations owing to the signal to noise problem, which might be caused by the nutation motion of the spin axis of the nucleus. If the CN rich jet is located in a different location to the H_2O jets, or the CN to H_2O ratio is different in places, then the exposure to solar radiation may cause differences in the production rate of these gases by the nutation motion. As it is well known that the photodissociation time constant of water is as long as one day at 1 AU, such an abrupt change is not understandable. But, if some of the hydrogen can be produced from some material which is easily dissociated such as an organic substance ejected from the nucleus, this phenomenon can be attributed to the spinning motion. Other candidates have also been proposed by Shimizu (1987). Such organic material candidates have been observed by the PUMA mass spectrometer on board the Vega-1 spacecraft (Kissel & Krueger 1987).

4 COMETOCENTRIC DISTRIBUTION OF HYDROGEN IN THE COMA

The imaging of the hydrogen coma became almost meaningless when the spacecraft approached the coma closely. Therefore, the UVI on board Suisei was switched to the photometric mode from the imaging mode. The photometry was carried out by pointing the imager to the direction of $\alpha_o = 11h\ 20m$ and $\delta_o = 5°\ S$ where there were no bright UV stars in the field of view of the photometer. This photometry observation was carried out during about 8 hours up to the time of closest approach. After that, Suisei carried out the plasma measurements as scheduled, owing to the exclusiveness of the two experiments. From this observation, the brightness distribution of hydrogen Lyman-α intensity inside the coma was measured, which can be converted to a density distribution for hydrogen atoms inside the coma. The observed result is shown in Fig. 4. The cometocentric distribution of

to the UVI, due to the small solar elongation angles. As the abscissa of this figure is limited to 52.9 hours, a phenomenon which appears in parallel to the coordinate of this figure can be designated as periodic, with a recurrence time of 52.9 hours. The triangles show the time of appearance of the activity peaks. Therefore, predominant activity happens quite regularly at 52.9 hour intervals. During the middle of November 1985 and the middle of March 1986, the position of the peak moves because of the change in the true anomaly of the comet. Before and after this period, the periodicity is excellent, indicating that the comet nucleus spins continuously during perihelion passage. The same spin rate of the nucleus, 2.2 days, is also found by Vega (Sagdeev *et al*. 1986) and IUE (Weaver *et al*. 1986) as well as by ground-based observations (Sekanina & Larson 1986). Another period such as

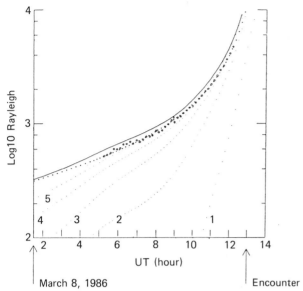

$$Q_{20}(t) = \begin{cases} Q(1 - e^{-T/\tau_{H_2O}})e^{-(t-t_0-T)/\tau_{H_2O}} & \text{for } t > t_0 + T \\ Q[1 - e^{-(t-t_0)/\tau_{H_2O}}] & \text{for } t_0 + T \geqslant t \geqslant t_0 \\ 0 & \text{for } t_0 \geqslant t \end{cases}$$

$$Q_8(t) = \begin{cases} \dfrac{Q}{(\tau_{H_2O} - \tau_{OH})}[\tau_{OH}(1 - e^{T/\tau_{OH}})e^{-(t-t_0)/\tau_{OH}} \\ \qquad - \tau_{H_2O}(1 - e^{T/\tau_{H_2O}})e^{-(t-t_0)/\tau_{H_2O}} & \text{for } t > t_0 + T \\ \dfrac{Q}{\tau_{H_2O} - \tau_{OH}}[(\tau_{H_2O} - \tau_{OH}) + \tau_{OH}e^{-(t-t_0)/\tau_{OH}} \\ \qquad - \tau_{H_2O}e^{-(t-t_0)/\tau_{H_2O}}] & \text{for } t_0 + T \geqslant t \geqslant t_0 \\ 0 & \text{for } t_0 \geqslant t \end{cases}$$

where t : present time

t_0 : time of generation of jet

T : duration of jet

Q : production rate of H_2O by jet (molecules s^{-1})

τ_{H2O} : characteristic time of photodissociation of H_2O

τ_{OH} : characteristic time of photodissociation of OH

Fig. 4 — Cometocentric distribution of the Lyman-α intensity. (open circles) The asterisks represent calculated values as described in the text. The dotted curves labelled 1–5 represent the intensities caused by outbursts of one to five ejections, respectively. Solid curve show the ultimate distribution calculated by superimposing many outbursts of the same magnitude.

hydrogen atoms in the coma can be considered to be inversely proportional to the square of the cometocentric distance if the coma expands freely without any decay of species, but this is found to be untrue. Haser proposed a two-step exponential production–decay model, which suggested uniform outward expansion with specific characteristic scale lengths for H_2O and OH (Haser 1957). In the present analysis, the time constants of photodissociation of H_2O and OH, as well as the effect of periodic ejection of source material by the spinning motion of the nucleus, have been taken into account. A simple model of hydrogen distribution including these effects is introduced here.

In this model, the parent molecules of hydrogen, H_2O, and OH, are assumed to be produced from a point source owing to the low ejection velocity from the nucleus. The production functions of hydrogen from H_2O and OH are defined as $Q_{20}(t)$ and $Q_8(t)$ respectively. The number 20 and 8 express the typical velocities of hydrogen atoms produced by the photo-dissociation of H_2O and OH by the solar ultraviolet radiation in kilometres per second.

The following formulae are then introduced:

By using these production functions at the origin, $Q_{20}(t)$ and $Q_8(t)$, the hydrogen density at cometocentric distance R at time t, $n(t, R)$ can be derived as

$$n(t,R) = \frac{Q_{20}(t - R/v_{20})}{4\pi R^2 v_{20}}e^{-R/\tau_H v_{20}} + \frac{Q_8(t - R/v_8)}{4\pi R^2 v_8}e^{-R/\tau_H v_8}$$

where v_{20} and v_8 are the velocities of hydrogen atoms produced by the photodissociation of H_2O and OH amounting to 20 and 8 km s^{-1} respectively, and τ_H is the lifetime of a hydrogen atom which will be ionized by the solar wind and ultraviolet radiation.

By integrating the above formula along an adequate line of sight, the line density can be calculated. Multiplying this value the factor of resonance scattering, g, of the solar Lyman-α line, one can compare the calculated intensity with that of the observed brightness if the line density is not optically thick Fig. 4 shows the relation between observed intensity versus time at the time of encounter of Suisei and comet Halley. Using the Haser model, by taking into account the relative positions of Suisei and the nucleus and the production rate of water, which was predicted by a working group of the Inter Agency Consultative Group (IACG), the calculated values are also shown in Fig. 4. The observed values shown by small circles seem to correspond well with the calculated values.

In this figure, the dotted line '1' expresses the expected value of intensity by one ejection from a jet, and successive lines correspond to the intensity caused by the accumulation of hydrogen from several ejections, as shown by the numbers. The general feature of the cometocentric distribution of hydrogen atoms can be explained by both models. However, the observed distribution shows some characteristic fine structure. The observed Lyman-α intensity shows a small increase in several positions, suggesting local increases in the production rate of hydrogen atoms. These increases appear rather regularly in time or distance, and although they are not yet fully analyzed, are presumed to be caused by the spinning motion of the nucleus. As the characteristic time of appearance of a small increase is almost one hour, the separation distance of these increases is roughly 25 000 km. Thus, the described small change can be considered to be caused by propagation of the hydrogen atom excess produced by the outburst-like event as described previously. This will be discussed in more detail in the next section by combining it with the result of other observations.

5 BRIGHTNESS CHANGE ACCOMPANIED BY OUTBURST

A 58-hour photometry mode observation of the hydrogen coma was carried out from 0300 UT on 21 March to 1300 UT on 23 March 1986. It completely covered one rotation of the nucleus of the comet. The observed result is shown in Fig. 5. Two processed curves are shown: one is the time variation of the Lyman-α intensity averaged over all pixels of the obtained histogram and the other is that obtained by summation of the 100 brightest pixels. Again, one can see that the brightness is highly variable, probably suggesting the existence of rapidly dissociated hydrogen atoms from ejected material from the nucleus. In the middle of the observation period, the field of view of the UVI was slightly adjusted to catch comet Halley in the best position. The solar elongation was, therefore, reset, and the background intensity was reduced, especially in the data for the averaged value

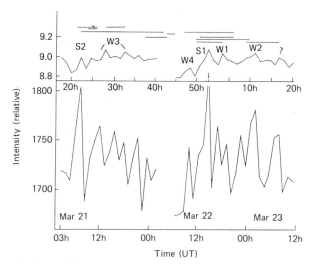

Fig. 5 — Brightness variation of the hydrogen coma during one rotation of the nucleus, obtained by photometric mode observation.

of all pixels. Two strong sources (S1 and S2) and four weak sources (W1, W2, W3, and W4) have been identified by many recurrent observations which are indicated by horizontal lines at the top of the figure. If the occurrence time of the strongest source, S1, is taken as the time origin, 0 h, corresponding to the spin of the nucleus, S2 occurs at 22 h and W1, W2, W3 and W4 occur at around 3 h, 11 h, 30 h, and 48 h, respectively. Between 33 h and 48 h of cometary time, the activity of the comet is much less than at other times during each rotation. At around 17h cometary time, another source may exist, but it is not confirmed. From the result of Giotto's HMC, seven jets have been identified. (Keller *et al.* 1986). Thus, the intensity increases observed by the Suisei UVI are presumed to correspond to the visibly identified jets, although there might exist many other sources which were not seen by Giotto. In this case, rapidly dissociated hydrogen is again necessary to explain these phenomena.

If the excess hydrogen thus produced travels with a certain velocity, expanding outwards, the small increase of brightness observed in the cometocentric distribution of brightness described in the previous section can be interpreted as follows.

By differentiating the data shown in Fig. 5 with time, the time of appearance of a peak can be determined. Then the position of this peak in cometocentric distance can be calculated as shown in the second column of Table 1. These peaks can be identified as corresponding with the outburst sources described in the previous section, by considering the relative time differences of each peak and the time of occurrence of each burst as shown in the first column of Table 1. It has been recognized that there appear, sometimes, shell-like structures in the images of the hydrogen coma. One image which was taken at 1209 UT on 25 February 1986, when the activity of the hydrogen coma was almost in phase with that at closest approach, 1306 UT on 8 March 1986, showed such a shell structure. A comparison between the shell structure in this image and the peak structure described above has been carried out (Kaneda *et al.* 1986b). For this comparison, the observed peak position should be corrected, because the group of peaks was obtained over several hours, whereas the image was taken in two seconds. Therefore, the peak position should be corrected by adopting an adequate expansion speed for the hydrogen atoms in this Lyman-α intensity increase. It was found that the structure showed excellent coincidence with peak phenomena as shown in the third and fourth columns of Table 1, if the expansion velocity of hydrogen atoms was assumed to be around 11 km s^{-1}. The possible candidate of rapidly dissociating material for the hydrogen, the CH radical, is quite favourable, with its short dissociation time of about 100 seconds and resulting velocity of about 11 km s^{-1}. (Kaneda *et al.* 1986b). This radical could be produced by various organic materials found by Vega.

Table 1 — Comparison of the positions of hydrogen intensity increases obtained by photometry and imaging modes. Distances from the nucleus shown in the table are expressed with unity of 10^4 km

Symbol of source	Photometry peak position (observed)	Photometry shell position (corrected)	Image shell position (observed)
S2	no data	–	32
?	50	55	54
–	60	68	70
W2	75	85	84
W1	92	105	105
S1	110	125	–
W4	120	136	–
W3	180	204	–

6 THE PRODUCTION RATE OF WATER VERSUS HELIOCENTRIC DISTANCE

The activity of a comet is usually expressed by the water production rate from the nucleus. The chemical abundance of a comet nucleus has been modelled. The main constituent, a particularly volatile one, is water ice, as proposed by Whipple (1950) as the reasonable primordial remnant of the Solar System. The activity has been measured from the ground by observing the OH line brightness. When spaceborne instruments became available, the hydrogen coma was observed directly by means of sounding rockets as well as satellites and it was found to be extremely large compared with the visible one. The prediction of the water production rate of comet Halley 1986 was carried out by members of the IACG working group before the closest encounters of the 'Halley armada' in March 1986 (Divine *et al.* 1986).

As one of the main objectives of the Suisei spacecraft, the water production rate derived from its ultraviolet observations at various heliocentric distances of comet Halley, is shown in Fig. 6. The method used in the present derivation is to model the hydrogen cloud as the summation of six jets, two strong ones and four weak ones, with a relative water production rate of 2:1, and to calculate the expected brightness contour maps for various water production rates. An example of a calculated contour map of Lyman-α brightness is shown in Fig. 7 (Kaneda *et al.* 1987). These calculated contours were compared with the observed brightness distribution, and permitted determination of the water production rate at the nucleus. Circles and crosses in Fig. 6 are the water production rates corresponding to pre-perihelion and post-perihelion passage of the comet, respectively. The dashed lines show the rate predicted by the IACG working group described above, multiplied by a factor of two. G and V show the values obtained by Giotto (Krankowsky *et al.* 1986) and Vega (Gringauz *et al.* 1986), respectively. Data from DE-1 (Craven *et al.* 1986) and IUE (Festou *et al.* 1986) are also superimposed.

From this figure, the water production rate is much higher during the post-perihelion phase than pre-perihelion. Since the observed water production rate shows a variation by factor of $2 \sim 3$, it is not easy to

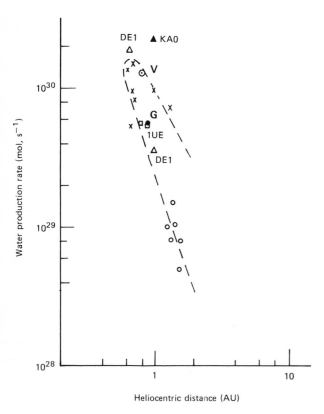

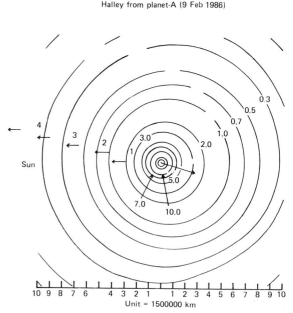

Fig. 7 — Calculated contour map of the Lyman-α brightness of the hydrogen coma viewed from Suisei for comet Halley on 9 February 1986. Intensity is expressed in kR.

Fig. 6 — The water production history of Comet Halley during the 1985/86 apparition. The dashed curve is that predicted by IACG, after multiplication by a factor of two. Other observed results (Giotto(G), Vega(V), IUE(□), DE-1(△), KAO(▲)) are also superimposed. Open circles and crosses represent UVI data obtained in the pre-perihelion and post-perihelion phases, respectively.

determine the general features of cometary activity for the pre-perihelion and post-perihelion phases. Observed results for Comet Halley 1986 show a large pre-perihelion and post-perihelion asymmetry of more than five at a heliocentric distance of 1 AU and less. Even 10 or more is reported by KAO observers. (Weaver *et al.* 1986). The increase of the water production rate near perihelion seems to be quite large. Qualitatively speaking, the water production rate at first increases rather slowly around the time of perihelion passage, and then increases rapidly by about four times, reaching maximum at around 0.8 AU post-perihelion, but decreases with a similar slope pre-perihelion. The observed values show much more variety than the IACG model curve, particularly near perihelion passage. Thus, the change of activity of the comet near perihelion passage is quite complex, and it probably depends to a great extent on the surface condition of the cometary nucleus.

7 CONCLUDING REMARKS

The *in situ* observations of comet Halley in 1986 by many spacecraft provided much new information for cometary science, including the results obtained by the UVI onboard Suisei.

The hydrogen coma was found to be very active and variable, which had not been expected. From its strong 'breathing', the spin period of the cometary nucleus was determined as 2.2 days. This was the first result, when the comet was still far beyond the Earth's orbit. The period was continuously recorded throughout the period of UVI observation.

The hydrogen atom distribution versus cometocentric distance was obtained by traversing inside the hydrogen coma. The averaged distribution is interpreted by a summation of the hydrogen dissociated from water molecules ejected by several jets, whose existence was confirmed by UVI. These jets were observed also by the cameras on board Giotto and Vega. Several small increases of hydrogen intensity superimposed on the average distribution were clarified.

The seven jet sources described were identified by continuous observation in the photometric mode.

To explain these phenomena — rapid changes in brightness and local increases of hydrogen atom density inside the coma — a second source of hydrogen, which must be rapidly photodissociated, was inferred. The second source is not necessarily abundant, but it is definitely necessary. The source is presumably organic material such as CH which is one of the parent molecules, and is easily dissociated from the various organic materials found by other spacecraft.

The time variation in the water production rate from the comet was also determined by UVI observation. The time variation is more complex than expected from the IACG model. There is a large asymmetry in the pre-perihelion and post-perihelion phases, suggesting that the water production rate is controlled not only by the heliocentric distance but also by the thermal history of the non-uniform nucleus surface during the perihelion passage of the comet.

From the hydrogen Lyman-α observations, various phenomena have been clarified. It must be emphasized that the analysis of these phenomena was supported by the results obtained by other spacecraft. Otherwise, it would have been very difficult to come to the above conclusions.

REFERENCES

Craven, J.D., Frank, L.A., Rairden R.L., & Dvorsky L.R.; (1986) The hydrogen coma of Comet Halley before perihelion: Preliminary observation with Dynamics Explore 1, *Geophys. Res. Letters* **13** 873–876

Divine, N., Fechtig, H., Gombosi, T.I., Hanner, M.S., Keller, H.U., Larson, S.M., Mendis, D.A., Newburn Jr., R.L., Reinhard, R., Sekanina, Z., & Yeomans, D.K. (1986) The Comet Halley dust and gas environment, *Space Sci. Rev.*

Festou, M.C., Feldman, P.D., A'Hearn, M.F., Arpigny, C., Cosmovici, C.B., Danks, A.C., McFadden, L.A., Gilmozzi, R., Patriarchi, P., Tozzi, G.P., Wallis, M.K., & Weaver, H.A.; (1986) IUE observations of Comet Halley during the VeGa and Giotto encounters. *Nature* **321** 361–363

Gringauz, K.I., Gombosi, T.I., Remizov, A.P., Apáthy, I., Szemerey, I., Verigin, M.I., Denchikova, L.I., Dyachkov, A.V., Keppler, E., Klimenko, I.N., Richter, A.K., Somogyi, A.J., Szegö, K., Szendro, S., Tátrallyay, M., Varga, A., & Vladimirova, G.A.; (1986) *Nature* **321** 282–285

Haser L.; Distribution d'intensite dans la tête d'une comete. *Bull. Acad. Roy. Belgique, Classe des Sciences* **43** 740–750

Krankowsky, D., Laemmerzahl, P., Herrwerth, I., Dolder, U.,

Herrmann, U., Schulte, W., Berthelier, J.J., Illiano, J.M., Hodges, R.R., & Hoffman, J.H.; (1986) *In situ* gas and ion measurements at Comet Halley. *Nature* **321** 326–329

Kaneda, E., Hirao, K., Takagi, M., Ashihara, O., Itoh, T., & Shimizu, M.; (1986a). Strong breathing of the hydrogen coma of Comet Halley. *Nature* **320** 140–141

Kaneda, E., Hirao, K., Shimizu, M., & Ashihara, O.; (1986b) Activity of Comet Halley observed in the ultraviolet *Geophys. Res. Letters.* **13** 833–836

Kaneda, E., Hirao, K., Ashihara, O., Shimizu, M., Itoh, T., & Oda, M.; (1987) Lyman alpha observations of Comet Halley by Suisei. *Special Report of ISAS.* **19** 111–124. (in Japanese)

Keller, H.U., & Lillie, C.F.; (1974) The scale length of OH and the production rate of H and OH in Comet Bennett (1970). *Astron. Astrophys.* **34** 187–196

Keller, H.U., & Lillie, C.F.; (1978) Hydrogen and hydroxyl production rate of comet Tago-Sato-Kosaka (1969). *Astron. Astrophys.* **62** 143–147

Keller, H.U., & Meier, R.R.; (1976) A cometary hydrogen model for arbitrary observational geometry. *Astron. Astrophys.* **52** 272–281

Keller, H.U., & Meier, R.R.; (1980) On the Lα isophotoes of Comet West (1976). *Astron. Astrophys.* **81** 210–214

Keller, H.U., Arpigny, C., Barbieri, C., Bonnet, R.M., Cazes, S., Coradini, M., Cosmovici, C.B., Delamere, W.A., Huebner, W.F., Hughs, D.W., Jamar, C., Malaise, D., Reitsema, H.J., Schmidt, H.U., Schmidt, W.K.H., Seige, P., Whipple, F.L., & Wilhelm, K.; (1986) First Halley multicolour camera imaging results from Giotto. *Nature* **321** 320–326

Kissel, J., & Krueger, F.R.; (1987) The organic component in dust from Comet Halley as measured by the PUMA mass spectrometer on board VeGa 1. *Nature* **326** 755–760

Mendis D.A., Opal, C.B., Keller, H.U, Page, T.L., & Carruthers, G.R. (1976). Hydrogen production rate from Lyman-α images of Comet Kohoutek (1973). *Astron. Astrophys.* **52** 283–290

Sagdeev, R.Z., Szabó, F., Avanesov, G.E., Cruvellier, P., Szabó, L., Szegö, K., Abergel, A., Balazs, A., Barinov, I.V., Bertaux, J.L., Blamont, J., Detaille, M., Demarelis, E., Dul'nev, G.N., Endroczy, G., Gardos, M., Kanyo, M., Kostenko, V.I., Krasikov, V.A., Nguyen-Trong, T., Nyitrai, Z., Reny, I., Rusznyak, P., Shamis, V.A., Smith, B., Sukhanov, K.G., Szabó, F., Szalai, S., Tarnopolsky, V.I., Toth, I., Tsukanova, G., Valnicek, B.I., Varhalmi, L., Zaiko, Yu.K., Zatsepin, S.I., Ziman, Ya.L., Zsenei, M., & Zhukov B.S.; (1986) Television observations of Comet Halley from VeGa spacecraft. *Nature* **321** 262–266

Samarasinha, N.H., A'Hearn, M.F., Hoban, S., & Kinglesmith, D.A.; (1986) CN jets of Comet P/Halley — rotational properties. *Proc. 20th ESLAB symposium on the exploration of Halley's Comet.* **1** 487–491

Sekanina, Z., & Larson, S.M.; (1986) Dust jets in Comet Halley observed by Giotto and from the ground. *Nature* **321** 357–361

Shimizu M. (1987); Halley's environment observed by Suisei, *ESA SP-278* 229

Weaver, H.A., Mumma, M.J., Larson, H.P., & Davis, D.S.; (1986) Post-perihelion observations of water in Comet Halley. *Nature* **324** 441–444

15

Jet and shell structures in the cometary coma: modelling and observations

N. I. Kömle

1 INTRODUCTION

One of the most conspicuous features of a cometary atmosphere is its inhomogeneous structure and its time-dependent behaviour. In this review, I shall point out what we have learnt from observations, during the recent comet Halley return, about two important structural elements of a cometary atmosphere: 'jets' and 'shells'. The latter are also sometimes termed as 'haloes'. Both features are seen in neutral gas and in dust observations.

It is not intended to give a complete survey of observations here. Rather, the gas-dynamical aspects of the currently existing theoretical models are emphasized, and only some representative observational material is shown for illustration. Further information regarding the large-scale structure of the cometary dust distribution can be found in the reviews by L. Masonne and Z. Sekanina in Volume 2 of this work.

2 BASIC PROPERTIES OF STATIONARY GAS JETS

2.1 Nozzle flow into vacuum

Although jet-like structures in the coma of comet Halley have been observed to consist both of dusty and of gaseous components, it is beyond any doubt that the driving agent must be a locally increased gas pressure. Such a pressure increase may be produced both by enhanced sublimation on the nucleus surface and by outgassing of sub-surface gas reservoirs through narrow holes or cracks. The flow which

develops from such a situation may be quite similar to rocket exhaust plumes. Therefore I shall first describe a simple analytic model of gas outflow from a reservoir through a nozzle (or thin hole) into a vacuum (Boettcher & Legge 1980, 1981, Koppenwallner et al. 1986). Although rocket and satellite nozzles are much smaller than the proposed 'comet nozzles', the formulae describing the model are largely independent of the absolute size of the nozzle, and can likewise be used to describe the cometary jets close to the nucleus.

Figs. 1 and 2 illustrate the different flow regimes and their basic geometry. If we are dealing with the outflow of a gas at rest in a reservoir through a thin hole or nozzle into vacuum, there exists always a transition region which separates the continuum (collision-dominated) flow regime from the free molecular flow further away from the nozzle. Within the continuum flow region two different flow regimes are identified: an isentropic core expansion region extending to a certain angle θ_o and a non-isentropic boundary layer expansion region, where the gas properties are influenced by the friction between the nozzle wall and the gas. This region extends to a larger angle θ_{lim}. The structure of the flow field is uniquely determined (a) by the geometry parameters r^* (nozzle throat radius), r_E (nozzle exit radius), θ_E (nozzle exit angle), and (b) by the temperature T_o and the pressure p_o of the gas in the reservoir. Further, the adiabatic index γ, the molecular weight μ, and the viscosity

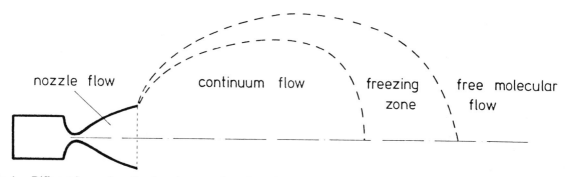

nozzle flow   continuum flow   freezing zone   free molecular flow

Fig. 1 — Different flow regimes forming when a gas flows through a nozzle from a reservoir into vacuum (taken from Boettcher & Legge 1980).

coefficient λ of the gas enter the formulae. R_g is the universal gas constant.

Then for the isentropic core the outflow velocity at the nozzle throat is,

$$u^* = \sqrt{\left(\frac{2\gamma}{\gamma+1} \frac{R_g}{\mu} T_o\right)} \qquad (1)$$

The maximum flow velocity u_{max} is reached in the isentropic core for $r \to \infty$, and is given by the relation,

$$u_{max} = \sqrt{\frac{2\gamma}{\gamma-1} \frac{R_g}{\mu} T_o} \qquad (2)$$

From r^*/r_E it is easy to calculate the corresponding nozzle area ratio,

$$\frac{A^*}{A_E} = \left(\frac{r^*}{r_E}\right)^2 \qquad (3)$$

The Mach number at the nozzle exit (M_E) is obtained by solving the following implicit equation for M_E:

$$\frac{A^*}{A_E} = \left(\frac{\gamma+1}{2}\right)^{\frac{\gamma+1}{2(\gamma-1)}} M_E \left(1 + \frac{\gamma-1}{2} M_E^2\right)^{-\frac{\gamma+1}{2(\gamma-1)}} \qquad (4)$$

Next the Prandtl–Meyer angle ν must be calculated for the Mach numbers $M = M_E$ and $M = \infty$:

$$\nu_E = \sqrt{\left(\frac{\gamma+1}{\gamma-1}\right)} \arctan\sqrt{\left(\frac{\gamma-1}{\gamma+1}(M_E^2-1)\right)} - \arccos\frac{1}{M_E} \qquad (5a)$$

$$\nu_\infty = \frac{\pi}{2}\left(\sqrt{\left(\frac{\gamma+1}{\gamma-1}\right)} - 1\right) \qquad (5b)$$

The limiting expansion angle of the jet is then given by,

$$\theta_{lim} = \nu_\infty - \nu_E + \theta_E \qquad (6)$$

From the given values for p_o and T_o the corresponding density ρ_o in the reservoir is also known via the ideal gas law. The density at the nozzle throat (ρ^*) follows from the isentropic relationship,

$$\frac{\rho^*}{\rho_o} = \left(1 + \frac{\gamma-1}{2}\right)^{-\frac{1}{\gamma-1}} \qquad (7)$$

keeping in mind that the Mach number at the nozzle throat is equal to 1.

To obtain the hydrodynamic flow variables in an arbitrary point (r, θ) of the axisymmetric free flow field outside the nozzle one can make a separation of variables for the density ρ. As the nozzle throat gas density ρ^* and the limiting expansion angle θ_{lim} are now known, ρ may be expressed in the following analytic form:

$$\rho(r,\theta) = \rho^* \cdot A_p \left(\frac{r^*}{r+r_E}\right)^2 \qquad (8)$$

$$\begin{cases} \left[\cos\left(\frac{\pi\theta}{2\theta_{lim}}\right)\right]^{\frac{2}{\gamma-1}} & \text{for } \theta < \theta_o \\ e^{-c_p(\theta-\theta_o)} & \text{for } \theta_o < \theta < \theta_{lim} \end{cases}$$

The constants A_p, θ_o and c_p are derived from mass conservation considerations, and depend also on the viscosity of the gas that determines the boundary layer thickness. For the complete expressions see Boettcher & Legge (1980). Once the constants are determined, the density ρ for an arbitrary point in the flow field is calculated from (8), and the corresponding Mach number in the isentropic core is given by

$$M^2 = \frac{2}{\gamma - 1}\left[\left(\frac{\rho}{\rho_o}\right)^{-(\gamma - 1)} - 1\right].$$ (9)

The flow variables normalized to their values in the reservoir can now be expressed very conveniently in terms of the Mach number. Because, by definition, the flow velocity in the reservoir is zero, the sound speed in the reservoir, defined by $c_o = (\gamma p_o/\rho_o)^{1/2}$, is used as a normalizing factor for the flow velocity u. For completeness we give here the relations for ρ, p, T and u as a function of the Mach number for the flow in the isentropic core:

$$\frac{\rho}{\rho_o} = \left(1 + \frac{\gamma - 1}{2}M^2\right)^{-\frac{1}{\gamma - 1}}$$ (10a)

$$\frac{p}{p_o} = \left(1 + \frac{\gamma - 1}{2}M^2\right)^{-\frac{\gamma}{\gamma - 1}}$$ (10b)

$$\frac{T}{T_o} = \left(1 + \frac{\gamma - 1}{2}M^2\right)^{-1}$$ (10c)

$$\frac{u}{c_o} = M \cdot \left(1 + \frac{\gamma - 1}{2}M^2\right)^{-1/2}$$ (10d)

The flow in the boundary layer can be described by similar formulae, the main difference being that the reservoir values ρ_o, p_o, T_o, and c_o must be replaced by their corresponding values at the nozzle exit close to the nozzle wall (see Fig. 2). These expressions will not be described in detail here, because the mass flux through the boundary layer gives only a minor contribution compared to the flow in the isentropic core, and it is not expected that these effects play a major role in connection with the jets seen at comet Halley.

The essential properties of a gas outflow from a nozzle into vacuum are now summarized as follows:

(a) As the gas leaves the nozzle exit, it expands laterally within a rather short distance and forms then a cone-like structure with essentially straight streamlines.

(b) Depending upon the adiabatic index γ, the limiting expansion angle θ_{lim} and the limiting

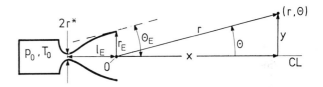

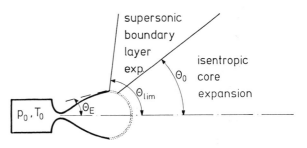

Fig. 2 — Geometry and model parameters for a supersonic flow from a reservoir into vacuum (taken from Boettcher & Legge 1980).

velocity u_{max} are practically reached between 10 and 100 nozzle radii r_E (Dankert 1985). In Fig. 3 the limiting expansion angle θ_{lim} for gases with various γ-values is plotted as a function of M_E. As can be seen from this plot, gases with large γ produce narrow cones, while gases with small γ expand into a wide solid angle and — in extreme cases — can even fill the whole space around the

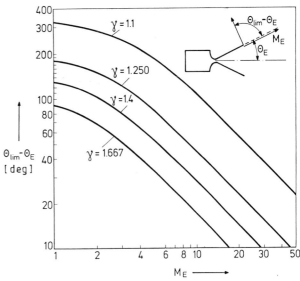

Fig. 3 — Limiting expansion angle θ_{lim} as a function of the Mach number at the nozzle exit (M_E) and the adiabatic index γ (adapted from Boettcher & Legge, 1980).

nozzle. Water vapour, with $\gamma = 1.33$, has a limiting expansion angle $\theta_{\text{lim}} \approx 150°$ if it expands through a narrow hole into vacuum.

(c) The gas density ρ decreases approximately as $1/(r + r_E)^2$ from the nozzle exit outward along any constant angle θ. Note that this expansion law implies a $1/r^2$ behaviour for $r \gg r_E$.

Thus the overall appearance of a gas jet expanding from a reservoir into vacuum would be a gas cone with maximum density at the centreline and a gradual density decrease toward the boundaries. The flow lines become radial close to the nozzle exit. I recall that the formulae given in this section are valid only in the continuum domain of a free jet flow. An approximative model of the flow in the transition region and the free molecular regime is described in Boettcher & Legge (1981). However, it appears unlikely that in the jets observed close to the nucleus of comet Halley the gas is in the free molecular regime, and therefore it is not necessary to treat this case here in more detail.

2.2 Nozzle flow into a background gas

Some additional features must be taken into account if a jet expands not into a vacuum but into a low-pressure background gas, which is dense enough to be treated as continuum. In this case the jet is confined by the external pressure and close to the jet boundary a shock system forms which — because of its shape — is called the 'barrel shock'. Another typical structure that may exist in such a flow field is the so-called 'Mach disk', where the supersonic plume expansion flow becomes again subsonic by a shock transition. The position of the Mach disk depends only upon the ratio between the reservoir pressure p_0 and the background pressure p_1, and is given by the expression (Ashkenas & Sherman 1966),

$$x_M = 2r^* \cdot 0.67 \sqrt{\frac{p_0}{p_1}} \qquad (11)$$

It should be noted that relation (11) is independent of the adiabatic index γ, and its validity was experimentally confirmed for a very wide range of pressure ratios ($15 \leq p_0/p_1 \leq 17\,000$). A very convenient method to calculate the structure of such flow fields numerically is the method of characteristics (see Dettleff 1981).

A simple analytical model predicting the shape of a jet boundary in the case of a comet has been given by Sagdeev et al. (1985). The geometry used in their model is illustrated in Fig. 4. Utilizing a spherically symmetric adiabatic gas expansion outside the jet, they find for the ratio of gas pressures inside and outside the jet the formula,

$$\frac{P_{\text{inside}}}{P_{\text{outside}}} \sim \frac{r^{2\gamma}}{(r^2\theta^2)^{\frac{2\gamma}{\gamma+1}}} \qquad (12)$$

Because there must be pressure equilibrium along the jet boundary $P_{\text{inside}} = P_{\text{outside}}$ must hold there. This gives for the expansion angle of the jet as a function of the radial distance from the jet source

$$\theta_{\text{boundary}} \sim r^{-\frac{\gamma-1}{2}} \qquad (13)$$

Such a simple model cannot, however, tell us anything about the density and pressure distribution inside the gas jet.

2.3 Numerical models

More sophisticated two-dimensional (axisymmetric) numerical calculations have recently been performed by Kitamura (1986, 1987) for a dusty gas jet and by Kömle & Ip (1987) for gas jets with different interaction possibilities. The main result from Kitamura's (1986) calculations is that a narrow gas/dust jet emanating from the cometary surface develops into a broad cone with increased density close to the boundary and a 'hole' along the jet axis. Only one dust particle size is used (0.65 μm), and it is found that

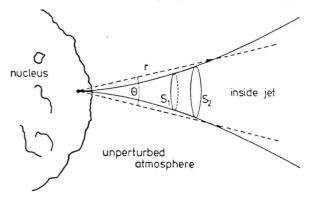

Fig. 4 — Geometry of the analytic jet model proposed by Sagdeev et al. (1985) (taken from Sagdeev et al. 1985).

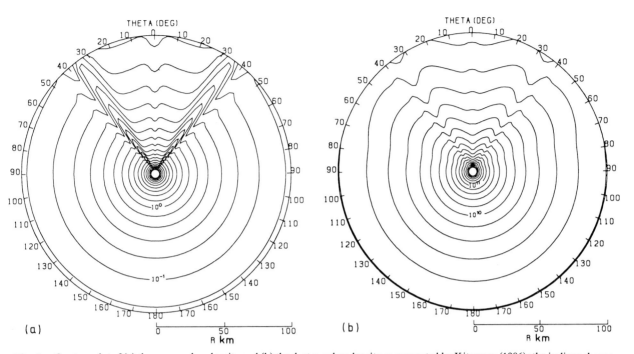

Fig. 5 — Contour plot of (a) the gas number density and (b) the dust number density as computed by Kitamura (1986); the isolines show a dusty gas jet of 10° half-width superimposed on a spherical symmetric outflow from the cometary nucleus. The symmetry axis of the jet corresponds to $\theta = 0$ (taken from Kitamura 1986).

these dust particles also move toward the boundary of the jet, leading to a density decrease close to the jet axis (Fig. 5a, b). The increased gas density along the boundary of the jet is seen even more clearly if one uses narrower source distributions, as in Kömle & Ip (1987).

From the gasdynamical point of view the behaviour of the density as described above is not surprising. It corresponds to the expansion of a free jet against a lower pressure background gas. The density increase close to the jet boundary is due to the interaction of the expanding jet gas with the background gas. It has nothing to do with the presence of the dust, because it is also seen in pure gasdynamic models. This view is confirmed by a calculation which assumed a 'ringlike' density distribution at the comet surface. In this case one can see a focusing effect due to the counterstreaming of the gas from different sides (Fig. 6).

In the second part of Kitamura's investigation (Kitamura 1987) the case of 'isolated jets' was treated numerically, assuming that gas and dust is ejected only from an isolated 'active spot', while there is no emission at all from the surroundings. In this case the

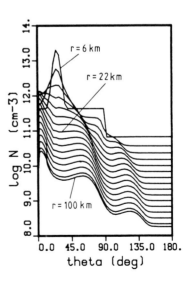

Fig. 6 — Gas number density profiles resulting from a 'ring-like' outgassing from the nucleus (at $r = 6$ km). Close to $\theta = 0$ one can see the increase of the density caused by the flowing of the gas toward the Sun-comet line, which is assumed to be the axis of symmetry in this axisymmetric calculation (taken from Kömle & Ip 1987).

density increase along the jet boundary does not occur, because there is no interaction with an ambient background medium. The gas flow field resembles that predicted by the analytic model of an isentropic isentropic nozzle flow into vacuum as described before (Boettcher & Legge 1980), although the interaction with dust particles (heating and drag effects) modifies the gas flow field somewhat. A more detailed comparison between these numerical results and the isentropic analytic model is still to be carried out.

Another interesting case treated in Kitamura's (1987) paper is a model, where the gas flow from the nucleus is assumed to be isotropic while the dust ejection is confined to an 'active spot' of 10° width around the nucleus (Fig. 7a, b). The result is that the

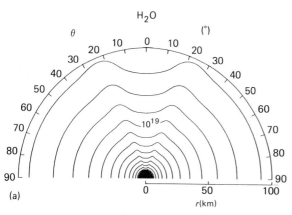

(a)

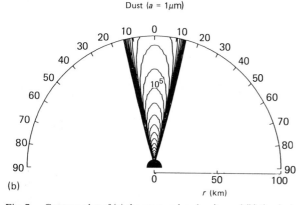

(b)

Fig. 7 — Contour plot of (a) the gas number density and (b) the dust number density (size 1 μm) as computed by Kitamura (1987). It is assumed that the gas emission from the nucleus is isotropic, while the dust comes only from an 'active spot' of 10° angular width (taken from Kitamura 1987).

dust jet remains almost confined to the original angular width of the active spot on the nucleus, and shows a very sharp boundary to the surrounding gas within 100 km from the nucleus. The gas density is depressed within the active area, owing to the lateral pressure gradient originating from the heating of the gas by the (hotter) dust.

3 WHICH TYPES OF JET STRUCTURES ARE SEEN AT COMET HALLEY?

In this section some representative observations of cometary jets are presented and critically discussed in the light of the theoretical models.

3.1 Near-nucleus jets

Fig. 8 is a composite picture produced by a combination of several frames recorded by the Halley Multicolour Camera (HMC) during the encounter of Giotto with comet Halley (Reitsema et al. 1986). It shows two major activity centres on the cometary nucleus, which are believed to be the source regions of the dust jets emanating into the sunward pointing hemisphere. The best resolution, namely $\approx$ 80 m, is achieved in the upper left corner of the nucleus (Keller et al. 1987; see also the review by H.U. Keller in Volume 2. Although only the dust is visible on the picture, it is quite clear that it must be driven either by sublimating gas from an active surface area or by outgassing from submerged gas reservoirs. As pointed out by Huebner et al. (1986), the diameter of the active region at the northern part of the nucleus is about 3 km. Inside this area several 'bright spots' are seen where the dust (and gas?) outflow may be even more enhanced. Their size is estimated to be only a few hundred metres, close to the resolution limit of the HMC. A reconstruction of the dust jets and their source regions on the nucleus is difficult because brightness differences on the image can be caused both by surface structure on the nucleus (different illumination and albedo) and by scattering from dust particles in the coma. A first (straightforward) interpretation would be that there is only one small active spot of $\approx$ 300 m diameter which expands within 1–2 km above the surface laterally to form the apparent source region of $\approx$ 3 km diameter mentioned above. However, in this case one would expect

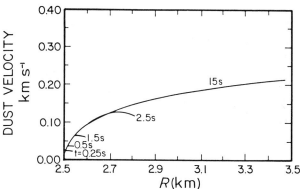

Fig. 9 — Dust velocity profile obtained with the time-dependent, spherically symmetric gas–dust-interaction model of Gombosi *et al.* (1985). The curve labelled with '15 s' corresponds approximately to the steady state solution (taken from Gombosi *et al.* 1985).

Fig. 8 — Giotto Halley multicolour camera composite image showing the nucleus and the dust jets of Comet Halley. North is at the top of the picture and the Sun is to the left and 15° behind the image plane (Courtesy H.U. Keller, MPAE, Lindau).

that the source region should be much brighter than the rest because of the strong lateral expansion close to the surface, which is not observed. As pointed out by Thomas & Keller (1987), the most probable interpretation of the structure of the northern activity region is that most of the dust comes from a distributed emission throughout an area of ≈ 3 km diameter. The fine structure of the dust jet may, however, be caused by local activity enhancements inside the active area. It is, however, not obvious from the HMC images whether or not the source regions of these fine dust jets coincide with the 'bright spots' mentioned above (N. Thomas, personal communication). Anyway, what we observe as one activity centre is most probably a conglomerate of several gas/dust sources of a few hundred metres or less in diameter, supplemented by somewhat weaker emission from the area in between.

It should be noted that — according to Thomas & Keller (1987) — the column density along a line crossing the single jets is inside the jets only about 10% above the column density of the overall dust emission coming from the northern activity area. The lateral density of this overall emission can roughly be fitted by a 'cos θ'-function. The fine dust jets themselves appear to remain collimated within an angle of 10° out to 100 km or more from the nucleus.

An important finding reported by Keller *et al.* (1986) is that the intensity profiles along the jets follow closely a $1/R$ law, indicating that the dust particle density along the jet decreases as $1/R^2$ from the brightest point outward. This means that the optically effective dust particles (diameter ≈ 1 μm) have already reached their terminal velocity very close to the surface ($\leq$ 1 km). This finding is in reasonable agreement with the results from the spherically symmetric gas-dust-interaction model of Gombosi *et al.* (1985) which predicts that the dust reaches almost its terminal velocity within 1 km from the nucleus (Fig. 9). On the other hand, the axisymmetric dusty jet

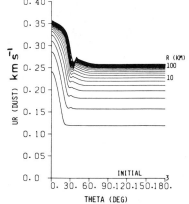

Fig. 10 — Dust velocity profiles along the angle θ (see Fig. 5 for definition) for different radial distances from the nucleus as calculated by Kitamura's (1986) model (taken from Kitamura 1986).

model of Kitamura (1986) gives a larger acceleration scale length for the dust. According to it, there should be considerable dust acceleration both inside and outside the jet in the region between the surface and at least 7 km above it (Fig. 10). This is, however, not seen in the Giotto/HMC pictures.

It seems to be clear from the facts reviewed above that the dust acceleration region lies below the resolution limit of even the closest HMC images. If we assume that this acceleration region corresponds to the length scale of lateral expansion from the source regions, then we can see on the images only the 'far field' of the expansion flow described by the simple analytic jet model (Fig. 2), namely straight streamlines, a $1/R^2$ density distribution in a radial direction, and a $\cos \theta$-density distribution in a lateral direction.

Thus the overall structure of the large dust jet emanating from the northern activity region is approximately consistent with the hydrodynamic description of a distributed gas outflow from a limited area that is 'made visible' by the embedded dust. However, the formation and collimation of the superimposed fine dust jets need further explanation.

Another feature which might be an indication for gasdynamic interaction between the jets is the curved structure visible in the southern part of Fig. 8. From laboratory experiments it is known that similar structures can be formed if two nozzles of different

strength interact with each other (Dankert & Koppenwallner 1984, Dankert 1986).

Fig. 11 shows the sunward-directed dust coma as seen on 9 March 1986 by the TVS camera aboard Vega-2 (Smith *et al.* 1986). Again, the straight appearance of the dust jet structures on a scale of several tens of kilometres (best seen on the processed picture, Fig. 11b) is obvious.

3.2 Large scale jets and anisotropies

Aside from the near-nucleus jets described before, plenty of large-scale structures have been observed at Halley that are usually also termed as 'jets' in the literature, although their nature might be quite different from a 'jet' in the hydrodynamic sense as described in the previous sections. It may be more correct to speak of 'inhomogeneities' or 'anisotropies' at least for some of the phenomena in question. Fig. 12 shows spectral scans from a region of 5 000 km × 4 700 km on the sunward side of the coma, which were obtained with the TKS-instrument aboard Vega-2 (Gogoshev *et al.* 1986). They show a 'spiky' structure which has been interpreted as the signature of gaseous jets extending several thousand km away from the nucleus. The spike is seen most clearly in the monochromatic image obtained in OH (panel A).

Large-scale curved dust features frequently ob-

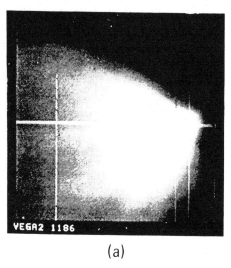

(a)

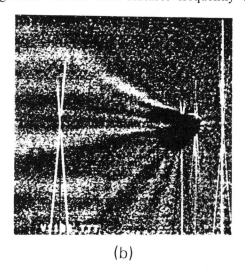

(b)

Fig. 11 — Sunward directed dust coma as seen by the TVS camera aboard Vega-2 on 9 March 1986 from a distance of 11 200 km from the nucleus; the field of view is 100 km; (a) original frame; (b) shift-differentiated version of the same frame to enhance the jet structure (taken from Smith *et al.* 1986).

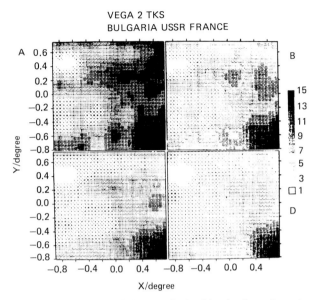

VEGA 2 TKS
BULGARIA USSR FRANCE

Fig. 12 — Monochromatic images obtained by the three channel spectrophotometer aboard VEGA 2 in the spectral bands OH (panel A), CN (panel B), C_2 (panel C) and in the continuum (panel D). The size of every image is 5000 km × 4700 km. The intensity maximum in the lower left corner corresponds to the near-nucleus region (taken from Gogoshev et al. 1986).

served in the outer coma both during the 1910 and during the 1986 apparition are described in more detail in the review by Z. Sekanina in Volume 2 of this work. An essentially new phenomenon observed for the first time during the 1985/86 apparition was the existence of long-lasting spiral-shaped gas-jets seen in the light of CN and C_2, most distinctly in CN (A'Hearn et al. 1986 a, b). These structures had an extension of about 70 000 km away from the nucleus, and their development could be followed over several months. An example is shown in Fig. 13. To explain the persistence of the structure out to these large distances, A'Hearn et al. suggest that the gas observed is released from sub-micrometre-sized dust particles on their way through the coma. They argue that, if the CN originated from parent molecules evaporating from the nucleus surface, one would need an unrealistically narrow beam of parent molecules to produce the observed structure of the CN-jets. Their argument is, however, valid only, if the main production region of the CN-molecules lies outside the collision-dominated flow regime, which may extend to $\approx 10^4$ km. If it lies inside, the CN-radicals would follow the bulk flow of the gas, and initial anisotropies can easily

be maintained out to large distances (see also the discussion in the following section). Thus there still exists the possibility that the observed gaseous jets originate from parent molecules evaporating from the nucleus and being dissociated inside $\approx 10^4$ km, although the explanation offered by A'Hearn et al. could likewise be correct.

Finally, I shall address some observations obtained from Earth-based facilities that indicate the existence of a large-scale density anisotropy between the sunward and the anti-sunward part of the coma. The observations reported by Watanabe et al. (1986) show rather significant differences between the disturbed dust coma and the disturbed gas coma (C2) during the outburst event of 12 December 1985. While the dust coma is concentrated toward the sunward-directed hemisphere (which might be a continuation of the sunward directed dust fans seen on the 'close up' images from Giotto/HMC and Vega/TVS) the isodensity lines of the C_2-coma look more circular (see also the reviews by T. Watanabe in Volume 2 of this work).

Another observation which indicates a large-scale

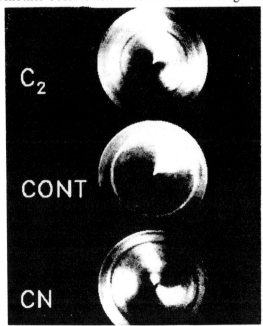

Fig. 13 — Images of comet Halley taken from Perth Observatory (Australia) on 23 April 1986 in the light of C_2, continuum and CN. The radius of the images (with the comet nucleus at the centre) is 10^5 km. The pictures show gas and dust 'jet' structures extending more than 60 000 km from the nucleus (taken from A'Hearn et al. 1986).

outgassing anisotropy of the water vapour itself has been reported by Weaver *et al.* (1986) and Larson *et al.* (1986). They measured the infrared spectrum of comet Halley on several days both pre-perihelion and post-perihelion with the aid of a Fourier-transform-spectrometer from aboard the Kuiper Airborne Observatory (KAO). During one observation run (on 25 April 1986) they measured the intensity of the H_2O-lines at positions offset from the photometric nucleus in sunward and in tailward direction, and found a ratio of about 3:1 between the column density on the sunward and tailward side of the comet. Simultaneous observations from other observatories indicate that around the time of this observation the comet was in a very active phase, and showed a major outburst (Rettig *et al.* 1987). Thus it remains unclear if the observed anisotropy of the water vapour content in Halley's coma occurs only during phases of enhanced activity (outbursts), or whether it exists also during more quiet phases.

A further independent indication for the existence of a large scale anisotropy of the gas flow is the measured asymmetry of the spectral line profiles both for the H_2O-lines (Larson *et al.* 1986) and for HCN-lines (Schloerb *et al.* 1986). In the following section the problem of large-scale anisotropies will be discussed further in connection with 'expanding halo' structures.

4 EXPANDING HALOES

4.1 Observations

Expanding halo structures were already frequently observed during Halley's apparition in 1910. Bobrovnikoff (1931) writes in his famous article on comet Halley: 'A peculiar feature of Halley's comet was the constant formation of spherical haloes. These haloes constituted envelopes around the nucleus.' And further, with respect to the unsteady nature of this phenomenon, Bobrovnikoff notes: 'The expulsion of matter from the nucleus and the formation of the haloes was an intermittent process. The haloes often had sharp inner and outer boundaries and in some cases actually formed rings with darker space on both sides . . . several haloes existed simultaneously.'

When interpreting these results from today's point of view we must keep in mind that Bobrovnikoff's photographic material consisted exclusively of photographs in integrated light, and therefore always show a mixture of gas and dust structures. He concluded, however, from the available cometary spectra that a large amount of the gaseous emission in the haloes was due to the CN-radical. From this one could expect that the expanding shells should be much better visible when observed through a narrowband CN-filter. And indeed the observations of Schlosser *et al.* (1986) show again striking evidence for the existence of expanding shells during the 1985/86 apparition. Because they are taken through a CN-filter they are much less contaminated by dust-scattered light, and should really show the distribution of the CN-emission along the line of sight. In Fig. 14 one of these photographs obtained at ESO (Chile) is reproduced. It shows two expanding shells surrounding an essentially spherical coma region. The average expansion velocities of the shells were found to be approximately 1 km s^{-1}. Because the observations covered two months post-perihelion (17 February–17 April 1986) it seems unlikely that the spherical appearance is only a projection effect. If the shell structures would have been restricted to the sunward-pointing side of the coma ('half-shells') this fact should have shown up in

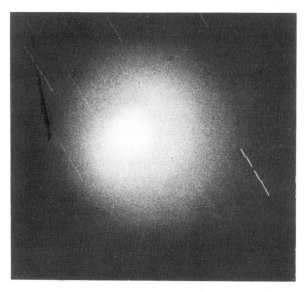

Fig. 14 — Expanding gas shells at comet Halley observed on 8 April 1986 at ESO (Chile) through a CN filter; the horizontal diameter of the image corresponds to approximately 10^5 km (courtesy R. Schulz, Bochum University, FRG).

the observations at least during the second half of March, when the angle between the Sun-comet and the Earth-comet line was more than 60°. The tailward distortion of the shells is most probably caused by a radiation pressure effect, while the different orientations of the main axes may indicate rotation of the activity centres on the nucleus.

4.2 The physical nature of the CN-haloes

The hydrodynamic formation process of the expanding haloes has up to now been treated only within the frame of spherically symmetric time-dependent models (Ip 1981). Such a model implies that the comet periodically changes its gas and dust output, whereby the period is identified with the rotation period of the nucleus. Spherical shells of the type predicted by these models can be generated only if either

(a) the original anisotropic outgassing (sunward facing outward flow) of the parent molecules is effectively transformed to a spherically symmetric flow in the innermost coma (i.e. by lateral jet expansion) or,

(b) the outflow of the parent molecules of the CN-radical from the nucleus is approximately spherically symmetric from the very beginning.

In both cases spherical gas shells of a certain thickness (proportional to the duration of the outburst) would be generated which might have the observed approximately spherical appearance when viewed along any line of sight. Assumption (b) would be difficult to understand if the outgassing variations are indeed caused by spin modulation of certain activity centres on the nucleus surface. So it remains to be investigated whether or not assumption (a) is a reasonable one. Let us assume an outgassing anisotropy of the bulk neutral gas from the nucleus of the form

$$F(\theta) = (F_{noon} - F_{night}) (\cos \theta)^{\alpha} + F_{night} \quad (14)$$

with $F_{night} \gg F_{noon}$ and $\alpha = 1$ or 2. Because the inner coma is a collision-dominated region the gas will expand hydrodynamically, starting approximately with sonic velocity close to the nucleus surface. Kömle & Ip (1987) have recently performed a hydrodynamic model calculation assuming outflow of H_2O gas from the nucleus, using the anisotropic boundary condition

(equation 14). The result was that in any case considerable deviations from spherical symmetry were maintained out to 10^4 km, which was estimated to be the extension of the collision zone. Thus it must be concluded that hydrodynamic lateral expansion of a mainly sunward directed gas outflow is inefficient for producing a spherically symmetric neutral gas halo. Because inside the collision-dominated flow regime the parent molecules of CN and the CN-radicals themselves should follow the bulk flow of the neutral gas, we expect that the angular distribution of the CN-molecules at the outer edge of the collision zone ($\approx 10^4$ km) is still rather anisotropic and forms a sunward-directed fan. Radio observations of the HCN-molecule in comet Halley (one possible parent molecule of CN) confirm the existence of an outgassing anisotropy in the case of this species (Schloerb et al. 1986). The basic question that remains to be answered is now: how can such an anisotropic CN (or HCN) cloud produce the observed, approximately spherical haloes?

One possibility which may be considered is that the excess energy of the photolytic reaction creating the CN-molecule is large enough that the molecules gain random velocities of several kilometres per second. If a considerable part of the CN production region lies outside the collisional zone, these random velocities would be maintained, and the whole space around the comet could be filled up with molecules, because the bulk flow speed is typically ≤ 1 km s^{-1}. If this scenario is true, the expanding shells would in reality be filled-up spheres of CN-radicals whose shell-like appearance is solely due to the simultaneous existence of overlapping spheres with different diameters. There is, however, an argument against this explanation. As already mentioned, the measured expansion velocities of the haloes are typically around ≈ 1 km s^{-1}, i.e. they are the same as the outflow velocities of the bulk gas from the nucleus. This means either that the main production region of CN lies inside the collision zone or that the reaction producing the CN has negligible excess energy. In both cases a transformation from a sunward-directed fan into a spherical, nucleus-centred shell appears to be impossible.

I now consider a further possibility to explain the spherical appearance of the CN-shells. In connection with an attempt to find a reasonable rotation- and nutation-period of Halley's nucleus, Wilhelm (1987) suggested that the cometary nucleus might be a not completely rigid body. Then, because of its irregular shape, there is the possibility that it is squeezed periodically owing to inertial forces. This squeezing could trigger outgassing activity from sub-surface reservoirs on the whole comet, not necessarily restricted to the sunward side. If the CN-haloes are produced during such squeezing periods of the whole comet, both their spherical shape and their periodic appearance could be naturally explained.

5 MODELLING OF OUTBURST EVENTS

Finally, some results from the most recent model calculations (Kömle & Ip 1987) are shown, which illustrate the propagation of a disturbance originating close to the subsolar point of the cometary surface. It is assumed that the gas outflow rate is locally increased during a period of 30 minutes: Compared to the undisturbed background flow (which is assumed to be spherically symmetric), the outgassing rate at the subsolar point is taken to be 75 times larger and to fall off in a Gaussian profile with a half-width of 10°.

Fig. 15* shows a meridional cut of the density distribution in the coma caused by such an outburst out to a distance of 10 000 km from the nucleus at different time levels. The quantity plotted is the relative density, i.e. at each point the density is divided by its 'undisturbed' value. The first picture of the sequence (Fig. 15a) is taken at the time when the outburst has ceased on the surface. It shows the typical density increase along the jet boundary as discussed in section 2. After the disturbance on the surface has stopped the neutral gas cloud expands and drifts through the sunward-directed coma. The increased density along the original jet boundary has now transformed into a drifting ring-like structure disconnected from the nucleus which is visible in Figs. 15b and 15c. In the last frame of the sequence (Fig. 15d) the internal density structure of the cloud has almost disappeared, but the whole cloud continues to expand and to drift in the sunward direction. It forms a kind of shell restricted to one half-sphere, but can never develop into a full circumnuclear shell.

---

* to be found in the colour illustration section in the centre of the book.

6 CONCLUDING REMARKS

At the time of writing, it is still not possible to give a consistent view regarding the structure of the gas and dust coma of comet Halley. The main controversy seems to be how strong the anisotropy between dayside and nightside gas and dust ejection really is. On the one hand, the Giotto/HMC and the Vega/TVS pictures clearly show the prefered sunward ejection of dust, mainly in the form of jets. However, Vega/TVS has also detected a weak nightside-directed dust jet (Smith et al. 1986). An anisotropy in the H_2O-column density of about 3:1 between nightside and dayside of the comet has been found by KAO infrared observations reported by Weaver et al. (1986).

On the other hand, the in situ measurements obtained by the neutral gas experiment (NMS) aboard Giotto show a density profile $n(R) \sim (1/R^2).e^{-R/v\tau}$ (τ = lifetime of H_2O molecules, v = bulk outflow velocity) which would be consistent with a spherically symmetric outgassing from the nucleus (Krankowsky et al. 1986). Further it has been shown by numerical simulation (Kömle & Ip 1986) that an initial outgassing anisotropy should be maintained out to at least 10 000 km from the nucleus. Thus the NMS results suggest a more isotropic outgassing profile. The spherical appearance of the CN-haloes points toward the same direction. If gas ejection is restricted to the sunward side, the formation of nucleus-centred haloes is difficult to explain, as discussed in section 4. A short sunward directed gas outburst results in a drifting torus-shaped cloud, but can never develop into a spherical halo. Long-lasting outbursts produce spiral jets due to the rotation of the nucleus. These features have been frequently observed for the dust (Sekanina, 1986) but also for the gas in CN and C_2 filter observations (A'Hearn et al. 1986 a, b).

The discrepancies discussed above might at least partly be resolved if one assumes that strong anisotropies of the gas density distribution occur only during periods of enhanced activity (outbursts). However, this assumption cannot be true for the dust production because at the time of Giotto encounter there was no unusual activity detected by ground-based observations. Clearly, more detailed comparisons between the various spacecraft experiments and ground-based observations are necessary to clarify the questions outlined above.

REFERENCES

A'Hearn, M.F., Birch, P.V., & Klingelsmith III, D.A. (1986a) *Gaseous jets in Comet P/Halley*, ESA-SP-250 **I** 483–486

A'Hearn, M.F., Hoban, S., Birch, P.V., Bowers, C., Martin, R., & Klingelsmith III, D.A. (1986b) Cyanogen jets in Comet P/Halley, *Nature* **324** 649–651

Ashkenas, H., & Sherman, F.S. (1966) The structure and utilization of supersonic free jets in low density wind tunnels, In: *Rarified gas dynamics* **2** 84–105

Bobrovnikoff, N.T. (1931) Halley's comet in its apparition of 1909–1911, *Publ. Lick Observatory* **XVII** Part II

Boettcher, R.-D., & Legge, H. (1980) *A study of rocket exhaust plumes and impingement effects on spacecraft surfaces: II. Plume profile analysis*
Part 1: *Continuum plume modelling*, DFVLR internal report IB 251–80 A 29 (1980)
Part 3: *Rarefaction effects*, DFVLR internal report IB 222–81 A 19 (1981)

Dankert, C. (1985) *Experimental verification of rocket exhaust plumes and impingement effects on spacecraft surfaces. Work package I: Plume model testing, Part 2: Velocity measurement*, DFVLR internal report IB 222–84 A 47

Dankert, C. (1986) Flow in the interaction region of two parallel free jets, In: Boffi, V., & Cercignoni, C., (eds.), *Proceedings of the 15th International Symposium on Rarefied Gas Dynamics, June 16–20, 1986 Grado/Italy*, 486–494

Dankert, C., & Koppenwallner, G. (1984) Experimental study of the interaction between two rarefied free jets, In: H. Oguchi (ed.), *Proceedings of the 14th International Symposium on Rarefied Gas Dynamics, July 16–20, 1984 Tsukuba Science City/Japan*, 477–484

Dettleff, G. (1981) *A study of rocket exhaust plumes and impingement effects on spacecraft surfaces: II. Plume profile analysis: Method of characteristics computer program for axisymmetric free jets, Part I: Modification and revision of a programme written by Vick, Andrews et al.*, DVFLR internal report IB 222–81 A 01

Gogoshev, M., Gogosheva, T.S., Sargoichev, S., Palazov, K., Georgiev, A., Kostadinov, I., Kanev, K., Spasov, S., Werner, R., Mendev, I., Moreels, G., Clairemidi, J., Mougin, B., Parisot, J.P., Zucconi, J.M., Festou, M., Lepage, J.P., Runavot, J., Bertaux, J.L., Blamont, J.E., Herse, M., Krasnopolsky, V.A., Moroz, V.I., Krysko, A.A., Troshin, V.S., Barke, V.V., Jegulev, V.S., Sanko, N.F., Tomashova, G.V., Parshev, V.A., Tkachuk, A.Yu., Novikov, B.S., Perminov, V.G., Sulakov, I.I., & Fedorov, O.S. (1985) Vega project — three channel spectrophotometer for the Halley's Comet investigation, *Adv. Space Res.* **5** (12) 133–136

Gombosi, T.I., Cravens, T.E., & Nagy, A.F. (1985) Time-dependent dusty gasdynamical flow near cometary nuclei, *Astrophys. J.* **293** 328–341

Huebner, W.F., Keller, H.U., Wilhelm, K., Whipple, F.L., Delamere, W.A., Reitsema, H.J., & Schmidt, H.U. (1986) *Dust-gas interaction deduced from Halley Multicolour Camera Observations*, ESA-SP-250 **II** 363–364

Ip, W.-H. (1981) Expanding haloes in cometary comae, *Nature* **289** 269–271

Keller, H.U., Delamere, W.A., Huebner, W.F., Reitsema, H., Schmidt, H.U., Schmidt, W.K.H., Whipple, F.L., & Wilhelm, K. (1986) *Dust activity of Comet Halley's nucleus*, ESA-SP-250 **II** 359–362

Keller, H.U., Delamere, W.A., Huebner, W.F., Reitsema, H.J., Schmidt, H.U., Whipple, F.L., Wilhelm, K., Kurdt, W., Kramm, R., Thomas, N., Arpigny, C., Barbieri, C., Bonnet, R.M., Cazes, S., Coradini, M., Cosmovici, C.B., Hughes, D.W., Jamar, C., Malaise, D., Schmidt, W.K.H., Seige, P. (1987) *Comet Halley's nucleus and its activity*, *Astron. Astrophys* **187** 807–823

Kitamura, Y. (1986) Axisymmetric dusty gas jet in the inner coma of a comet *Icarus* **66** 241–257

Kitamura, Y. (1987) *Axisymmetric dusty gas jet in the inner coma of a comet: II. The case of isolated jets* *Icarus* **72** 555–567

Kömle, N.I., & Ip, W.-H. (1987) *Anisotropic non-stationary gas flow dynamics in the coma of Comet Halley*, *Astron. Astrophys.* **187** 405–410

Kömle, N.I., & Ip, W.-H. (1987) A model of the anisotropic structure of the neutral gas coma of a comet, In: *Proceedings of the Symp. on 'Diversity and Similarity of Comets', Brussels, April 6–9*, ESA SP-278 247–254

Koppenwallner, G., Boettcher, R.-D., Dettleff, G., & Legge, H. (1986) *Rocket exhaust plume flow into space*, ESA SP-265, 83–88

Krankowsky, D., Lämmerzahl, P., Herrwerth, I., Woweries, J., Eberhardt, P., Dolder, V., Hermann, V., Schulte, W., Berthelier, J.J., Illiano, I.M, Hodges, R.R., & Hoffmann, J.H. (1986) *In situ gas and ion measurements at Comet Halley*, *Nature* **321** 326–329

Larson, H.P., Davis, D.S., Mumma, M.J., & Weaver, H.A. (1986) *Velocity-resolved observations of water in Comet Halley*, ESA SP-250 **I** 335–340

Reitsema, H.J., Delamere, W.A., Huebner, W.F., Keller, H.U., Schmidt, W.K.H., Wilhelm, K., Schmidt, H.U., & Whipple, F.L. (1986) *Nucleus morphology of Comet Halley*, ESA-SP-250 **II** 351–354

Rettig, T.W., Kern, J.R., Ruchti, R., Baumbaugh, B., Baumbaugh, A.E., Knickerbocker, K.L., & Dawe, J. (1987) Observations of the coma of Comet Halley and the outburst of March 24–25 (UT) 1986 In: *Proceedings of the Symp. on 'Diversity and Similarity of Comets', Brussels April 6–9 1987*, ESA SP-278 265–269

Sagdeev, R.Z., Shapiro, V.D., Shevchenko, V.I., & Szegö, K. (1985) *Jet formation in comets*, KFKI-1985–89 internal report

Schloerb, F.P., Kinzel, W.M., Swade, D.A., & Irvine, M. (1986) *HCN production from Comet Halley*, ESA SP-250 **I** 577–581

Schlosser, W., Schulz, R., & Kozcet, P. (1986) *The cyan shells of Comet P/Halley*, ESA SP-250 **III** 495–498

Sekanina, Z. (1986) *Dust environment of Comet Halley*, ESA SP-250 **II** 131–143

Smith, B., Szegö, K., Larson, S., Merenyi, E., Toth, I., Sagdeev, R.Z., Avanesov, G.A., Krasikov, V.A., Shamis, V.A., & Tarnapolski, V.I. (1986) *The spatial distribution of dust jets seen at Vega-2 fly-by*, ESA SP-250 **I** 327–332

Thomas, N., & Keller, H.U. (1987) Comet P/Halley's near-nucleus jet activity, In: *Proceedings of the Symp. on 'Diversity and Similarity of Comets', Brussels April 6–9 1987* ESA SP-278 337–342

Watanabe, J., Kawakami, H., Tomita, K., Kinoshita, H., Nakamura, T., & Kozai, J. *On the outburst of Comet Halley on December 12, 1985*, ESA SP-250 **III** 267–272 (1986)

Weaver, H.A., Mumma, M.J., Larson, H.P., & Davis, D.S., Post-perihelion observations of water in Comet Halley, *Nature* **324** 441–444 (1986)

Wilhelm, K. (1987) Rotation and precession of Comet Halley *Nature* **327** 27–30

16

Variations of the gaseous output of the nucleus of comet Halley

M.C. Festou

1 INTRODUCTION

The magnitude and rate of variation of the gaseous output of a comet nucleus is closely related to the thermodynamical properties of its surface and to the intensity of the incoming solar flux. The classical thermal equilibrium equation for a comet nucleus (e.g. Huebner & Whipple 1976 or Houpis 1989) shows that while part of the incident solar energy is directly reflected into space, a very large fraction of the energy is absorbed and both heats the nucleus and is used to vapourize the volatile component of its surface layers. Infrared observations have shown that the mean geometric albedo of comet nuclei is very small. Each time that important parameter has been measured, a value of less than about 5% has been found and, as a consequence, most of the light that reaches a comet nucleus is absorbed. Far from the Sun, equilibrium temperatures for nucleus cores are very low, of the order of 150 K or less, depending on the thermal conductivity of the nucleus and on the heliocentric distance of the comet (Klinger 1983). No sublimation occurs and the captured energy is transferred towards the nucleus interior whose temperature is then fixed by the amount of radiation freed at infared wavelengths. Closer to the Sun, an important fraction of the solar energy is employed to evaporate volatile matter and the vapourizing surface temperature can not then be much higher than about 200 K (e.g. Weissman & Kieffer 1981). If a comet could be assimilated to an asteroid, subsolar temperatures in the range 100–400 K would be obtained; the lower values would apply to comets situated at large heliocentric distances while the larger could apply only to comets moving on orbits taking them inside the Earth's orbit.

The recent exploration of Halley's comet confirmed the validity of this scheme and has shown that the nucleus of that comet is not entirely covered by ices (Keller *et al.* 1986) and that it has dark areas which are quite hot ($T \approx 400$ K; Combes *et al.* 1986; Emerich *et al.* 1987). Dust (and consequently gases) was observed escaping from at least three distinct regions at the time of encounter with the Giotto spacecraft (Keller *et al.* 1987a); these emissive areas are mostly located in the northern hemisphere (Keller *et al.* 1987b). They are now commonly referred to as 'active areas', since they free large amounts of gas while the dark regions seem to be quiescent. These active areas are probably the coldest regions of the nucleus. Solid particles are expelled from the bright spots by collisions with surrounding molecules leaving the nucleus surface at a velocity of a few hundred metres per second.

One of the questions yet to be answered is to know at which point in the orbit the outgassing begins. If comets were made of a single chemically simple species, the knowledge of such an important parameter should theoretically allow to infer some of the chemical and thermodynamical properties of the species that controls the evaporation of cometary matter, a difficult question that Levin (1943) first attempted to answer long ago without success. Even though there is some evidence that the outgassing rate of comets becomes significant only below 3 AU from the Sun pre-perihelion (e.g. Delsemme 1982), we know that reality is not that simple and that a comet nucleus must consist of a complex mix of various chemical species that, in all likelihood, is not homogeneous on a small scale. One of the main reasons to observe comet Halley as early as 1981, was to attempt to detect the first signs of its activity. As this activity showed some kind of permanency after February 1985 (Wyckoff *et al.* 1985), the observing effort was directed towards the study of the nature of the outgassing law. Were all gases released in a constant ratio independent of the heliocentric distance? Was the production of dust particles running in parallel with that of the main gases? When the comet came closer to the Sun, more detailed activity surveys were conducted in order to evaluate the activity of the nucleus on a short time scale. Others were conducted at the highest possible spatial and temporal resolutions to study the distribution of dust particles (were active areas and large scale structures linked, were dark areas outgassing? Was the existence of active areas permanent or not?).

The *in situ* exploration of Halley's comet has provided a wealth of new information. The many new observations of these rapidly changing bodies allows us to build a fairly complete phenomenological picture of how a comet develops. The monitoring of the nucleus production of matter, a very difficult undertaking, is one of the tools at our disposal towards understanding in detail how that happens.

2 THE ONSET OF ACTIVITY

Since dark areas are found to be quite hot, it is clear that they cannot release the observed gases. The gas production per unit area would be extremely large, and the total outgassing surface would have to be very small to account for the observed quantities, in contradiction to what has been observed, i.e. that the active areas represent about 10% of the total surface of the nucleus. Standard thermal models predict total surfaces of that order of magnitude only if the surface temperature does not rise much above 200 K. Outgassing in those regions will begin when their temperature is high enough to free significant

amounts of gas by sublimation. Since comet comae are mostly made of water dissociation products, comet nuclei should become active when they are closer to the Sun than about 3 AU, the distance from the Sun at which the vapour pressure above an icy surface begins to have an appreciable value. Then, the production rate of gases should reach its maximum value at the time of perihelion passage, and one should observe a reverse series of events when the comet is on its way toward its aphelion.

This simple picture does not adequately describe the actual observations. Although, generally, outgassing does become obvious below about 3–2 AU from the Sun, comet nuclei have been observed to be active, often erratically, at much larger heliocentric distances (Roemer 1976, West 1982), and in the case of Halley's comet at least at 6 AU from the Sun pre-perihelion. The comet was still active at the end of 1987, although it was situated at more than 5 AU from the Sun ($m_v \approx 15$; spectral features due to neutral gases were easily detected, Cochran & Barker 1987, Wisniewski *et al.* 1987, Wyckoff *et al.* 1987). Observations made at 8.5 AU from the Sun in 1988 revealed well defined inner and outer comae (West & Jorgensen 1989). The characteristics of the former can be interpreted as a clue that outgassing was still taking place at that distance from the Sun. Observations in early 1989 indicated an object at 10.1 AU from the Sun that could have ceased to be active if the ring structure that was then observed was a remnant of the outer coma observed in 1988 (West 1989). Since we know the bulk composition of the volatile fraction of comet nuclei, i.e. that H_2O and CO are the two most abundant species, the fact that a comet nucleus can be active so far from the Sun tells us that either we do not understand how comet nuclei free their volatiles, or that comet nuclei must be inhomogeneous on a scale large enough so that the most volatile species can evaporate *independently* of the surrounding dominant water molecules. If this is true, one of the most simple mechanisms which could explain nucleus activity so far from the Sun is the building up of mechanical pressure inside pockets of CO (its thermodynamical properties would play a key rôle if it is not trapped in clathrates or hydrates; other minor species such as CO_2 or H_2CO can be present as well in those pockets), until that pressure is large enough to break the icy lattice (Festou *et al.* 1986a; Feldman *et al.* 1986; Prialnik & Bar-Nun 1988). Interestingly enough, the same process could explain the outbursts occurring at much smaller heliocentric distances. However, since outbursts do not occur in every comet, admitting the existence of a single process for comet outbursts, implies that volatile species are either chemically trapped or are underabundant in comets which do not undergo activity bursts. The other mechanism that is often proposed to explain the occurrence of bursts is the crystallization of amorphous ice (Larson *et al.* 1989; Prialnik & Bar-Nun 1988), an exothermic process that could be triggered by the above-mentioned simple downward transport of solar energy.

Early observations of comet Halley indicate that the comet was active very far from the Sun (reviews are given by Meech *et al.* 1986, and Festou *et al.* 1986a). Brightness fluctuations were observed first on a time scale of a few hours (West & Pedersen 1983), which strongly suggested that the comet surface could be outbursting. Nevertheless, the comet remained stellar in appearance until September 1984, when a diffuse coma was detected for the first time (Spinrad *et al.* 1984). That coma was not found to be systematically present in subsequent observations, an indication that it was produced by a source of matter erratically turning on and off. The continuous decrease of the minimum magnitude of the comet (Meech *et al.* 1986; Festou *et al.* 1986a) was an indication that, despite the fact that the comet had a stellar appearance most of the time, its nucleus was probably surrounded at times by a cloud of rapidly evaporating particles. Both groups of authors estimated that the comet was active at least at 6 and 8 AU from the Sun pre-perihelion, respectively. In February 1985, the spectral signatures of CN (Wyckoff *et al.* 1985; Cochran & Barker 1986) and OI (Spinrad *et al.* 1986) were detected for the first time while the comet was 5.4 AU from the Sun. The detection of OH by its UV emission failed in April 1985 (Festou *et al.* 1986b) but was detected at radio wavelengths three months later when the comet was at a heliocentric distance of 4.3 AU (Gérard *et al.* 1986; Schloerb *et al.* 1986b). Today, the question of the origin of the brightness fluctuations of the comet is not definitely settled. It is

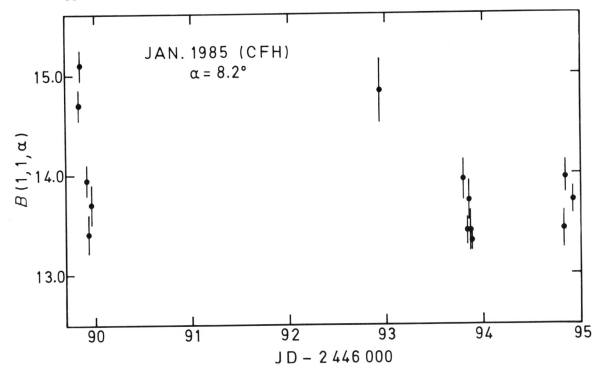

Fig. 1 — Brightness variations of Halley's comet in January 1985 (from Festou *et al.*, 1986a)

probably related to the rotational properties of the nucleus, a yet very controversial issue (Belton *et al.* 1986, Festou *et al.* 1987, Parisot *et* al. 1987, Peale & Lissauer 1987, 1989, West & Jorgensen 1989).

3 LONG TERM ACTIVITY OF THE NUCLEUS

The activity of the comet, as measured by the gas production rates of the main coma species or by the rapidity at which the inner dust coma structures changed, has been the object of numerous studies. It appeared to vary greatly from one day to another and even from one hour to the next, making Halley's comet an exceptionally variable object. Note that this finding is not an artefact due to the major observing activity deployed in 1985/86. A recent campaign, of similar amplitude, was organized to observe comet Wilson (19861) with the International Ultraviolet Explorer (IUE) in 1986/87: the gas production of that comet was found to vary less than 3–4% on a time scale from hours to days (Roettger *et al.* 1989).

3.1 The visual observations

More than 3 000 visual estimates of Halley's comet magnitude have been collected, mostly by amateur astronomers (Green & Morris 1986). The light curve of the comet departs from that of a solid reflecting body possibly as far as 11–12 AU pre-perihelion and certainly below 6 AU. Table 1 gives the values taken by the parameters H_o and n in equation (1) which approximates the total magnitude of the comet as a function of both its heliocentric (r) and its geocentric distance (Δ):

$$m_T = H_o + 5 \log \Delta + 2.5 \, n \log r \qquad (1)$$

The gross structure of the visual light curve is characterized by four distinct phases (Fig. 2). First, from early 1983 until January 1985, was a period during which activity was detected at times and during which the coma had no permanent existence. Then the coma was spectroscopically detected and a steady increase in the total brightness of the comet was observed until mid-November 1985, as noted by Keith (1986) and Fischer (1986). This can also be seen in the analyses of Marcus (1986) and Hasegawa (1986). Then, effects that limited the release of matter by the nucleus were in action until the end of February

1986 (m_T does not obey equation (1) during that period: was that due to seasonal effects, to mantle development that limited the outgassing surface, or to another phenomenon that limited the increase of the nucleus temperature?). From the beginning of March 1986 until at least mid-June 1986 (the comet was then lost by most visual observers), the light curve showed a uniform and fairly slow brightness decrease, on top of which a short time scale activity of large amplitude was present. From what was observed far from perihelion in 1987/88 (West & Jorgensen 1989; West 1989), one may expect the comet to have returned to its quiescent phase of activity at the end of 1988 or at the beginning of 1989. A phase of erratic activity of the kind observed at the end of 1984 occurred in 1988. When the complete light curve has been constructed, it will be interesting to observe whether or not a turning point, similar to that observed at the end of 1985 (Fig. 2), will be found in 1986 or later in the light curve.

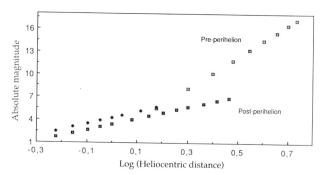

Fig. 2 — Absolute magnitude of comet Halley in 1985/86 derived from visual observations. Note the change in the slope of the curve pre-perihelion when $r \approx 2.0$ AU. Short-term variations have been removed.

3.2 Monochromatic observations

Monochromatic observations of comets have been widely used in the past 15 years as a powerful means of assessing the gas production rate of coma species. Modelling strongly affects the derivation of these gas productions from brightness measurements, though. Let us first consider the production of the dominant species, water. In Table 1, parameters Q_o and n' that allow us to compute the total water production rate according to equation (2) were derived from OH emission observations (unfortunately, direct measurements of the water production are more limited, Weaver *et al.* 1987). Note that the linear fit to observations shown here is only an approximation that ignores short time scale variations and is simply intended to approximate the gross structure of the production curve:

$$\log Q(H_2O) = \log Q_o + n' \log r \qquad (2)$$

Table 1 — Magnitudes and water production rates of comet Halley in 1985/86

Period	H_o[1]	n[1]	Log Q_o	n
Jan. 85 ($r \approx 5.4$ AU) > Dec. 2, 85 ($r = 1.5$ AU)	1.8	8.4		
Sep. 12, 85 ($r = 2.59$ AU) > Dec. 31, 85 ($r = 1.01$ AU)			29.60[2]	-2.47[2]
Dec. 3, 85 ($r = 1.5$ AU) > Jan. 28, 86 ($r = 0.6$ AU)	4.3	3.2		
Mar. 9, 86 ($r = 0.84$ AU) > Jul. 2, 86 ($r = 2.18$ AU)			29.68[2]	-1.18[2]
Mar. 1, 86 ($r = 0.6$ AU) > Jun. 13, 86 ($r = 2.9$ AU)	3.4	3.0		
July 1985 ($r \approx 3.4$ AU) – > July 1986 ($r \approx 3$ AU)			≈ 29.3[3]	-2[3]
Nov. 30, 85 ($r = 1.5$ AU) – > Jan. 25, 86 ($r = 0.68$ AU)			29.66[4]	-2.30[4]
Feb. 20, 86 ($r = $ AU) – > April 2, 86 ($r = $ AU)			29.74[4]	-1.62[4]

[1] Green & Morris 1986.
[2] From Feldman *et al.* 1986.
[3] From Fig. of Schloerb *et al.* 1986.
[4] 0.5 Q_H. Craven & Frank 1987.

It is clear that the total brightness of the comet increased much more rapidly than would be expected from the rate at which the gaseous output of the nucleus increased, a clear indication that most of the recorded light was due to the component reflected by dust grains and not to the gas coma. This latter assumption is confirmed by the results of Tokunaga *et al.* (1986), who showed that the dust production decreased as $r^{-3.17}$ against $r^{-3.20}$ for the pre-perihelion variation of the magnitude as indicated by Green & Morris (1986). The pre/post perihelion asymmetry noticed in the 1910 apparition was again observed: gas and dust productions were higher post-perihelion than pre-perihelion at similar heliocentric distances by a factor of about three when the comet was at about 1 AU from the Sun (e.g. Boyarchuk *et al.* 1986). From the analysis of visual observations, Marcus (1986), showed convincingly that the comet behaved similarly in 1910 and in 1986. The perihelion asymmetry was much greater at large heliocentric distances (e.g.

Wyckoff *et al.* 1987). In contrast to what was observed in the visual light curve, the gas production curve slope did not change in November 1985. Was this magnitude change induced by the appearance after mid-November 1985 of new emissive areas having different dust to gas contents? This seems unlikely, since, one should then have simultaneously observed a modification of the gas production rate corresponding to a likely change of the surface of those emissive areas. Alternative explanations should be sought, for example a mechanism to explain an accumulation of dust in the inner coma during the few months around perihelion time, such as the fractionation of large grains into smaller ones.

Radio observations of the OH 1665 MHz and 1667 MHz transitions provide results somewhat different than those obtained with the IUE. Both Schloerb *et al.* (1986a), and Gérard *et al.* (1986) found that log Q (OH) and the heliocentric magnitude were linearly correlated up to heliocentric distances of at least 2 AU, and that no significant pre/postperihelion asymmetry was present. Why UV and radio observations of the OH emission do not provide similar results is a question that still awaits an answer (see the discussions in Schloerb *et al.* 1986b, and in references 1, 3 and 6 of that paper; according to Schloerb 1987, UV measurements should be less affected by modelling uncertainties). Results derived from HCN radio observations by Schloerb *et al.* (1986b), are very similar to those obtained from the observation of the UV emission of OH, i.e. they show a marked perihelion asymmetry. This surprising result might be an indication that the OH 18 cm emission mechanism could be erroneously modelled while the simpler HCN emission would not (in particular, the quenching of the hyperfine OH levels is delicate to estimate). Radio and visual observations suggest that the strong activity increase observed around 3 AU preperihelion corresponds to the water 'turn-on' that vapourization theories predict. The gas production reached its maximum in the second half of February or at the beginning of March 1986, and Stewart (1987), indicated a water production rate of the order of $1–2 \times 10^{30}$ molecules per second for that period. The total loss of water was according to Stewart (1987), about 400 million tonnes, or a layer 30 m thick out of the 36 km² active areas supposed to be of density 0.4.

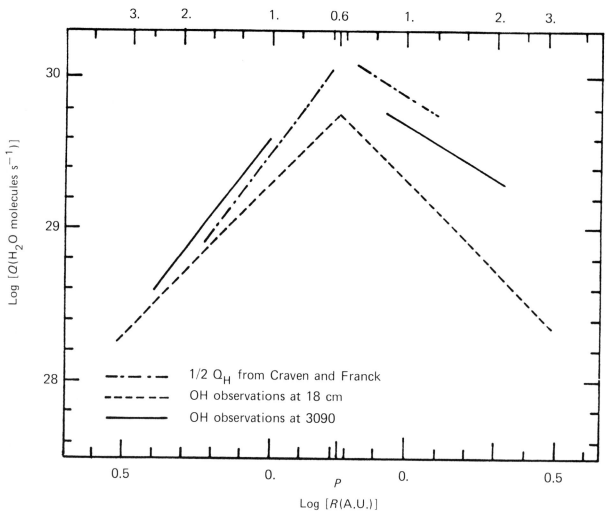

Fig. 3 — Long-term variation of the gaseous output of Halley's comet (data from Table 1). Note that differences between the gas production laws reflect either model uncertainties or a changing composition of the matter ejected by the nucleus.

Let us turn now to the minor species gas production curves. A few ground based observers monitored the production of CN, C_3, C_2, NH_2 over almost the entire apparition. Those observations are very interesting since they do not suffer from the lack of time resolution exhibited by the UV and some of the radio observations. The extensive data set from which the 7.4 day periodicity was found (Schleicher & Millis 1986) should be published soon (Millis *et al.* 1987). These latter observations reveal that the mean Q(CN) varied as $r^{-3.1}$ pre-perihelion (2.2–0.66 AU) and as $r^{-2.2}$ post-perihelion (0.95–2.65 AU). Apparently, the mean water and minor species production rates did not follow the same law (see Table 1). The OH to C_2 and C_2 to C_3 gas production rates were found to vary from 200 to 800 and from 15 to 5 in the post-perihelion phase, respectively. If those ratios are confirmed by other studies, this would imply a change in the chemical composition of the comet material. In our opinion, all those results are very preliminary, and one should be certain that the conversion of the observed emissions into gas production rates were not too much affected by modelling uncertainties and/or nucleus activity changes before coming to any firm conclusion (e.g. compare the above figures with those given by Cochran & Barker 1986).

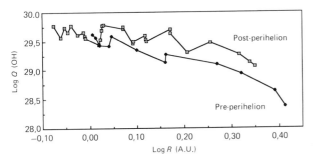

Fig. 4 — Water production as derived from IUE observations (Feldman *et al.*, 1987). The pre/post perihelion asymmetry was about 3 in the range 1–2.5 AU, with variations due to temporal variability of the comet on a time scale of one day.

3.3 Seasonal effects or else?

The work by Weissman (1987), about the origin of the pre/post-perihelion asymmetry shows that even though a seasonal effect is an assumption which qualitatively explains the observations, it however fails to explain both the short-term variations and the magnitude of the perihelion asymmetry. This is possibly due to the fact that the theory does not take into account the complex rotational motion of the nucleus and the fact that the outgassing is localized in few spots.

In opposition, Houpis (1989) indicates that the perihelion asymmetry can probably be attributed to mantle development and thermal lag. Conversion of the water production rates given in Table 1 into emissive area surfaces indicates that active areas had a total surface that first decreased when the comet was moving toward its perihelion ($\approx 70 \text{ km}^2$ in November 1985 according to standard thermal models of the nucleus) to reach 36 km² at the time of the Giotto encounter with comet Halley (Keller *et al.* 1987), then increased when the comet was receding from the Sun ($\approx 100 \text{ km}^2$ in April 1986), in qualitative agreement with Houpis' explanation. Still, a detailed quantitative model is urgently needed to fully validate that idea. As a matter of fact, one would like to understand why a similar general behaviour is observed in comets whose dust production is very low and whose exponent *n* of the production law differs from 2, the only value which does not correspond to a strong variation of the outgassing surface (most of the time, *n* has a value between 3 and 4). It seems that the seasonal effect could be appropriate to explain

observations of periodic comets (n ≫ 2) while new comets (n ≈ 2) would rather require models in which mantles develop. Comet Halley might exhibit a seasonal effect because of its long past as a periodic comet and behave simultaneously as new comets because of its high dust to gas ratio and could consequently develop a mantle.

4 SHORT-TERM FLUCTUATIONS

In addition to the general behaviour of the comet described above, brightness variations of large amplitude were observed on very short time scales, of the order of one day or even a few hours, and sometimes much less (Rettig *et al.* 1986, Larson *et al.* 1986, 1989, Feldman *et al.* 1987, Goraya *et al.* 1987, McFadden *et al.* 1987). This was reported by visual observers as well as by astronomers conducting photometric studies. Among the most extensive data sets available to study this phenomenon are those of radio astronomers (Mirabel *et al.* 1986, Gérard *et al.* 1986, Schloerb *et al.* 1986b, Galt 1987), some of UV observers (Festou *et al.* 1986c, Feldman *et al.* 1987; McFadden *et al.* 1987) and those of optical observers (Cochran & Barker 1986, Spinrad *et al.* 1986, Sterken *et al.* 1986, Millis *et al.* 1987, Schleicher *et al.* 1987).

4.1 Spinning motions and variation of the gas and dust productions

The discovery by Schleicher & Millis (1986), that a periodicity of ≈ 7.4 days was present in their photometric observations was first interpreted in terms of a single rotation period of the nucleus in 7.4 days. However, not only is that periodicity not always found by other observers, (Leibovitz & Brosch 1986, found evidence for a 52 h period, Schlosser *et al.* 1986, 1989, did not find that the times of formation of the cyanogen shells were following any regular pattern, even though the average time interval between the formation of two consecutive shells was about 50 h, while the Pioneer Venus Orbiter data (Stewart 1987) favour the 7.4 day periodicity), but those authors found during the analysis of their complete data set that such a periodicity was not always obviously present (Schleicher *et al.* 1987). Consequently, since the March and April 1986 data showed good indications that the periodicity was real (Fig. 5),

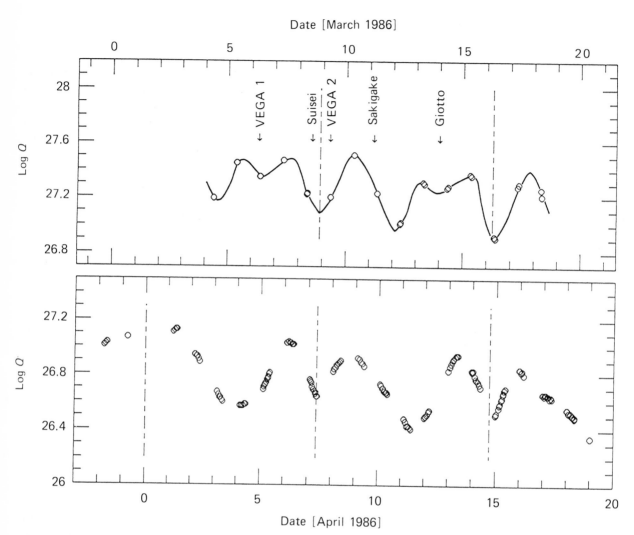

Fig. 5 — Short term variations of the C$_2$ production rates in March and April 1986 (courtesy of Schleicher & Millis, 1986). A periodicity of about 7.4 days is clearly present. The complete data set, extended over a period of a few months around perihelion, show that this regular pattern slowly evolves with time, suggesting the existence of a marked seasonal effect.

production rate estimates were rebinned by Millis *et al.* 1987, in monthly intervals assuming the existence of a 7.4 day periodicity: they found then that once the secular variation had been removed, gas production *minima* and *maxima* of the monthly sets were approximately in phase.

From the works of Festou *et al.* (1987), Parisot *et al.* (1987), Peale & Lissauer (1987, 1989) and Szego *et al.* (1988), it appears that a complex rotational motion of the nucleus (rotation, precession and nutation) coupled to seasonal effects (orbital motion of the comet)

should allow one to completely explain the observations and allow the derivation of the various periods of the nucleus motions. At the present time, only the 2.2 day period seems likely to correspond to a directly detected physical motion (spin around the small axis or precession about that axis, Keller & Thomas 1988) while the 7.4 day period might result from the combination of slower spinning motions with the above fast one (Festou *et al.* 1987, Szego *et al.* 1988). If the periodic phenomenon found by Schleicher *et al.* (1986), is associated with rotation of the nucleus, then the rotational state of Halley's comet nucleus is not affected by the reaction force induced by the outgassing of the surface.

4.2 Did outbursts occur?

If we call an 'outburst' any event during which the gas (or dust) production increases significantly and rapidly compared to any of the rotational motions of the nucleus, numerous outbursts occurred in comet Halley. The best documented events are those of 15 and 21 November 1985 (Meech & Jewitt 1986), 12 December 1985 (Watanabe *et al.* 1986), 18 March 1986 (Feldman *et al.* 1986, Hanner *et al.* 1987), 24–25 March 1986 (Rettig *et al.* 1986, 1987, Hanner *et al.* 1987, Weaver *et al.* 1986, 1987, McFadden *et al.* 1987, Larson *et al.* 1989). The appearance of cyanogen shells reported by Schlosser *et al.* 1986, 1989, are possibly events of a similar nature, since the discontinuous aspect of the gas production is then obvious. If one calls an 'outburst' an event of explosive nature, then only the events of 18 March 1986 (Feldman *et al.* 1986), that of 24 March 1986 (Larson *et al.* 1989) and perhaps the early ones of 1983–85 (West & Pedersen 1983; Festou *et al.* 1986a) deserve that name.

In any case, if the above mentioned events were solely the effect of the spinning of the nucleus (which would bring active areas into sunlight), one should have observed them continuously, which was far from being the case. Only the strongest March events were accompanied by a documented modification of the composition of the gaseous output of the nucleus. One of the aims of the yet to be published detailed modelling of the evolution of gas production rates should be to tell us whether the observed gas production variations are compatible with the assumption that the ejected matter had a constant composition or if they were not simply the effect of the existence of various dissociation time constants among parent species. One should note that the 500 m scale morphological structures recently discovered with the Giotto camera by Keller *et al.* (1987b), may further complicate nucleus activity modelling if active areas are not restricted to terrains which would be fairly flat relative to those of an ideal tri-axis ellipsoid.

5 SOURCES OF MATTER

Since the Giotto camera (HMC) imaged the nucleus in the dominant continuum light, the presence of dust jets arising from bright spots is not an absolute and definitive clue that no gas was diffusing through the dust matrix. However, the good agreement which exists between the area of the calculated outgassing surface and the observed one is an indication that production of gas in dark areas should have been small, if not negligible. This is in agreement with the latest results provided by the analysis of Giotto images which indicate that dust particles were carried *across* the nucleus by the gas (Keller *et al.* 1987b) and not away from dark areas. Consequently, outgassing should occur only in the active areas, likely to be the only parts of the nucleus where volatile matter is directly exposed to incoming solar photons. Such a scheme qualitatively explains the observed daily variations of the activity of the nucleus. However, if the existence of outgassing through the deep layers of dark areas were established, then it would be an indication that dust particles in those regions stick together sufficiently to resist the dragging forces due to the diffusing gas.

There is no strong evidence that outgassing occurs on the dark side of the nucleus (Keller and Thomas 1989), which means that the fraction of the solar energy which is stored in the core is small as compared to that employed to vapourize the upper layers. However, transport by collisions and other mechanisms combine to populate the anti-Sun side with parent molecules (e.g. Magee-Sauer *et al.* 1989,) indicate that [OI] 630 nm emission does not show any significant sunward/antisunward asymmetry; the direct measurement of the H_2O density profile by Krankowsky *et al.*, 1986, indicate a R^{-2} distribution – the comet was likely to be close to a quiescent state at that time (Festou *et al.* 1986c). The rôle of collisions in efficiently redistributing the parent molecules in all directions, even though the matter is known to leave the nucleus surface in a very anisotropic manner, is demonstrated by coma models in which time dependency of the source strength has been included. Time dependency appears to be the dominant factor in determining the shape of the observed intensity profiles or isophotes (e.g. Festou *et al.* 1989). A similar model applied to data collected on Vega 1 (NGE instrument, Hsieh *et al.* 1987) would probably produce an even more significant result than that reported by Festou *et al.* (1989).

Solid grains contribute significantly to the production of coma species: CN jets have been observed by A'Hearn *et al.* (1986), and Rees *et al.* (1986), suspected that CN could be emitted from grains. A C_2 jet has been noticed by Watanabe *et al.* (1986). We should note here that those jets are not true gas jets but rather gas associated to dust jets. A very important question regarding the structure of the coma is whether or not true gas jets exist. As we said above, collisions should redistribute in space molecules that were anisotropically released by the nucleus. Combi (1987), claims that jets of gas could have two origins, the evaporation of CHON particles and the background gas which could 'remember' as far as 10^5 km from the nucleus its original anisotropy of ejection. A way to discriminate between the two kinds of jets is to examine their spatial structure which should be very different. Let us mention, finally, a few observations that require distributed sources to be understood. The mass spectrometer measurements made on board the Giotto spacecraft indicate that CO molecules could be released by CHON particules or a short lived parent (Eberhardt *et al.*, 1986, 1987). Observations by Woods *et al.* (1986, 1987), of the spatial distribution of the CI 165.7 nm emission imply a distributed source as well. Note that if CO is the parent of the observed carbon atoms, one is facing a severe difficulty: the total lifetime of CO is very long to account for the characteristics of the observed spatial distribution. An alternative solution could be production via the electron recombination of CO^+ ions, a process that is more likely than the dissociation of an abundant undetected parent. However, this scheme requires the rapid formation of CO^+ ions, a task that cannot be accomplished by mere photoionization of CO molecules. There is some indication that formadehyde could also be produced in an extended region (Snyder *et al.* (1989), although that conclusion can be questioned according to Mumma & Reuter (1989). The case of the radio observations of the OH 18 cm emission by de Pater *et al.* (1986), is puzzling and have not been confirmed by any other team of radio astronomers having observed the comet at the same

time: the VLA observations indicate that the OH emission arises from clumps clustering around the centre of the coma. Those observations could be another indication that the emission mechanism of the OH radio lines is more complex than generally believed.

6 CONCLUDING REMARKS

The gaseous output of a comet nucleus is determined by the thermodynamical properties of its surface and by the insolation conditions of that surface. Since comet matter is not a chemically pure species and since the nucleus is not entirely covered by frozen volatiles, there is no hope of determining the nature of this matter by a study of the shape of the gas production curve. Even though the outgassing significantly increases near 3 AU pre-perihelion, demonstrating thus that water is a key factor in determining the amount of molecules freed close to the Sun, the continuous monitoring of comet Halley's activity over the entire apparition clearly shows that some molecules can escape in an unknown manner, sometimes during outbursts, at all distances from the Sun both pre- and post-perihelion. What triggers that activity is not well established. The apparition of Halley's comet allowed us to emphasize the extremely important role played by the insolation conditions of the evaporating surfaces. As a matter of fact, it appears possible to explain the observed short time scale variations as well as the long-term variations of the nucleus output of Halley's comet simply by taking into account the complex rotational motions of its nucleus, the details of which are still not completely understood. Yet, it remains to explain how erratic phases occur and whether they are related to those spinning motions.

Key information about the homogeneity of the nucleus ices is provided by the comparison of the shape of the gas production curves of the various coma species: if it is confirmed, beyond modelling uncertainties, that the relative outgassing rates have varied during the apparition, it will have been shown for the first time that a comet nucleus is inhomogeneous. Such an inhomogeneity would exist on a centimetre scale (the depth removed in a few hours of nucleus activity) if compositional variations were observed during outbursts. Compositional differences associated with the secular trend of the gas production curves would imply differences in composition from one active area to another or within an active area on a scale of the order of one to ten metres. The comparison of absolute species production rates is potentially very diagnostic of the overall composition of the nucleus. As an example, if the difference between the H and OH production rates is significant (see Table 1), another important source of hydrogen atoms exists in the coma besides water molecules, and it has not been identified. CHON particles are an obvious candidate, but one should refrain from concluding too rapidly given the simplicity of coma models that are currently in use to derive coma abundances (isotropy of the ejection of matter, steady state gas production, optical depth effects are neglected, constant parent molecules velocities, etc.).

Images from the Giotto camera (HMC) strongly suggest that dark areas are not outgassing. The current modelling of the production curve of the comet should tell us if the production goes towards zero at times or if some gas diffusion through the dust mantle is necessary to provide a background source that accounts for the observed curves (no well marked gas production minima, (e.g. Festou *et al.* 1986c, McFadden *et al.* 1987, Schleicher *et al.* 1986). The absence or presence of such outgasssing provides information on the dust bulk coherence (resistance to an eventual induced mechanical pressure) and on the dust thermal conductivity (is the temperature gradient towards the inner core of the nucleus large enough to protect the snows buried below the surface?). An answer to such a question may have to wait the first drilling of a comet surface.

Active areas represent about 10% of the total surface of the nucleus. Studies of the light curve and of the shape of some dust structures of faint periodic comets show that these objects have their activity governed by one main spot which occupies a small fraction of the total surface of their nuclei. In that respect, Halley's comet seems to be an old comet. This leads us to the natural question: what is the evolutionary path between new and old comets? Or, what is the lifetime of an active area? It seems difficult to assume that the potato shape of the nucleus is not of pristine nature, for that would require an unexplainable highly asymmetric wearing of the surface. In that context, the large dust mantle of

comets implied by the existence of large dark areas would be formed in the early years of the comet's life. Then, considering the size of nucleus structures (that result from the nucleus activity) and the outgassing potential of active areas (the shaping agent), the likely number of orbital revolutions accomplished by the comet cannot be much larger than about a thousand, at the present level of nucleus activity. Another possibility would be that active areas do not maintain their potential during more than an equivalent time interval. Let us conclude by mentioning that if the size of active areas truly changes by factors of two to three during a single apparition, as suggested by the simple conversion of the measured water production rates into emissive surfaces, the above speculations are very fragile and that much will be learned by observing, during one of the planned space missions to a comet, the development of active areas as a nucleus moves along in its orbit.

REFERENCES

A'Hearn, M.F., Hoban, S., Birch, P.V., Bowers, C., Martin, R., & Klinglesmith, D.A. (1986) *Nature* **324** 649

Belton, M.J.S., Wehinger, P., Wyckoff, S., & Spinrad, H. (1986) *ESA SP* **250, I**, 599

Boyarchuk, A.A., Grinin, V.P., Zvereva, A.M., & Sheibet, A.I. (1986), *ESA SP* **250**, 193

Cochran, A., & Barker, E.S., (1986) *ESA SP* **250, I**, 439.

Cochran, A., & Barker, E.S., (1987) *IAC Circ.* **4353**

Combes, M., Moroz, V.I., Crifo, J.F., Lamarre, J.M., Chara, J., Sanko, N.F., Soufflot, A., Bibring, J.P., Cazes, S., Coron, N., Crovisier, J., Emerich, C., Enerenaz, T., Gispert, R., Grigoryev, A.V., Guyott, G., Krasnopolsy, V.A., Nikolsky Yu. V., & Rocard F. (1986) *Nature* **321**, 266.

Combi, M.R. (1987) *Icarus* **71** 178

Craven, J.D., & Frank, L.A. (1987) *Astron. Astrophys.* **187** 351

Delsemme, A.H. (1982) In *Comets*, L.L. Wilkening ed., The University of Arizona Press, Tucson, AZ

De Pater, I., Palmer, P., Snyder, L.E., & Ip, W.H. (1986) *ESA SP* **250, I**, 409

Eberhardt, P., Krankowsky, D., Schulte, W., Dolder, U., Lämmerzahl, P., Berthelier, J.J., Woweries, J., Stubbermann, U., Hodges, R.R., Hoffman, J.H., & Illiano, J.M. (1986) *ESA SP* **250, I**, 383

Eberhardt, P., Krankowsky, D., Schulte, W., Dolder, U., Lämmerzahl, P., Berthelier, J.J., Woweries, J., Stubbermann, U., Hodges, R.R., Hoffman, J.H., & Illiano, J.M. (1987) *Astron. Astrophys.* **187**, 481

Emerich, C. *et al.* (1987) *ESA SP* **278**, 703

Feldman, P.D., A'Hearn, M.F., Festou, M.C., McFadden, L.A., Weaver, H.A., & Woods, T.N. (1986) *Nature* **324** 433

Feldman, P.D., Festou, M.C., A'Hearn, M.F., Arpigny, C., Butterworth, P.S., Cosmovici, C.B., Danks, A.C., Gilmozzi,

R., Jackson, W.M., McFadden, L.A., Patriarchi, P., Schleicher, D.G., Tozzi, G.P., Wallis, M.K., Weaver, H.A., & Woods, T.N. (1987) *Astron. Astrophys.* **187**, 325.

Festou, M.C., Lechacheux, J., Kohl-Moreira, J.L., Encrenez, T., Baudrand, J., Combes, M., Despiau, R., Laques, P., Le Fevre, O., Lemonnier, J.L., Lelievre, G., Mathez, G., Pierre, M., & Vidal, J.L. (1986a) *Astron. Astrophys.* **169** 336

Festou, M.C., Arpigny, C., Bertaux, J.-L., Carey, W., Danks, A.C., Gilmozzi, R., Hughes, D.W., Ip, W.-H., Patriarchi, P., Tozzi, G.P., Wallis, M.K. & Zarnecki, J. (1986b) *Astron. Astrophys.* **155** L17

Festou, M.C., Feldman, P.D., A'Hearn, M.F., Arpigny, C., Cosmovici, C.B., Danks, A.C., McFadden, L.A., Gilmozzi, R., Patriarchi, P., Tozzi, G.P., Wallis, M.K., & Weaver, H.A., (1986c) *Nature* **321** 361

Festou, M.C., Drossart, P., Lecacheux, J., Encrenaz, T., Puel, F., & Kohl-Moreira, J.L. (1987) *Astron. Astrophys.* **187** 575

Festou, M.C., Tozzi, G.P., Falciani, R., Smaldone, L.A., Felenbok, P., & Zucconi, J.-M. (1989) *Astron. Astrophys.*, in press

Fischer, D. (1986) *ESA SP 250*, **III**, 303

Galt, J. (1987) *Astron, J.* **93** 747

Gérard, E. Bockelée-Morvan, D., Bourgois, G., Colom, P., & Crovisier, J. (1986), *ESA SP 250*, **I**, 589

Goraya, P.S., Wahab Uddin, & Srivastava, R.K. (1987) *Earth, Moon and Planets* **38**, 53

Green, D.W.E., & Morris, C.S. (1986). *ESA SP 250*, **I**, 613

Hanner, M.S., Tokunaga, A.T., Golisch, W.F., Griep, D.M., & Kaminski, C.D. (1987) *Astron. Astrophys.* **187** 653

Hasegawa, I. (1986) *ESA SP 250*, **III**, 255

Hsieh, K.C., Curtiz, C.C., Fan, C.Y., Hunten, D.M., Ip, W.-H., Keppler, E., Richter, A.K., Umlauft, G., Afonin, V.V., Ero, J., & Somogyi, A.J. (1986) *Astron. Astrophys.* **187**, 375.

Houpis, H.L.F. (1989), chapter 'Models of cometary nuclei', in Volume 2 of this work

Huebner, W.F., & Whipple, F.L. (1976) *Ann. Rev. of Astron. and Astrophys.*, 143

Kaneda, E., Tagaki, M., Hirao, K., Shimizu, M., & Ashihara, O. (1986) *ESA SP 250*, **I**, 397

Keith, G. (1986) *The Astronomer* **22**, 196

Keller, H.U., Arpigny, C., Barbieri, C., Bonnet, R.M., Cazes, S., Coradini, M., Cosmovici, C.B., Delamere, W.A., Huebner, W.F., Hughes, D.W., Jamar, C., Malaise, D. Reitsema, H.J., Schmidt, H.U., Schmidt, W.K., Seige, P., Whipple, F.L., & Wilhelm, K. (1986) *Nature* **321** 320

Keller, H.U., Delamere, W.A., Huebner, W.F., Reitsema, H.J., Schmidt, H.U., Whipple, F.L., Whilhelm, K., Curdt, W., Kramm, R., Thomas, N., Arpigny, C., Barbieri, C., Bonnet, R.M., Cazes, S., Coradini, M., Cosmovici, C.B., Hughes, D. W., Jamar, C., Malaise, D., Schmidt, K., Schmidt, W.K.H. & Seige, P. (1987a) *Astron. and Astrophys.* **187** 807

Keller, H.U., Kramm, R., Thomas, N., Craubner, H., Milkusch, E., & Schwarz, G. (1987b) *B.A.A.S.* **19**, 3, 878

Keller, H.U., & Thomas, N. (1988) *Nature*, **333**, 146

Keller, H.U., & Thomas, N. (1989) *Astron. Astrophys.*, in press

Klinger, J. (1983) *Icarus* **55** 169

Krankowsky, D., Lämmerzahl, P., Herrwerth, I., Woweries, J., Eberhardt, P., Dolder, U., Herrmann, U., Schulte, W., Berthelier, J.J., Illiano, J.M., Hodges, R.R., & Hoffmann, J.H. (1986) *Nature* **326**

Larson, H.P., Davis, S.D., Mumma, M.J., & Weaver, H.A. (1986) *ESA SP 250*, **I**, 335

Larson, H.P., Hu, H.Y., Mumma, M.J., & Weaver, H.A. (1989)

Icarus, in press

Leibovitz, E.M., & Brosch, N. (1986) *ESA SP 250*, **I**, 605

Levin, B.J. (1943) *C.R. Acad. Sci. USSR (NS)* **38**, 82

Magee-Sauer, K., Roesler, F.L., Scherb, F., & Harlander, J. (1989) *Icarus* **76**, 89

McFadden, L.A., A'Hearn, M.F., Feldman, P.D., Roettger, E.E., Edsall, D.M., & Butterworth, P.S. (1987) *Astron. Astrophys.* **187**, 333

Marcus, J. (1986) *ESA SP 250*, **III**, 307

Meech, K.J., Jewitt, D.C., & Ricker, G.R. (1986) *Icarus* **66** 561

Meech, K.J., & Jewitt, D.C. (1986) *ESA SP 250*, **I**, 553

Mirabel, I.F., Bajaja, E., Arnal, E.M., Cerosimo, J.C., Colomb, F.R., Martin, M.C., Mazzaro, J., Morras, R., Poppel, W.G.L., Silva, A.L., & Boroakoff, V. (1986) *ESA SP 250*, **I**, 595

Millis, R.L., Schleicher, D.G., Birch, P.V., Martin, R., & A'Hearn, M.F., (1987) *B.A.A.S.* **19**, 3, 880

Mumma, M.J., & Reuter (1989) *Astrophys. J. Lett.*, in press

Parisot, J.P., Puel, F., & Festou, M.C. (1987) *B.A.A.S.* **19**, 3, paper 12.26

Peale, S.J., & Lissauer, J. (1987) *B.A.A.S.* **19**, 3, 879

Peale, S.J., & Lissauer, J. (1989) *Astrophys, J.*, in press

Prialnik, D., & Bar-Nun, A. (1988) *Icarus* **74**, 272

Rees, D., Meredith, N.P., & Wallis, M.K. (1986) *ESA SP 250*, **I**, 493

Rettig, T., Ruchti, R., Baumbaugh, D., Baumbaugh, A., Knicker-bocker, K., & Dawe, J. (1986) *ESA SP 250*, **III**, 93

Rettig, T.W., Hern, J.R., Ruchti, R., Baumbaugh, B., Baumbaugh, A.E., Knickerbocker, K.L., & Dawe, J. (1987) *ESA SP 278*, 265

Roemer, E. (1976) *NASA SP 393*, 380

Roettger, E.E., Feldman, P.D., A'Hearn, M.F., Festou, M.C., McFadden, L.A., & Gilmozzi, R. (1989) *Icarus*, in press

Schleicher, D.G., & Millis, R.L. (1986) *Nature*, **324**, 646

Schleicher, D.G., Millis, R.L., Tholen, D.J., Hammel, H.B., Piscitelli, J.R., Lark, N., Birch, P.V., & Martin, R. (1987) *B.A.A.S.* **19**, 3, 879

Schloerb, P.F., Claussen, M.J., & Tacconi-Garman, L. (1986a) *ESA SP 250*, **I**, 583

Schloerb, P.F., Kinzel, V.M., Swade, D.A., & Irvine, W.M. (1986b) *ESA SP 250*, **I**, 577

Schloerb, P.F. (1987) In *Workshop on Cometary Radio Astronomy*, in press

Schlosser, W., Schultz, R., & Koczet, P. (1986) *ESA SP 250*, **III**, 495

Schlosser, W., Schultz, R., & Koczet, P. (1989) *Astron. Astrophys.* **214**, 375

Snyder, L.E., Palmer, P., & de Pater, I. (1989) *Astron. J.*, in press

Spinrad, H., McCarthy, P.J., & Strauss, M.A. (1986) *ESA SP 250*, **I**, 437

Spinrad, H., Djorgovsky, S., & Belton, M.J.S. (1984) *IAU ;Circ.* **3996**

Sterken, C., Manfroid, J., & Arpigny, C. (1986) *ESA SP 250*, **I**, 445

Stewart, A.I.F. (1987) *Astron. Astrophys.* **187** 369

Szego, K., Kondor, A., Larson, S., Merenyi, E., Sagdeev, R.Z., & Toth, B.A. (1988) to appear in *Adv. Space Res.*

Tokunaga, A.T., Golisch, W.F., Griep, D.M., Kaminski, C.D., & Hanner, M.S. (1986) *Astron. J.* **92**, 1183

Watanabe, J., Kawakami, H., Tomita, K., Kinoshita, H., Nakamura, T., & Kozai, Y. (1986) *ESA SP 250*, **III**, 267

Weaver, H.A., Mumma, M.J., Larson, H.P. & Davis, D.S. (1986) *Nature* **324**, 441

Weaver, H.A., Mumma, M.J., & Larson, H.P. (1987) *Astron. Astrophys.* **187**, 411

Weissman, P.R. (1986) *ESA SP* **250, III**, 517

Weissman, P.R. (1987) *Astron. Astrophys*, **187** 873

Weissman, P.R., & Kieffer, H.H. (1981) *Icarus* **47**, 302

West, R., (1982) In: *The need for coordinated ground-based observations of Halley's Comet*. ESO publication, P. Véron, M.C. Festou & K. Kjär (eds), 123

West, R. (1989) *IAU Circ.* **4712**

West, R.M., & Pedersen, H. (1983) *Astron. Astrophys.* **121**, L11

West, R., & Jorgensen, H.E., (1989) *Astron. Astrophys.* in press

Wisniewski, W., Levy, D., & Fink, U. (1987) *IAU Circ.* **4372**

Woods, T.N., Feldman, P.D., & Dymond, K.F. (1986) *ESA SP* **250, I**, 431

Woods, T.N., Feldman, P.D., & Dymond, K.F. (1987) *Astron. Astrophys.* **187**, 380

Wyckoff, S., Wagner, R.M., Wehinger, P.A., Schleicher, D.G., & Festou, M.C. (1985) *Nature* **316**, 241

Wyckoff, S., Wehinger, P.A., Ferro, T., Tegler, S., Theobald, J., Womack, M., & Peterson, B. (1987) *B.A.A.S.* **19**, 3, paper 17.25

17

Rocket and satellite observations of the ultraviolet emissions of comet Halley

Paul D. Feldman

INTRODUCTION

The advent of the space age and the resulting ability to observe celestial objects from above the Earth's atmosphere, which obscures all but a small part of the electromagnetic spectrum to ground-based observers, has had as significant an impact on our understanding of comets as the *in situ* measurements made by the recent several space probes to comet Halley. The vacuum ultraviolet region of the spectrum, defined at the long wavelength end by the atmospheric cut-off at ~ 300 nm and which extends to the soft X-ray region at a few tens of nanometres contains the resonance transitions of the most abundant chemical elements in the universe (e.g. H, He, C, O, N, S) as well as those of many of the simple molecules made of these elements. For comets, this region is particularly important for a number of reasons: the radiation observed from comets, other than the component of sunlight *reflected* by cometary dust grains, is mainly the result of resonance scattering or fluorescence of solar radiation by atoms, ions, and molecules in the gaseous coma of the comet. In addition, the dissociation products of the primary constituent of cometary ice, H_2O, all fluoresce in the ultraviolet (water molecules themselves do not fluoresce strongly at all), as do CO, CO_2 and their dissociation and ionization products. The gaseous species which comprise the visible coma, such as CN, C_2, C_3 and NH, are all dissociation products of polyatomic molecules which exist as impurities (totalling less than 1%) in the water ice. Ultraviolet observations of nearly three dozen comets, made

during the past two decades, have shown water ice to be the dominant constituent, confirming the 'icy-conglomerate' model of Whipple (1950, 1951). In turn, the *in situ* mass spectrometer measurements at comet Halley (Krankowsky *et al.* 1986) have confirmed this picture, at least for one comet.

The first ultraviolet observations of comets were made serendipitously in 1970 by two orbiting observatories, OAO-2 and OGO-5, which detected the extended hydrogen envelope surrounding comet Bennett (1970 II) (Bertaux *et al.* 1973, Code *et al.* 1972). Ultraviolet spectra and direct hydrogen Lyman-α images were first obtained for two bright comets, Kohoutek (1973 XII) and West (1976 VI), using instruments carried above the atmosphere aboard sounding rockets, in 1974 and 1976 (Feldman & Brune 1976, Opal *et al.* 1974). However, routine ultraviolet observations of comets became possible only with the launch of the International Ultraviolet Explorer (IUE) satellite observatory in January 1978, a resource which continues in operation today. The results of the sounding rocket and early IUE work which provided the corroboration of water ice as the dominant cometary constituent were reviewed by Feldman (1983). A description of the IUE observations of comets was recently given by Festou & Feldman (1987).

For the 1985/86 apparition of Halley, in addition to campaigns of IUE observations and sounding rocket launches, several other Earth-orbital programmes were planned, based on utilization of the Space Shuttle as the observing platform. One of these, an autonomous payload (known as SPARTAN) from the University of Colorado, was aboard the *Challenger* when it exploded on 28 January 1986. The other payload, a much more ambitious observatory comprising three ultraviolet telescopes, ASTRO-1, had its scheduled launch date, 6 March 1986, deferred by over four years as a result of the *Challenger* accident. The ASTRO-1 payload instruments provided measurement capabilities in terms of spectral range, spectral and spatial resolution, and polarimetry, not previously available for cometary observations, together with a significant improvement in sensitivity over that of IUE. The extension of the spectral range below 115 nm (the short-wavelength limit of IUE) would have made possible the detection of the noble gases He, Ne, and Ar, if they were present in any reasonable abundance, in the cometary ice (Feldman & Davidsen 1983).

Despite the loss of Shuttle-based observations, a large number of ultraviolet observations of comet Halley were made, as summarized in Table 1. A number of valuable observations were made from spacecraft which were designed for completely different scientific objectives. Notable in this regard are the observations made by the ultraviolet spectrometer on Pioneer Venus Orbiter, as Halley was well placed for observation at perihelion from Venus, but not from Earth. In the sections below, the initial results from each of the instruments listed in Table 1 will be given. The principal point of comparison is the derivation of the water vapourization rate, a quantity common to all of the ultraviolet experiments. Note that this discussion is limited only to the *remote* ultraviolet observations of comet Halley, and thus excludes the data from the Suisei Lyman-α imager and the Vega TKS experiment, both of which were part of the spacecraft flotilla to Halley.

2 SPECTROSCOPY

The ultraviolet spectrum of comet Halley was monitored with the International Ultraviolet Explorer (IUE) between 12 September 1985 and 8 July 1986 (r < 2.6 AU) pre-perihelion and post-perihelion) at regular time intervals except for a two-month period

around the time of perihelion. The first spectrum of comet Halley ever recorded from space was a long-wavelength IUE exposure taken on 12 September 1985 (Feldman *et al*. 1987). Typical of the period following perihelion is the exposure from 11 March 1986, shown in Fig. 1, which displays all of the known neutral cometary emission features in this spectral range as well as a very strong component of sunlight reflected by particulate grains in the coma. Comparison with the measured solar spectrum (Mount & Rottman 1981), shown as the dotted curve in Fig. 1, indicates that the reflected spectrum is more reddened (7% per 10 nm) than in the visible (Jewitt & Meech 1986), and is in accord with the theoretical predictions of Lamy & Perrin (1986) based on the *in situ* determination of the grain size distribution. The strong solar component tends to mask the emission from the CO_2^+ doublet near 289 nm. However, when a properly reddened solar spectrum is subtracted from a spectrum such as in Fig. 1, the resultant CO_2^+ emission suggests an abundance relative to OH comparable to that found in other comets (Festou *et al*. 1982).

As with the long-wavelength spectra, the short-wavelength spectrum (Fig. 2) does not show any previously undetected cometary emission features. In fact, the apparent weakness of OI, CI, and SI, relative to those emissions in the spectrum of comet Bradfield (1979 X) (Weaver *et al*. 1981), reflects lower fluorescence efficiencies and photodissociation rates owing to the lower solar UV fluxes characteristic of solar minimum as compared with 1980 which was the year of solar maximum.

A similar, less extensive programme of observations of ultraviolet spectra was made by the Soviet ASTRON orbiting observatory (Boyarchuk *et al*. 1986, 1987). This programme included three observation dates in December 1985, three in April 1986, and one in June 1986. The spectra in the wavelength range 190–350 nm are qualitatively identical to those obtained with the IUE except for the presence of a feature at ~219 nm which is identified as CO+. However, this identification is questionable as the first negative system of CO+ would exhibit two other detectable features at nearby wavelengths, as seen in the spectrum of comet West (1976 VI) (Feldman & Brune 1976). The longest IUE exposures do not show

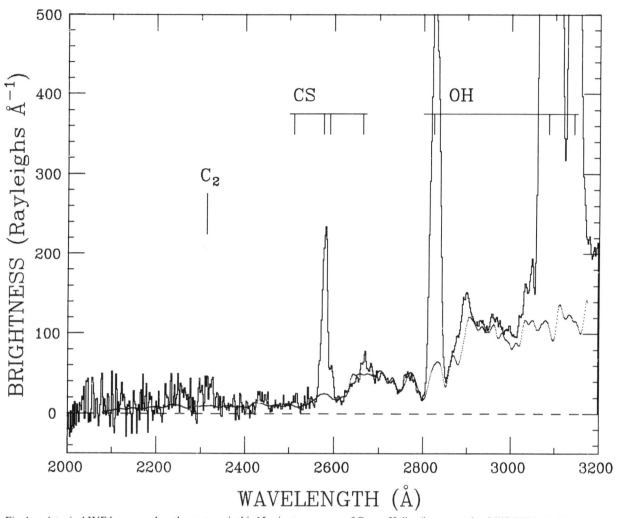

Fig. 1 — A typical IUE long-wavelength spectrum is this 15-minute exposure of Comet Halley (image number LWP 7773) obtained on 11 March 1986. (from Feldman *et al.* 1987). Note, 1 Å = 10^{-10} m or 0.1 nanometres (nm).

this feature which would easily have been detected at the brightness level reported by Boyarchuk *et al.* It is possible that this feature is a result of a transient outburst, similar to that observed on 18 March 1986 by Feldman *et al.* (1986). However, it should be noted that the CO_2^+ emission at 289 nm appears to have the same strength relative to the continuum in both the IUE and ASTRON spectra. This problem remains to be resolved.

While the satellite observatories offer the unique possibility of permitting observations to be made over an extended period, and consequently, over a large range of heliocentric distances, the use of sounding

rockets as a platform for cometary ultraviolet observations also has some unique advantages despite the short observing time above the atmosphere (~ 5 minutes). These advantages include the ability to observe a comet at relatively small ($\sim 20°$) elongation angles from the Sun by using the Earth's limb as a solar occultation disk, and the use of instruments specifically designed for cometary observations and often employing state-of-the-art ultraviolet technology. Both of these advantages were exploited in the sounding rocket experiments of Woods *et al.* (1986). The payload consisted of a Dall–Kirkham telescope, a Rowland circle spectrograph and a slit-jaw camera,

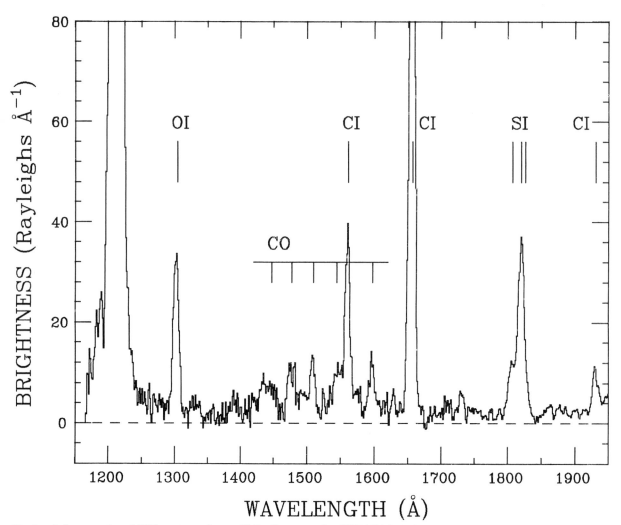

Fig. 2 — A short-wavelength IUE spectrum of comet Halley (image number SWP 27884) obtained on 9 March 1986. The exposure time was 45 minutes (from Feldman *et al.* 1987). Note, 1 Å = 10^{-10} m or 0.1 nm.

the last of which was used to transmit, in real time, an image of the comet to the ground for remote control of the telescope pointing. The spectrograph utilized an astigmatism-corrected concave grating to achieve a spatial resolution of $\sim 10''$ along a 7.7 arc-minute long slit, and a two-dimensional photon-counting resistive anode detector. The spectral resolution was 1.2 nm. The spatial resolution in the spectra obtained, an example of which from 26 February is shown in Fig. 3, was slightly degraded from the resolution of the spectrograph by short-term jitter in the pointing stability of the rocket payload. Nevertheless, both launches of this payload (a second flight, the day of the

Giotto encounter with Halley, was scheduled after the postponement of the ASTRO-1 mission) were completely successful and provided the first long-slit spectra of any comet in the vacuum ultraviolet.

The data of Fig. 3 can be reduced to the spectra shown in Fig. 4, from Woods *et al.* (1986). The combination of increased sensitivity, higher activity of the comet (as measured by its water production rate), and being closer to the Sun (on 26 February 1986) leads to a higher signal-to-noise ratio spectrum for these data (72.2 second exposure) as compared to the 45 minute exposure with IUE on 9 March 1986 shown in Fig. 2. In addition to the usual cometary

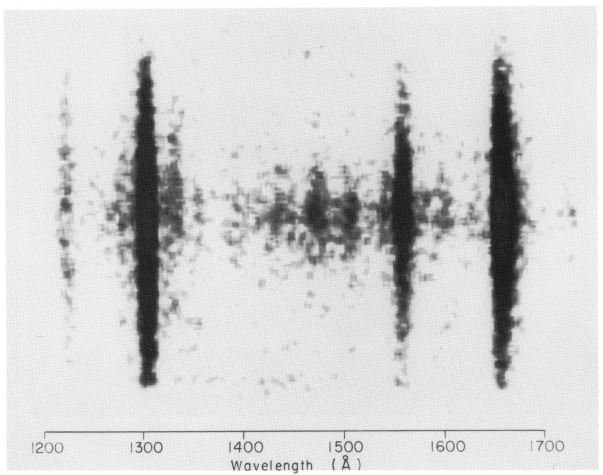

1200   1300   1400   1500   1600   1700
Wavelength (Å)

Fig. 3 — A photographic representation of the raw counts from the experiment of Woods *et al*. (1986) is shown for a 72.2 second exposure of comet Halley centred in the entrance slit of the spectrograph. The horizontal axis is the wavelength dispersion from 120 to 180 nm and the vertical axis is along the 7.7 arc-minute slit with the sunward direction pointing down. The three brightest lines are OI (130.4 nm), CI (156.1 nm) and CI (165.7 nm). HI (121.6 nm) appears as a weak feature owing to the heavy attenuation below 123 nm by a CaF_2 filter in front of the entrance slit. Note, 1 Å = 10^{-10} m or 0.1 nm.

features of HI, OI, and CI, Woods *et al*. have identified 11 bands of the CO fourth positive system as well as weak emissions of CI (128.0 nm), CII (133.5 nm), and OI (135.6 nm). The latter is of particular importance as a diagnostic of electron impact excitation in the inner coma, a mechanism which is needed to reconcile the spatial profiles of atomic carbon emission with those of CO, as the former cannot be matched solely by photodissociation alone (Woods *et al*. 1987). The abundance of CO derived from the rocket data confirm the mass 28 identification from the Giotto neutral mass spectrometer (Eberhardt *et al*. 1987) as being due to CO rather than N_2, and it being

the second most abundant molecule in the coma, after H_2O, at 15–20% relative to water. The rocket data are also consistent with the distributed source origin of a large fraction of the CO observed by Eberhardt *et al*.

For emission line spectra, such as those shown in Figs 2 and 4, an objective grating spectrograph provides, at its focal plane, monochromatic images of the comet in each of the spectral lines. As with the long slit spectra, these images contain the spatial distribution of the emitting species and provide a direct test of the photochemical models of the coma. Such images were obtained by one of the instruments onboard the sounding rocket payload of McCoy *et al*.

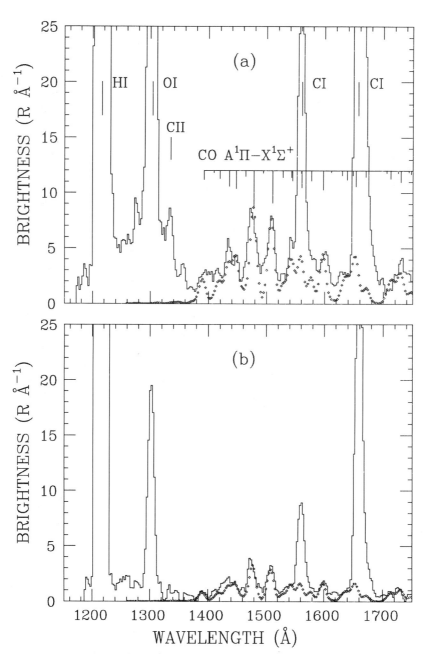

Fig. 4 — The central coma spectrum from the two rocket flights of Woods *et al.* (1986): (a) 26 February 1986 and (b) 13 March 1986. The synthetic CO spectrum (diamonds), convolved to the instrument resolution, is based on calculated g-factors of CO fluorescence by ultraviolet solar radiation. Note, 1 Å $= 10^{-10}$ m or 0.1 nm.

(1986). These data, taken with a spectrograph with a $12°$ square field of view and 1 arc-minute resolution, contain more of the extended atomic oxygen and carbon emissions than do the data of Woods *et al.* (1986) which were limited by the 7.7 arc-minute long slit of their instrument.

3 LYMAN-α IMAGING AND PHOTOMETRY

As water is the primary molecular constituent of the gaseous coma, its dissociation product, hydrogen, is the most abundant atomic species in the coma. Approximately 90% of the photodissociations of water result in H and OH, with the hydroxyl radical, OH, further broken into O and H atoms. The long lifetime of H, $\tau \approx 1 \times 10^6$ s at 1 AU (H atoms are both photoionized and ionized through charge exchange collisions with solar wind protons), coupled with mean excess velocities of dissociation of 20 and 8 km s^{-1}, give a scale length, vτ, ∼0.1 to 0.2 AU for a comet 1 AU from the Sun. The cometary atoms resonantly scatter the strong solar Lyman-α line at 121.6 nm producing a bright and very extended image of the comet at this wavelength. Such an image of Halley, from the sounding rocket experiment of McCoy *et al.* (1986) is shown in Fig. 5.* The linear scale is measured in units of Gm, or 10^6 km (the Sun's diameter is ∼1.5 Gm). The distortion of the isophotes in the anti-sunward direction results from the acceleration of the light hydrogen atoms by the radiation pressure of solar Lyman-α photons acting over the approximately ten-day lifetime of the atoms. From the successive displacements of the outer isophotes, McCoy *et al.* were able to derive a mean velocity of 8 km s^{-1} under the assumption of a simple radial outflow model.

Two spacecraft intended for other objectives were also able to obtain large-scale Lyman-α images of comet Halley. These were the Pioneer Venus Orbiter (PVO), in orbit around Venus since 1978, and the Dynamics Explorer-1 (DE-1) satellite, launched in 1981. Both of these are spin-stabilized spacecraft, unlike the observatory spacecraft which are three-axis stabilized and are capable of pointing at a given celestial target with an accuracy of a few arc-seconds. The ultraviolet spectrometer on PVO was designed to study the ultraviolet emissions from the upper atmosphere of Venus, and its wavelength range was similar to that of IUE. However, to obtain data on Halley, the grating wavelength was held fixed, at either HI (121.6 nm), OI (130.4 nm), or CI (165.7 nm), and the spin axis of the spacecraft was orientated so

* to be found in the colour illustration section in the centre of the book.

that the spectrometer field of view scanned through the calculated position of the comet on the sky. The effective field of view was 1°.4 × 2°.2, so that the atomic oxygen and carbon appeared point-like, while only the HI Lyman-α emission exhibited a spatial variation along the scan track. By keeping the spacecraft spin axis fixed in inertial space, the motion of the comet on the sky enabled the PVO scans to build up an image, in two dimensions, of the Lyman-α envelope. A complete image requires several days to acquire and consequently does not represent an instantaneous snapshot of the hydrogen distribution in the coma. Nevertheless, the continuous operation of the PVO ultraviolet spectrometer during the period around perihelion, when comet Halley was obscured to Earth-based viewers, provided a unique record of the evolution of the cometary activity during the time when the comet was closest to the Sun (Stewart 1987).

The DE-1 satellite, placed in a highly elliptical orbit about the Earth, carries an ultraviolet imaging photometer designed to observe the global distribution of the terrestrial aurora. As HI Lyman-α is brighter, by some two orders of magnitude, than any other emissions in the photometer pass-band, the spin-scan image, with a resolution of ∼17 arc-minutes, gives essentially a Lyman-α image of the comet (Craven *et al.* 1986). Fortunately, there is sufficient overlap between the PVO, DE-1 and the rocket experiment of McCoy *et al.* (1986) to make direct comparisons of the various Lyman-α observations possible. None of these experiments showed the rapid time fluctuations or point-like image quality reported from the Lyman-α imager on Suisei (Kaneda *et al.* 1986) and so comparison with this latter data set is not possible. It should be noted that while IUE also recorded the Lyman-α brightness of comet Halley, the small field of view of the IUE spectrographs, 10″ × 20″, translates into spatial scales of 10^3–10^4 km, considerably smaller than the Gm scale of the imaging experiments, again inhibiting a direct comparison of the data.

4 WATER PRODUCTION RATES

As noted above, all of the ultraviolet measurements have in common the ability to be used to derive the production rates of the parent molecules of the observed species. In the case of H_2O, all three dissociation products, H, O, and OH, are detectable in the ultraviolet, and it was shown fairly early that the relative brightnesses of the ultraviolet emissions from these three species were consistent with water being the dominant parent species being vapourized from the surface of the cometary nucleus (Feldman 1983). The *in situ* neutral mass spectrometer measurements of Krankowsky *et al.* (1986) confirmed this result for comet Halley, with water being 80% of the gaseous component of the coma. Moreover, the *in situ* measurements additionally confirmed two premises of the models used to derive the water production rate, Q_{H_2O}, from remote observations: (1) radial outflow, implying a R^{-2} variation in H_2O density, R being the distance from the nucleus; and (2) an outflow velocity of ~ 1 km s^{-1} for the parent molecules.

For the IUE and ASTRON measurements of OH using relatively small fields of view (10″ × 20″ and 1′ diameter, respectively, for IUE and ASTRON), the derivation of Q_{H_2O} from the observed surface brightness is generally achieved with the use of steady-state spherically symmetric models. Although the validity of these assumptions is clearly questionable in light of both *in situ* and remote observations of Halley, they provide the only means of inter-comparing observations made over a large range of heliocentric and geocentric distances. For the IUE data, the vectorial model of Festou (1981) and the OH fluorescence efficiencies of Schleicher and A'Hearn (1988), which depend on the comet's heliocentric velocity, are used to derive Q_{H_2O}. For the parent molecules, the outflow velocity was taken to be 1 km s^{-1}, while the photodissociation lifetimes of both H_2O and OH were taken from Festou (1981). Boyarchuk *et al.* (1986, 1987) used an independent model, with similar parameters, for the ASTRON data. The IUE results of Feldman *et al.* (1987), together with the ASTRON values of Q_{H_2O}, are shown in Fig. 6. The agreement between the two, for times that near-simultaneous

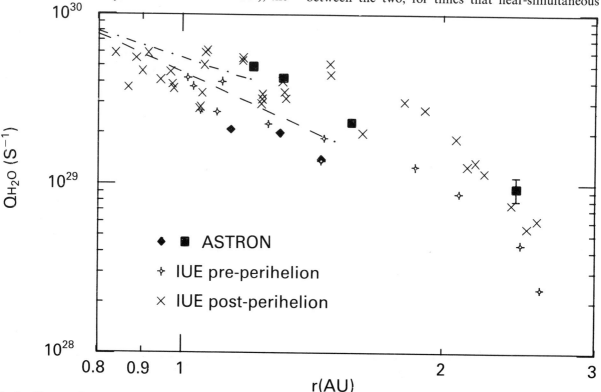

Fig. 6 — Water production rate for comet Halley as a function of heliocentric distance as derived from IUE and ASTRON observations of OH emission. The dashed and dot-dashed lines are the power-law fits, pre-perihelion and post-perihelion, respectively, to the Lyman-α data of Craven & Frank (1987).

observations were made, is excellent. One notes the large scatter in IUE data points, particularly for heliocentric distances, $r < 1.5$ AU. This scatter does not reflect instrumental uncertainties, but rather the short-term variability of the water production rate, and will be discussed further below. The large asymmetry in Q_{H_2O} between pre-perihelion and post-perihelion is evident in the figure, and mirrors the behavior of the visual light curve.

The derivation of Q_{H_2O} from the Lyman-α data is not quite as straightforward. One approach is to compare the observed images with detailed models, such as those of Meier & Keller (1985). However, as all of the Lyman-α data (except for that from IUE) are obtained with wide-field imaging instruments, at large distances from the nucleus ($R > 1$ Gm), the flow of hydrogen atoms is effectively radially outward, and the simple radial outflow model can be used instead. A

difficulty arises in that at large values of R, > 5 Gm, the cometary emission becomes blended with the sky background of geocoronal and interplanetary Lyman-α emission, and this background must be accurately known and subtracted from the image. A second problem arises from the uncertainty in the fluorescence efficiency for the transition, as the solar flux at the centre of the Lyman-α line is variable in time, both as a function of the 27-day solar rotation and 11-year solar activity cycle. Although the total Lyman-α flux from the Sun has been continuously monitored over the past few years by the Solar Mesosphere Explorer satellite, the flux at line centre has not. Various approaches to determining this value are used by McCoy *et al.* (1986), Craven & Frank (1987), and Stewart (1987), and the resulting values agree to within 25%. In general, there is fairly close agreement, in the derived hydrogen atom production rate,

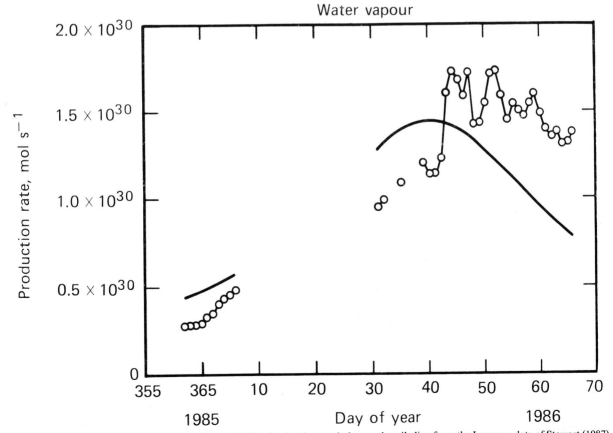

Fig. 7 — Apparent water production rate of comet Halley for the time period around perihelion from the Lyman-α data of Stewart (1987).

between the three sets of Lyman-α measurements. There is also good agreement with the values of Q_{H_2O} derived from the OH measurements, as indicated by the two lines in Fig. 6 which show the power law fit of Craven & Frank (1987) to their data.

The PVO data of Stewart (1987) provide unique information about the activity of comet Halley at perihelion, and the derived values of Q_{H_2O} are shown in Fig. 7. The solid curve in the figure is the behaviour expected if Q_{H_2O} varied simply as r^{-2}. That is clearly not the case here, where a large increase in activity occurs immediately after perihelion (day 40). The data after this point also exhibit fluctuations which follow, with about a one day delay, the shape of the visible light curve of comet Halley as determined from the ground-based observations of Millis & Schleicher (1986). In the case of the Lyman-α data, the amplitude of the variation is considerably smaller than that in the visible as a result of the long lifetime of the hydrogen atoms (5 to 10 days at the heliocentric distances of the observations) which tends to integrate the H-atom production rate over this period. The long lifetime also tends to enhance the 7.4-day sidereal period derived by Millis and Schleicher at the expense of the shorter time-scale structure (resembling, in some instances, a 2.2-day period) seen in their data.

There is, unfortunately, little overlap between the PVO and IUE measurements. Comparison of Figs 6 and 7 shows good agreement in Q_{H_2O} for the period at the end of December 1985. Between the last PVO observation, on 7 March 1986, and the first IUE post-perihelion observation on 9 March, there is an apparent decrease of about a factor of 2 in Q_{H_2O}. This decrease is also seen in the data of Craven & Frank (1987) and in the sounding rocket data of Woods *et al.* (1986). In the latter case, the two rockets were launched 15 days apart, exactly 2 periods in the light curve, and so observed the comet both times at the same phase. From the observed OI (130.4 nm) emission rate, Woods *et al.* derived a factor of three difference in Q_{H_2O}, which again is in good agreement with the PVO value for 26 February and the IUE value for 13 March. It is fair to say, in summary, that no serious discrepancies exist between any of the remote ultraviolet observations of H, O, and OH.

Several of the authors have used their data on Q_{H_2O} to estimate a water mass loss, Δm, for the entire

apparition of comet Halley. The results, as expected, are fairly similar: Feldman *et al.* (1987) find $\Delta m = 3 \times 10^{11}$ kg; Boyarchuk *et al.* (1986, 1987) give $\Delta m = 3.7 \times 10^{11}$ kg; and Stewart (1987) gives $\Delta m = 4 \times 10^{11}$ kg. Taking the mass of comet Halley to be $m \approx 1 \times 10^{14}$ kg (Rickman 1986), this gives $\Delta m/m$ in the range of $(3-4) \times 10^{-3}$. Allowing for the dust component and the other gaseous species, the total mass loss is thus of the order of 0.5% of the cometary mass.

Both the IUE and ASTRON data contain information about the production rates of the parents of other observed species. In particular, both Feldman *et al.* (1987) and Boyarchuk *et al.* (1986) give values for the production rate of the CS parent, CS_2. The latter give

$$\frac{Q_{CS_2}}{Q_{H_2O}} = (3.4 - 6.2) \times 10^{-4},$$

while the IUE data give a range of $(3.0-13.4) \times 10^{-4}$ for the same range of heliocentric distances. The apparent discrepancy results from the short-term variability of the comet, as discussed below, as the actual value obtained depends critically on the time of observation.

5 SHORT-TERM VARIABILITY

The variability of the activity of comet Halley noted above is vividly illustrated by the visible light curves obtained by Millis & Schleicher (1986) over many consecutive nights in March and April 1986. From these data, Millis & Schleicher derived a 7.4-day sidereal period for the light curve even though the data showed fine structure indicative of a shorter rotation period commensurate with the 2.2-day period claimed by other investigators. Light curve data are also obtained by the IUE Fine Error Sensor (FES), which is a slit-jaw camera used in the acquisition and tracking of the celestial targets being observed by the satellite. As a visible light photometer, the FES has an effective aperture of about 18″, and because the IUE is in a geosynchronous orbit, it can uniquely provide continuous monitoring of the visible brightness of a comet over extended observation periods. During the months of March and April 1986, when there were frequent IUE observing shifts scheduled, large variations in the visible brightness were detected from day

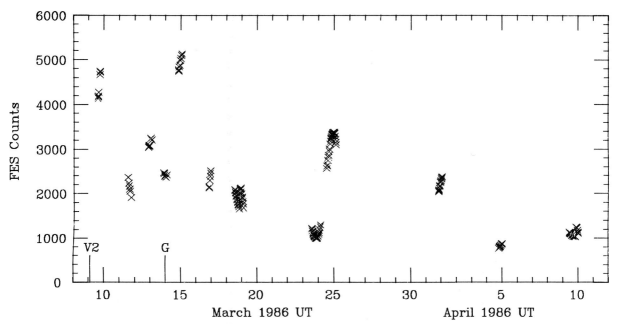

Fig. 8 — IUE Fine Error Sensor photometry for March and April 1986 (from Feldman *et al.* 1987). The times of the Vega-2 and Giotto encounters are indicated.

to day and also even within an 8- or 16-hour observing shift, as can be seen in Fig. 8 (Feldman *et al.* 1987) These data, while discontinuous, agree well, in both phase and amplitude, with the ground-based observations of Millis & Schleicher. Two features in these data bear noting.

Between 23 and 25 March 1986, IUE observations were made over a 40-hour period, all but eight of which were spent on comet Halley. The FES data, shown in the top panel of Fig. 9, exhibit a smooth, nearly sinusoidal variation with an apparent period of about 2 days. The lower panel of Fig. 9 shows that the ultraviolet emissions during the same period, from OH, CS, CO_2^+, and dust, all followed the visible light curve with the only difference being the amplitude of the variation (McFadden *et al.* 1987). In particular, OH, from the relatively long-lived water parent ($\tau \sim 1$ day), does not increase as rapidly as does CS (with a parent lifetime ~ 500 seconds) or the dust continuum. It is for this reason that there appear large fluctuations in the derived values of Q_{CS2}/Q_{H_2O}, noted above, and in the water production rate shown in Fig. 6. The large amplitude variations in both the visible light curve and the ultraviolet emissions continued through mid-

April 1986 and then diminished during the remainder of the IUE observing dates.

The other temporal event of note occurred on 18–19 March 1986 in the form of a transient outburst, of ~ 0.2 day duration, seen in the FES count rate superimposed on the more gradual light curve of Millis & Schleicher (1986). Accompanying this outburst was the detection of a large column abundance of CO_2^+ ions at a distance of 150 000 km from the nucleus, but without any increase in OH emission at the nucleus. This event has been interpreted by Feldman *et al.* (1986) as resulting from the eruption of a sub-surface pocket of volatile ices, which also produced an outward moving, extended clump of CO^+ ions detected from ground-based observations on 20 March 1986. Outbursts of this nature appear to be associated with many comets, most notably P/Schwassmann-Wachmann 1, but the frequency of occurrence on comet Halley must be relatively low.

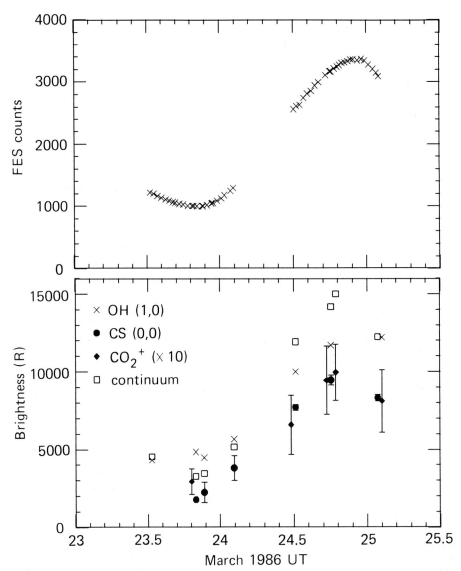

Fig. 9 — IUE FES lightcurve for the period 23–25 March 1986 together with the brightnesses of the principal ultraviolet emission features recorded during the same period.

6 CONCLUSION

Nearly complete monitoring of the production rate of the dominant cometary species, water, of comet Halley was provided by remote ultraviolet observations of the emissions of H, OH, and O by a variety of spacecraft instruments during a ten-month period centred on perihelion. The various data from these instruments give a consistent picture of the evolution of the gaseous coma during this period, and are in good agreement with the results derived from the *in situ* measurements made during the Giotto and Vega encounters with Halley. Both the short-term variability and the long-term trend, in which the water production rate reaches its maximum shortly after perihelion and remains higher on the outbound leg of the orbit than on the inbound leg, mirror the visual behaviour of the comet. Thus, these data provide the basis for normalizing the extensive set of ground-based observations of visible emissions of minor constituents, and for comparison with ultraviolet observations of some two dozen other comets.

ACKNOWLEDGEMENTS

The preparation of this review was partly supported by NASA grants NAG 5–619 and NSG-5393.

REFERENCES

Bertaux, J.L., Blamont, J.E., & Festou, M. (1973) Interpretation of hydrogen Lyman-alpha observations of Comets Bennett and Encke, *Astron. Astrophys.* **25** 415–430

Boyarchuk, A.A., Grinin, V.P., Zvereva, A.M., & Sheihet, A.I. (1986) Estimations of water production rate in the Halley Comet using the ultraviolet data obtained on the space station 'Astron', In: *Exploration of Halley's Comet*, ESA SP-250 **3** 193–194

Boyarchuk, A.A., Grinin, V.P., Sheikhet, A.I., & Zvereva, A.M. (1987) Pre- and post-perihelion Astron ultraviolet spectophotometry of Comet Halley: a comparative analysis, *Sov. Astron. Lett.* **13** 92–96

Code, A.D., Houck, T.E., & Lillie, C.F. (1972) Ultraviolet observations of comets, In: *The scientific results from Orbiting Astronomical Observatory (OAO-2)*, (ed. A.D. Code) NASA SP-310, 109–114

Craven, J.D., Frank, L.A., Rairden, R.L., & Dvorsky, M.R. (1986) The hydrogen coma of Comet Halley before perihelion: preliminary observations with Dynamics Explorer 1, *Geophys. Res. Lett.* **13** 873–876

Craven, J.D., & Frank, L.A. 1987, Atomic hydrogen production rates for P/Halley from observations with Dynamics Explorer 1, *Astron. Astrophys.* **187** 351–356

Eberhardt P. *et al.* (1987) The CO and N_2 abundance in Comet P/Halley, *Astron. Astrophys.* **187** 481–484

Feldman, P.D. (1983) Ultraviolet spectroscopy and the composition of cometary ice, *Science* **219** 347–354

Feldman, P.D., & Brune, W.H. (1976) Carbon production in Comet West (1975n), *Astrophys. J.* (*Letters*) **209** L145-L148

Feldman, P.D., & Davidsen, A.F. (1983) Planned observations of P/Halley with the Hopkins Ultraviolet Telescope on Space Shuttle, In: *Cometary Exploration III*, (ed. T.I. Gombosi) Hungarian Academy of Sciences, Budapest, 65–71

Feldman, P.D., A'Hearn, M.F., Festou, M.C., McFadden, L.A., Weaver, H.A., & Woods, T.N. (1986) Is CO_2 responsible for the outbursts of Comet Halley?, *Nature* **324** 433–436

Feldman, P.D. *et al.* (1987) IUE observations of Comet P/Halley: evolution of the ultraviolet spectrum between September 1985 and July 1986, *Astron. Astrophys.* **187** 325–328

Festou, M.C. (1981) The density distribution of neutral compounds in cometary atmospheres: I. Models and equations, *Astron. Astrophys.* **95** 69–79

Festou, M.C., Feldman, P.D., & Weaver, H.A. (1982) The ultraviolet bands of the CO_2^+ ion in comets, *Astrophys. J.* **256** 331–338

Festou, M.C., & Feldman, P.D. (1987) Comets, In: *Exploring the universe with the IUE satellite*, (ed. Y. Kondo) Reidel, Dordrecht, 101–118

Jewitt, D., & Meech, K.J. (1986) Cometary grain scattering versus wavelength, or, 'What color is comet dust?', *Astrophys. J.* **310** 937–952

Kaneda, E., Ashihara, O., Shimizu, M., Takagi, M., & Hirao, K. (1986) Observation of Comet Halley by the ultraviolet imager of Suisei, *Nature* **321** 297–299

Krankowsky, D. *et al.* (1986) *In situ* gas and ion measurements at Comet Halley, *Nature* **321** 326–329

Lamy, P.L., & Perrin, J.M. (1986) Comet Halley: Implications of the impact measurements for the optical properties of the dust, In: *Exploration of Halley's Comet*, ESA SP-250 **2** 65–68

McCoy, R.P., Opal, C.B., & Carruthers, G.R. (1986) Far-ultraviolet spectral images of Comet Halley from sounding rockets, *Nature* **324** 439–441

McFadden, L.A., A'Hearn, M.F., Feldman, P.D., Roettger, E.E., Edsall, D.M., & Butterworth, P.S. (1987) Activity of Comet P/Halley 23–25 March 1986: IUE observations, *Astron. Astrophys.* **187** 333–338

Meier, R.R., & Keller, H.U. (1985) Predictions of the hydrogen Lyman-α coma of Comet Halley, *Icarus* **62** 521–537

Millis, R.L., & Schleicher, D.G. (1986) Rotational period of Comet Halley, *Nature* **324** 646–649

Mount, G.H., & Rottman, G.J. (1981) The solar spectral irradiance 1200–3184 Å near solar maximum: July 15, 1980, *J. Geophys. Res.* **86** 9193–9198

Opal, C.B., Carruthers, O.R., Prinz, D.K., & Meier, R.R. (1974) Comet Kohoutek: ultraviolet images and spectrograms, *Science* **185** 702–705

Rickman, H. (1986) Masses and densities of Comets Halley and Kopff, In: *The comet nucleus sample return*, ESA SP-249 195–205

Schleicher, D.G., & A'Hearn, M.F. (1988) The fluorescence of cometary, OH *Astrophys. J.* **331** 1058–1077

Stewart, A.I.F. (1987) Pioneer Venus measurements of H, O, and C production in Comet P/Halley near perihelion, *Astron. Astrophys.* **187** 369–374

Weaver, H.A., Feldman, P.D., Festou, M.C., A'Hearn, M.F., & Keller, H.U. (1981), IUE observations of faint comets, *Icarus* **47** 449–463

Whipple, F.L. (1950) A comet model. I. The acceleration of Comet Encke, *Astrophys. J.* **111** 375–394

Whipple, F.L. (1951) A comet model. II. Physical relations for comets and meteors, *Astrophys. J.* **113** 464–474

Woods, T.N., Feldman, P.D., Dymond, K.F., & Sahnow, D.J. (1986), Rocket ultraviolet spectroscopy of Comet Halley and abundance of carbon monoxide and carbon, *Nature* **324** 436–438

Woods, T.N., Feldman, P.D., & Dymond, K.F. (1987), The atomic carbon distribution in the coma of Comet P/Halley, *Astron. Astrophys.* **187** 380–384

18

Evidence for the composition of ices in the nucleus of comet Halley

D. Krankowsky and P. Eberhardt

1 INTRODUCTION

Ices as main constituents in a cometary nucleus were conceptually introduced by Whipple (1950, 1951) in his 'dirty snowball' model, although the idea that comets are made of ice dates back to the last century (Laplace 1813, Hirn 1889). Whipple's model describes the nucleus as a fragile icy conglomerate of solar type material consisting of a mixture of stony material and ices composed of compounds of C, N, O, and H which exists in a solid form at very low temperatures ($\ll 100$ K). While Whipple's model specifically assumed water ice to be the major ice component, Delsemme & Swings (1952) suggested that other more volatile species can be present, either in the form of clathrate hydrates or ionic hydrates embedded in the water ice. The icy conglomerate model had a remarkable success in explaining in a quantitative or at least qualitative way many of the observed features of comets.

Before the space missions to comet Halley, all knowledge of the chemical nature of the volatile component in comets has been gained from spectra of cometary comae. Molecules, radicals, atoms, and ions identified spectroscopically are mostly products from dissociation and ionization of the gases released from the comet's nucleus. The composition of the volatile component in the nucleus has been estimated from the observed dissociation products by accounting for photo- and ion-chemistry in the coma. *Ad hoc* assumptions are made about the molecules present when the comets were formed with guidance from what is currently known about interstellar molecules and theories of the presolar nebula.

Evidence for the dominance of water ice in comets is obtained from the observed variation with heliocentric distance, of the brightness curves of 'new' comets (Delsemme 1975, and 1985) and from the nongravitational force on short-period comets (Delsemme 1972). Furthermore, the presence of the photodissociation products of water in the ultraviolet spectra of comets — atomic hydrogen, oxygen, and the hydroxyl radical — and their relative abundances and spatial variations, suggests the dominance of water ice (Weaver *et al.* 1981 a, b). Another indication for H_2O comes from the spectroscopical detection of H_2O^+ ions (Herzberg and Lew 1974) which are now frequently observed in comets from the ground. Indications for its presence in the comet Giacobini–Zinner were obtained when the International Cometary Explorer spacecraft flew past the comet in September 1985 (Ogilvie *et al.* 1986).

Although the accumulated evidence has made the presence and dominance of water in comets almost a fact, attempts at a direct detection have failed or gained marginal results at best. The tentative identification of water in comet Bradfield (1974b) by its 1.35 cm line at radio wavelength by Jackson *et al.* (1976) was challenged by Crovisier *et al.* (1981) and by Hollis *et al.* (1981). Searches for the 1.35 cm line in other comets gave negative results except for comet IRAS-Araki-Alcock (1983d) where apparently this line was detected by Altenhoff *et al.* (1983). Similarly, efforts to search for water ice were not conclusive. Reports of the detection of H_2O absorption at 3 μm (Campin *et al.* 1983, Hanner 1984) were questioned by A'Hearn *et al.* (1984). However, observational evidence for water ice in comet Kohoutek (1973XII) and in comet P/Schwassmann-Wachmann 1 was claimed by Crifo (1983) and Hartmann & Cruikshank (1983), respectively.

Despite the lack of conclusive direct observations of water in comets, vapourization of water ice has been accepted as a major process controlling the gas production. Some observations suggesting unusual abundances of CO and/or CO_2 (Wyckoff and Wehinger 1976, Miller 1980, A'Hearn & Cowan 1980) were taken as indications that more volatile species can play an important rôle for the gas production at large heliocentric distance. However, recently Delsemme (1985) argued that all new and quasi-new comets, similarly to the short-period comets, have their sublimation rate controlled by water ice or snow only.

2 COMA GAS COMPOSITION

The first definite detection of water vapour in a comet by remote observation was achieved on 22 and 24

December 1985 (Mumma *et al.* 1986), when nine infrared spectral lines of the ν_3 band (2.65 μm) were found in the coma of Halley's comet by a Fourier transform spectrometer onboard the NASA-Kuiper Airborne Observatory. Fig. 1 shows a spectrum measured on 24 December 1985. In addition to the Doppler-shifted cometary fluorescent emission lines, a lunar reflectance spectrum is displayed showing terrestrial absorptions by H_2O and CO_2. From these data and later observations production rates, spatial distribution, and outflow velocities of H_2O were derived (Weaver *et al.* 1986).

The first *in situ* measurement of water vapour in a cometary coma was obtained from the neutral mass spectrometer (NMS) experiment onboard the European Space Agency's Giotto spacecraft on 13 March 1986, when it flew past the nucleus of Halley's comet at a distance of about 600 km (Krankowsky *et al.* 1986). Fig. 2 shows an excerpt from a neutral gas mass spectrum measured by the instrument's mass analyzer, where the peak at 18 amu/e is due to cometary H_2O. The peak at 17 amu/e (where e is the charge of the electron) has contributions from cometary OH and OH produced by the dissociation of H_2O in the ion source as well as from the cometary NH_3. The 16 amu/e mass peak is made up of O, CH_4, and NH_2 from the coma gas and from O and NH_2 arising from dissociation of H_2O, OH, and NH_3 in the ion source.

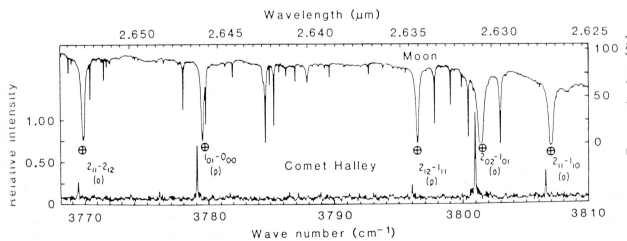

Fig. 1 — Part of infrared spectrum of comet Halley obtained on 24 December, 1985, from the Kuiper Airborne Observatory (KAO). The lower trace shows the emission spectrum from comet Halley, the upper one the lunar reflectance spectrum. The cometary H_2O lines are Doppler-shifted relative to the terrestrial absorption lines visible in the lunar spectrum. The labels give the quantum numbers and the ortho/para designation. Figure from Mumma *et al.* (1986).

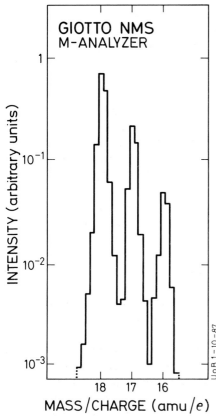

Fig. 2 — Mass spectrum of the gas in Halley's coma recorded on 14 March 1986, by the M-Analyzer of the neutral mass spectrometer on the Giotto spacecraft. Shown is the mass 16 to 18 amu/e range. The largest peak at mass 18 amu/e is due to cometary H_2O. For the molecules contributing to the other mass peaks see text. Figure adapted from Krankowsky et al. (1986).

In Fig. 3 the radial profile of the water density as measured by the NMS experiment on Giotto is shown. For the first time it was established that water is the most abundant volatile species in a comet, accounting for about 80% by volume of the coma gas at 1 000 km. Production rates for water and other gases were determined as well as the radial profiles of the gas outflow velocity and the gas temperature, and the actual photodestruction rate of water.

Other direct measurements of the water vapour in Halley's comet were obtained from the infrared instrument IKS on the Vega-1 spacecraft (Combes et al. 1986) on 6 March by observing the v_3 band at 2.65 μm and from the three-channel spectrometer on Vega-2 (Krasnopolsky et al. 1986a) on March 9 by utilizing the 1.38 μm band. The International Ultraviolet

Explorer (IUE) satellite observatory provided indirect monitoring of the water production of Halley in the period from September 1985 to July 1986, except for a two-month period between January and March when the solar elongation angle to the comet was too small. The H_2O gas production rates as a function of the heliocentric distance of Halley were derived from the fluorescent emission of OH (O–O band) (Festou et al. 1986, Feldmann et al. 1986).

Data from the NMS experiment on Giotto have established that H_2O is the most abundant volatile with approximately 80% by volume. It was also shown that carbon monoxide (CO) is the second most abundant gas in the coma of Halley when it is assumed that the peaks at 28 amu/e in the mass spectra are due to CO only (Eberhardt et al. 1986a). The upper limit given by these authors corresponds to a gas production rate for CO relative to H_2O of 15% at distances between 20 000 and 75 000 km. It drops to 7% at a distance of 1 000 km from the nucleus. This can be understood in terms of an extended source of CO in the coma (Eberhardt et al. 1987a) contributing about half the amount of the measured CO. Implications of the extended CO source will be discussed later.

Carbon monoxide was also measured by an

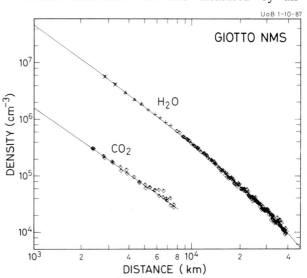

Fig. 3 — Water density in the coma of comet Halley measured in situ by the neutral mass spectrometer as function of the cometocentric distance. The curve shown for CO_2 must be considered as upper limit, as other molecules could also contribute to the mass 44 amu/e peak in the instrument (see text). Figure adapted from Krankowsky et al. (1986).

ultraviolet spectrometer on two sounding rockets (Woods *et al.* 1986). The data obtained on 13 March gave a CO gas production rate relative to H_2O of $17 \pm 4\%$ in very good agreement with the value from the NMS experiment, thus confirming the assumption that the 28 amu/e mass spectral peaks are mostly due to CO. The radial brightness profile from the rocket flight shown in Fig. 4 also seems to suggest the presence of a second source of CO, in addition to the nucleus, as can be seen from the bending of the brightness profile at decreasing distances. Marginal detections of CO were also reported by Combes *et al.* (1986) from the infrared experiment IKS aboard the Vega-1 spacecraft and by Festou *et al.* (1986) from IUE observations. Estimates for CO gas production rates relative to H_2O, ranging from 10 to 20%, are not in contradiction with the Giotto and rocket results cited above.

Carbon dioxide (CO_2) was detected for the first time in a comet by the IKS instrument on Vega-1

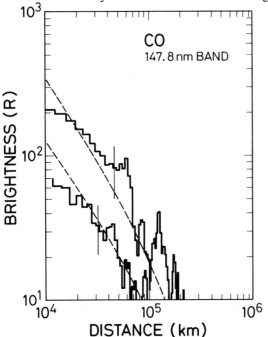

Fig. 4 — Distribution of the 147.8 nm CO band brightness as function of the distance from the nucleus of comet Halley. The upper and lower curves were obtained on 26 February and 13 March, 1986, respectively, with a UV spectrometer flown on sounding rockets. The dashed lines represent the expected emission profiles for a radial outflow model with CO production from the nucleus only. Figure from Woods *et al.* (1986).

(Combes *et al.* 1986, 1988) with an abundance of 2.7% relative to H_2O. From the radial density profile of the mass peak at 44 amu/e (Fig. 3), measured by the NMS experiment on Giotto, an upper limit of 3.5% relative to water was found, allowing for the possible contributions of other species such as for example CS and C_3H_8 to the 44 amu/e mass peak (Krankowsky *et al.* 1986).

Methane (CH_4) has been considered as another carbon-bearing candidate parent molecule in models of the comet formation. Originally suggested by Wurm (1943) for a scenario where comets formed in the neighbourhood of the giant planets, it is also included at various abundance levels in the recent models which attempt to link comets to the interstellar clouds (Mitchell *et al.* 1981, Biermann *et al.* 1982, Yamamoto *et al.* 1983). A search for CH_4 in the comet Halley on 20 March, 1986, utilizing the Fourier Transform Spectrometer on the NASA-Kuiper Airborne Observatory, resulted in an upper limit for the gas production rate of 4×10^{28} molecules s^{-1} (Drapatz *et al.* 1986) corresponding to an abundance of 4% relative to H_2O. The evaluation of the CH_4 abundance from the neutral gas spectra of the NMS experiment on Giotto is difficult, owing to the dissociation fragments from H_2O, NH_3 and heavier molecules which lead to interference in the mass range 12 to 16 amu/e. So far, only an upper-limit of 7% for the volume mixing ratio relative to water has been published (Krankowsky *et al.* 1986). An estimate for the CH_4 abundance can be obtained from the *in situ* ion composition measurements. Thus, from the relative abundance of ion mass peaks in the range 12 to 19 amu/e measured by the ion mass spectrometer (IMS) experiment, Allen *et al.* (1987) derived a CH_4 gas production rate of 2% relative to H_2O by modelling the photo- and ion-chemistry in the inner coma. Ground-based infrared spectroscopy of the v_3 band of CH_4 around 3.3 microns between March 15 and April 8 1986, have yielded estimates for the CH_4 production rate (Kawara *et al.* 1988). Assuming rotational temperatures of 50 to 200 K methane production rates of 0.2 to 1.2% relative to water have been derived.

Ammonia (NH_3), considered as one possible major nitrogen-bearing parent molecule in comets, was detected at radio-wavelength in comet IRAS Araki-

Alcock (1983d) (Altenhoff *et al.* 1983). From the NMS experiment on Giotto an upper limit for the volume mixing relative to H_2O of 10% has been reported (Krankowsky *et al.* 1986). Similarly, as for CH_4, improved estimates are obtained from the mass spectral data of the ion mass spectrometer on Giotto through modelling. Limits on the NH_3 production rate relative to H_2O from the IMS data inside the contact surface are 1 to 2% (Allen *et al.* 1987). This derivation of the NH_3 abundance has been critized by Marconi and Mendis (1988). Their main point is that by increasing the solar UV flux, responsible for ionizing H_2O, by a factor of ten over what Allen *et al.* (1987) assumed in their model, the NH_3 abundance required to fit the Giotto IMS ion composition data is drastically reduced. Such an increase in UV flux will also slightly decrease the photodestruction lifetime of H_2O. The decreased H_2O lifetime (4.4×10^4 s compared to the generally accepted value of 6.6×10^4 s for 0.9 AU from Huebner and Carpenter, 1979) as derived from a first interpretation of the Giotto NMS measurements (Krankowsky *et al.* 1986), was taken as circumstantial evidence for an elevated UV flux during the Giotto encounter. However, as shown in the following chapter when taking into account the radial variation of the measured outflow velocity (Lämmerzahl *et al.* 1987), the H_2O lifetime is 5.6×10^4 s, close to the expected value for quiet Sun conditions. From three independant spectrophotometric observations of NH_2 in the visible Wyckoff *et al.* (1988) derived a low NH_3 abundance relative to H_2O of 0.002 ± 0.001 for Halley. Such a low concentration of NH_3 is substantiated by a Fabry-Perot measurement of NH_2 which yielded a relative NH_3 abundance of 0.001 for Halley (Magee-Saur *et al.* 1988).

Molecular nitrogen (N_2) is invoked as an alternative form of nitrogen in models of comet formation from a gas mixture which contains interstellar molecules. Molecular nitrogen has not been found in cometary spectra, although the N_2^+ ion was observed (Fowler 1910, Swings & Page 1948). In a preliminary evaluation of neutral mass spectral peaks at 28 amu/e from the Giotto NMS experiment, Eberhardt *et al.* (1986a, 1987a) derived an upper limit of 10% for the N_2 gas production rate relative to H_2O. Mass spectra obtained at two electron energies in the electron impact source with different relative contributions of

N_2 and CO to mass 28 amu/e were taken into account. An improved upper limit for the N_2 abundance comes from ion composition data. From the ion spectrum obtained by the Giotto IMS experiment, in the distance interval from 1.1×10^5 to 1.7×10^5 km, an abundance of less than 10% for N_2^+ relative to CO^+ ions was deduced (Balsiger *et al.* 1986). From this limit and from the maximum value of 20% for the relative CO gas production rate from the rocket data measured on 26 February 1986 (Woods *et al.* 1986) it was concluded that the abundance of N_2 is less than 2% in comet Halley (Allen *et al.* 1987). Ground-based emission spectra of molecular ions in the visible provided for a still lower estimate of the N_2 abundance in comet Halley. From the measured column density ratio N_2^+/CO^+ and taking into account the measured CO/H_2O ratio Wyckoff and Theobald (1989) derived an N_2 abundance of 4×10^{-4} relative to H_2O.

The major gases released from the nucleus of Halley's comet are H_2O, CO and CO_2 representing about 90% of the volatile fraction in the nucleus. The remainder is composed of a number of species with abundances around a few percent. The infrared instrument IKS on Vega-1 detected in the 3.2 to 3.7 μm range spectral features which were attributed to saturated and unsaturated hydrocarbons and to formaldehyde (H_2CO) (Combes *et al.* 1986, 1988). The H_2CO production rate relative to H_2O was estimated to be ≤ 4%. Observations of the radio emission of H_2CO at 6 cm wavelength yielded a H_2CO production rate of about 1.5% relative to H_2O (Snyder *et al.* 1989). There is a clear indication in the data that H_2CO is also released from an extended source in the coma in addition to its release from the nucleus. Hydrogen cyanide (HCN) was detected as a minor species in comet Halley with the IRAM 30-metre millimetre-wave telescope at the tenth of a percent level (Bockelée-Morvan *et al.* 1986). The 28 amu/e peak in the ion spectra from the Giotto IMS experiment has been interpreted as being mainly due to protonated hydrogen cyanide H_2CN^+ (Ip 1988). An upper limit of 3×10^{-4} relative to H_2O was derived for the HCN abundance in comet Halley. This value is about 3 times lower than what is obtained from the IRAM observations. Searches for methyl cyanide (CH_3CN) and cyano-acetylene (HC_3N), as possible parents for the CN radical observed in Halley, resulted in upper limits only, which are of the order of 10^{-3} relative to H_2O. Table 1 summarizes parent molecules found in comet Halley.

A strong and broad emission emission feature at 3.2-3.5 μm detected by the IKS experiment on Vega-1 has been attributed to the C-H stretch in saturated und unsaturated hydrocarbons (Combes *et al.* 1986). For the origin of this emission hydrocarbons in the gas phase, polycyclic aromatic hydrocarbons (PAH), and organic mantles of small grains have been considered. If the observed emission is due to solids, then the carbon production rate has been estimated to be on the percent level or less relative to water (Combes *et al.* 1988). For gaseous fluorescence a similar production rate for carbon was originally quoted (Combes *et al.* 1986). However, in the meantime the authors have corrected upward this value by almost a factor of 100 (Combes *et al.* 1988). Such a high carbon production rate is not supported by the *in situ* gas and ion measurements in the coma of comet Halley.

Isotopic abundances from measurements in the volatile fraction of comet Halley have been reported so far for the elements hydrogen, carbon, nitrogen, oxygen, and sulphur. The $^{34}S/^{32}S$ ratio as determined from ion mass spectra in the inner coma was 0.045 ± 0.010 (Krankowsky *et al.* 1986) in agreement with the terrestrial ratio of 0.044. The mass spectrometric measurement of isotopic composition of light volatile elements in the coma is difficult, because the low-abundance isotopes are always heavier than the abundant isotopes, and can thus be masked by hydrids. Only for the ion H_3O^+ is the interference for the isotopic H_2DO^+ limited to a few possible species which can be corrected for with sufficient accuracy. Fig. 5 is an example of the 17 to 21 amu/e region of an ion spectrum recorded inside the contact surface by the NMS experiment on Giotto. From such spectra Eberhardt *et al.* (1987b) deduced a value of 0.0023 ± 0.0006 for the $^{18}O/^{16}O$ ratio in the water of Halley. While this figure is identical to the terrestrial ratio, the errors in the preliminary evaluation are too large for a meaningful comparison with observed meteoritic ^{16}O abundance variations. For the hydrogen in Halley's water Eberhardt *et al.* (1986b, 1987b) found an isotopic ratio of $0.6 \times 10^{-4} \leq D/H \leq 4.8 \times 10^{-4}$. In comparison Schleicher *et al.* (1986) obtained an upper

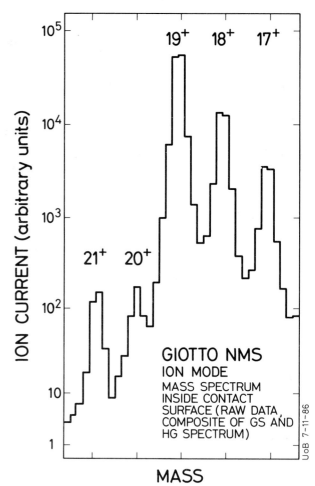

Fig. 5 — High-resolution mass spectrum of the ion composition-in the coma of comet Halley. Shown is the mass 16.5 to 21.5 amu/e range as measured in the ion mode by the M-Analyzer of the neutral mass spectrometer on the Giotto spacecraft. The dominant peak at mass 19 amu/e is due to H_3O^+ ions. The small peaks at masses 20 and 21 amu/e are due to H_2DO^+, $H_3{}^{18}O^+$ and other ions with one or more low-abundance isotopes. From these peaks the D/H and $^{18}O/^{16}O$ ratios were derived. Figure from Eberhardt *et al.* (1986b, 1987b).

limit of OD/OH $\leq 4 \times 10^{-4}$ for comet Halley from IUE observations. Emission lines of $^{13}C^{14}N$ resolved in ground-based spectra for the first time, have been utilized to determine the $^{12}C/^{13}C$ isotope ratio of CN in comet Halley (Wyckoff *et al.* 1989). A $^{12}C/^{13}C$ ratio of 65 ± 9 has been found which is 2.7σ lower than the bulk Solar System value of 89. For the nitrogen isotope ratio $^{14}N/^{15}N$ a 2σ lower bound of 200 has been estimated (Wyckoff *et al.* 1989) consistent with the bulk Solar System ratio of 272.

Table 1 — Abundances of probable parent molecules in the coma of comet Halley. For a detailed discussion, see text and quoted references.

Species	Gas Production Rate Relative to H_2O (by number)	Instrumental Technique
CO	0.07[1]	Giotto NMBS, gas spectra[5]
	$0.17 \cdots 0.20$[2]	Rocket UV experiment[6]
	0.15[2]	Giotto NMS, gas spectra[5]
CO_2	$\leqslant 0.035$	Giotto NMS, gas spectra[7]
	0.027	Vega IKS, IR spectra[8]
CH_4	$\leqslant 0.07$	Giotto NMS, gas spectra[7]
	$\leqslant 0.04$	KAO, IR spectra[9]
	≈ 0.02[3]	Giotto IMS, ion spectra[19]
	$0.002 \cdots 0.012$	IR spectra, ground based[14]
NH_3	$\leqslant 0.1$	Giotto NMS, gas spectra[7]
	$0.01 \cdots 0.02$[3]	Giotto IMS, ion spectra[10]
	0.002	Spectrophotometry[12]
	0.001	Fabry-Perot[13]
N_2	< 0.1	Giotto NMS, gas spectra[5]
	< 0.02[3]	Giotto IMS, ion spectra[10]
	≈ 0.0004	N_2^+ Spectrophotometry[4]
H_2CO	$\leqslant 0.04$	Vega IKS, IR spectra[8]
	0.015[2]	VLA, Radiowave[15]
HCN	≈ 0.001	IRAM telescope, millimetre spectra[11]
	≈ 0.0003[3]	Giotto IMS, ion spectra[16]

[1] Released from nucleus.
[2] Released from nucleus and extended source.
[3] Inferred from models of coma.
[4] Wyckoff and Theobald, (1989).
[5] Eberhardt et al. (1986b, 1987b).　　[11] Bockelée-Morvan et al. (1986)
[6] Woods et al. (1986).　　[12] Wyckoff et al. (1989).
[7] Krankowsky et al. (1986).　　[13] Magee-Saur et al. (1988).
[8] Combes et al. (1988).　　[14] Kawara et al. (1988).
[9] Drapatz et al. (1986).　　[15] Snyder et al. (1989).
[10] Allen et al. (1987).　　[16] Ip et al. (1989).

3 INFERENCE OF VOLATILES IN THE NUCLEUS

The chemical composition of the gas in the coma is different from the composition of the volatiles in the nucleus. Photoprocesses as well as the chemical reactions between neutrals, and between neutrals and ions, lead to substantial alterations of the molecular composition of the expanding gas in the coma. Expansion velocities are typically of the order of 1 km s^{-1}, and hence a molecule takes 3 hours to reach a distance of 10 000 km. For some species, lifetimes against photodissociation are much shorter. Optical composition measurements, depending on the strength of the investigated lines, often cover a substantial radial range. It is then necessary to make appropriate corrections for the change in the chemistry occurring over this distance range, in order to obtain the composition at the nuclear surface. These corrections are facilitated if radial profiles can be obtained which can be fitted to the theoretical models, such as the Haser model (Haser 1957, 1966), or the more sophisticated vectorial model (Combi & Delsemme 1980, Festou 1981). Measurements closer to the nucleus than 10^4 km are, however, often difficult, and the radial resolution obtainable is often only a few thousand kilometres (see Fig. 4).

The *in situ* measurements by the Giotto spacecraft give data to a minimum distance of approximately 1 000 km with a resolution of 250 km. This high spatial resolution facilitates the extrapolation of the molecular composition to the nuclear surface. The NMS also provided for the first time an accurate measurement of the gas expansion velocity profile (Lämmerzahl et al. 1986, 1987). Fig. 6 shows the profile of the radial velocities for gases obtained from the measured ram energy of the individual species. Spectral observations of HCN (Despois et al. 1986, Schloerb et al. 1986) and measurements of the time delays between associated dust and water vapour jets (Krasnopolsky et al. 1986b) give results agreeing with the average velocity obtained by the NMS, lacking, however, the radial resolution. In a Monte Carlo approach to study the nonadiabatic expansion in a cometary atmosphere Hodges (1989) showed that photolytic heating cannot explain the *in situ* measured increase of the H_2O radial expansion velocity. On the other hand Crifo (1989) argues that the release of latent heat by the formation of water clusters through recondensation in the expanding coma leads to radial expansion velocities 30% higher than measured by the Giotto NMS.

A detailed knowledge of the velocity profile is important for determining the lifetimes against photodestruction from the directly measured density profiles of the species. With an average velocity of 0.9 km s^{-1}, derived from a first analysis, Krankowsky et al. (1986) inferred from the data shown in Fig. 3 a lifetime of 4.4×10^4 s for the H_2O molecule. If the radial variation of the expansion velocity is taken into account, the resulting lifetime is distinctly higher. In Fig. 7 the quantity $\log [n(R) \times R^2 \times v(R)]$ is plotted as a function of the distance R from the nucleus where

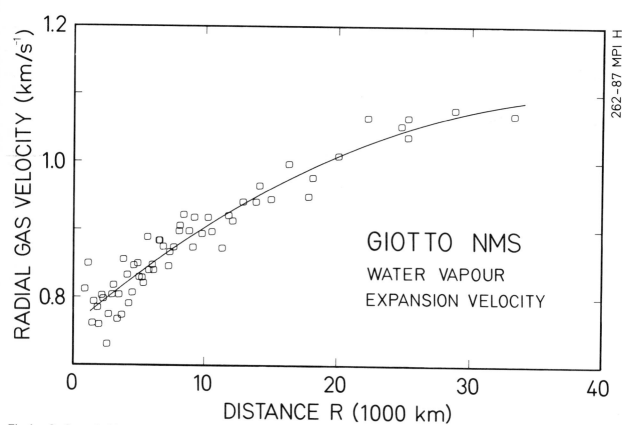

262–87 MPI H

Fig. 6 — Outflow velocities of the gases in the coma of comet Halley measured by the E-Analyzer of the Giotto neutral mass spectrometer. Figure adapted from Lämmerzahl *et al.* (1987). The solid curve represents a least squares polynomial fit to the data points where the coefficients are: $c_0 = 7.56 \times 10^{-1}$, $c_1 = 1.67 \times 10^{-5}$, $c_2 = -2.01 \times 10^{-10}$.

($n(R) = H_2O$ number density, $v(R) =$ expansion velocity). Only those data points from Fig. 3 are taken for which absolute calibration is available. The distance scale has been shifted 110 km closer to the nucleus in order to account for the changes in timing and in the flyby geometry which became known after the publication of the data in Fig. 3. The middle curve corresponds to photodestruction of H_2O with a lifetime of 5.6×10^4 s representing a least squares fit to the data points. The upper and the lower curve are models with H_2O lifetimes 20% lower and higher than the best fit value. The photodestruction lifetime of water vapour derived from the *in situ* measurements of the NMS experiment is 20% lower than the generally used value of 6.6×10^4 s at 0.9 AU (Hüebner and Carpenter, 1979). In a recent study of the photodissociation of water, Crovisier (1989) derived a dependence of the H_2O photodestruction rate on the solar

UV flux as characterized by the 10.7 cm solar index. The value of $F_{10.7} = 74.3$ on March 13, 1986 results in a theoretical lifetime of 6.2×10^4 s at 0.9 AU. The water lifetime of 5.6×10^4 s measured *in situ* by the Giotto NMS experiment is only 10% smaller, an agreement which is — in view of the uncertainties involved — quite satisfactory. The H_2O density profile in the distance range shown, is adequately described by the radial expansion of H_2O released from the nucleus, and does not indicate contributions from an extended source in the coma (see also section 4). Thus, the extrapolation of the measured H_2O abundance to the nucleus' surface involves only a small, well established correction.

The Giotto NMS made measurements in a distance of ≈ 1000 km from the nucleus. At such a far range, very reactive molecules or species with a lifetime much smaller than 1 000 s were not observable by the

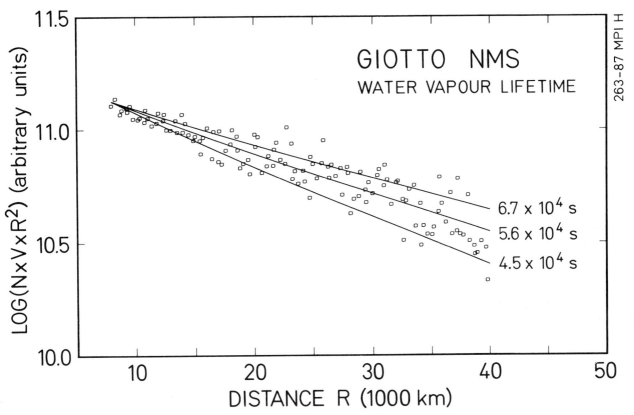

Fig. 7 — Comparison of the measured H_2O density with model calculations. The model used takes the radial dependence of the expansion velocity into account (see Fig. 6). The middle curve represents a least squares fit to the data points, corresponding to a photodestruction lifetime of 5.6×10^4 s for H_2O. The upper and lower curves assume 20% higher and lower lifetimes. Figure based on data published by Krankowsky *et al.* (1986)

instrument. This limitation can cause considerable uncertainty in deducing the chemical composition at the nucleus from the measurements in the coma. For instance, it will probably not be possible to decide whether S_2 (lifetime ≈ 350 s), discovered by A'Hearn *et al.* (1983) in Comet IRAS-Araki-Alcock, is a possible precursor of the abundant neutral and the ionized sulphur observed by the Giotto mission.

4 EXTENDED SOURCE OF GASES

Gas jets were discovered in the visible molecular emission bands of the radicals CN and C_2 in narrow band filtered images of the comet Halley Fig. 8)* by A'Hearn *et al.* (1986a, b). These jets, not present in the reflected continuum from the dust, extended to projected distances of more than 50 000 km, and

* to be found in the colour illustration section in the centre of the book.

persisted for several weeks. It was estimated that as much as half the amount of CN observed in the coma by Millis & Schleicher (1986) could be produced within the jets. A'Hearn and co-workers linked the observed gas jets to submicrometre dust particles, and argued that the radicals observed in the jets were produced by the photodissociation of the parent molecules on the grain surface or from the short-lived parents released from the grains into the gas phase. The mass loss rate of the dust required to support the observed amount of the radicals in jets was estimated to be only a small fraction of the total mass loss rate of the dust from the comet (McDonnell *et al.* 1986, Mazets *et al.* 1986). The CHON particles detected by the particle impact analyzer experiment on the Giotto and Vega spacecraft and composed primarily of C, H, O, and N, were suggested as candidate grains. Analyzed dust particles in the micrometre to submicrometre range were found to be composed essentially of two components, a silicate-like and a refractory organic fraction, in proportions varying from particle to particle (Kissel *et al.* 1986, Clark *et al.* 1986). A quantitative study of photosputtering from CHON grains and of the spatial development of trace gas jets as possible mechanisms for the gas jet formation showed the feasibility of both processes to produce gas jets (Combi 1987).

The radial density profile of CO obtained from measurements of the 28 amu/e peak in the mass spectrum of the NMS experiment on Giotto can be understood by the presence of an extended source for CO (Eberhardt *et al.* 1986a, 1987a). In Fig. 9 the intensity of the 28 amu/e peak times the square of the distance is shown as a function of the distance from the nucleus. Owing to the large photodestruction scale length of $\approx 7 \times 10^5$ km for CO the plotted quantity should be constant, if the number density of CO decreases only because of the radial expansion. It is evident from Fig. 9 that the signal grows exponentially, doubling within 10 000 km. At 20 000 km the signal becomes essentially constant. From these data the

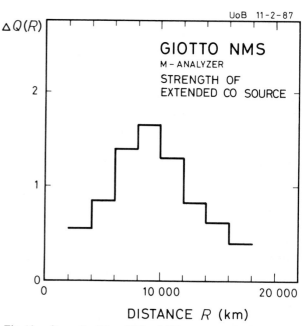

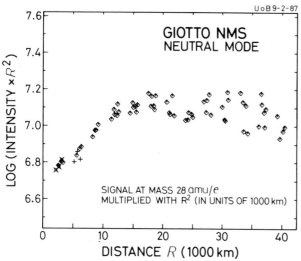

Fig. 9 — Evidence for extended CO source obtained from the Giotto neutral mass spectrometer. Shown is the ion current in the M-analyzer at mass 28 amu/e multiplied with the square of the distance from the nucleus. This value should be independent of the distance from the nucleus if no sources or sinks are present, and provided that the expansion velocity is constant. The increase by more than a factor of two observed for distances < 20 000 km shows that a parent molecule or dust grains must be decaying into fragments — presumably CO — contributing to the mass 28 amu/e signal. Figure from Eberhardt *et al.* (1986a, 1987a).

Fig. 10 — Strength of the additional CO source required to explain the radial increase observed for the 28 amu/e signal (see Fig. 9). The quantity $\triangle Q(R)$ (in arbitrary units) is the total amount of gas released in concentric shells with 2 000 km thickness. Figure from Eberhardt *et al.* (1987).

radial distribution of an extended source of CO was calculated in shells of 2 000 km thickness (Fig. 10). The maximum CO gas production was found at $\approx 9\,000$ km. Micrometre-sized dust particles of the CHON type were proposed as a source for the additional CO. It is interesting to note that Vaisberg *et al.* (1986) observed a distinct variation of the grain size distribution with distance from the nucleus. Small grains become overabundant, and the relative abundance of the medium-sized grains seems to decrease with increasing distance. The depletion starts closest to the nucleus ($R \leq 10\,000$ km), with grains in the 10^{-14} to 10^{-13} g range. This has been interpreted as the result of splitting of larger grains. Aggregate particles seem to lose their interstitial material and eventually break up into very small grains (Eberhardt *et al.* 1986a, 1987a). This process proceeds faster with smaller grains, and the lifetime of the aggregates should increase with their size. Once the break-up begins, additional surfaces of volatile material will be exposed, leading to an enhancement of gas production. The shape of the source function and the location of the maximum at $\approx 9\,000$ km (Fig. 10) is in qualitative agreement with the results of Vaisberg *et al.* (1986). If the observed CO enhancement is due to a release from grains, then photosputtering from grains is favoured as the mechanism of CN jet formation. Support for such a mechanism comes from the radio observation of an extended source of H_2CO between 10^3 and 10^4 km by Snyder *et al.* (1989).

The chemical nature of the material in the grains releasing CO, CN and H_2CO or the short-lived parent of these species is uncertain. The proposed photolyzed outer mantle of interstellar grains (Greenberg 1982) could be the source of evaporating organic molecules, including CO or a parent of it. In interstellar clouds CO is condensing directly on the grain surfaces as solid CO (Lacy *et al.* 1984), which requires temperatures as low as 17 K (Léger 1983). Laboratory irradiation with UV of CO, NH_3 and other ice mixtures has shown that molecules containing cyano ($C \equiv N$) groups are formed, and that these reaction products have much lower vapour pressure, and are not lost even at temperatures as high as 150 K (Lacy *et al.* 1984). From IR absorption spectra Lacy *et al.* conclude that XCN molecule formation also occurs on the interstellar grains. The observation of the extended CO and CN sources in the coma of comet Halley may thus be an indication for the presence of relatively unaltered interstellar grains in the nucleus.

Interpreting spectra obtained by the positive ion cluster composition analyzer (PICCA) experiment on the Giotto spacecraft, Huebner (1987) and Mitchell *et al.* (1987) proposed the existence of polymerized formaldehyde $((H_2CO)_n)$, also known as polyoxymethylene (POM), in dust grains of Halley. They argued that impact or photodissociation will disintegrate POM into compounds that consist of alternating methylene (CH_2) and atomic oxygen units, which were presumably detected as ions by the PICCA instrument (Korth *et al.* 1986, Mitchell *et al.* 1986). Decomposition of POM was proposed as the mechanism for the extended source of CO and as a source for H_2CO. The affinity of POM to silicates and graphite suggests POM as the interstitial glue of the CHON dust particles bonding submicrometre grains into larger particles which disintegrate when the glue has evaporated. This process, as discussed above, could then explain the observed change in the dust particle size distribution around 10 000 km distance from the

nucleus (Vaisberg *et al.* 1986). Other polymers containing atomic nitrogen or cyanide radical could represent the source for the observed CN jets. The mass ratio of POM to dust was estimated to be ≈ 0.02 at 10 000 km, assuming a dust to gas mass ratio of 0.3 (Mitchell *et al.* 1987). However, to account for the CO released from grains, a CO production rate relative to H_2O of $\approx 8\%$ is required (Eberhardt *et al.* 1987a). This number translates into a mass ratio of CO to dust of ≈ 0.3 (dust/gas ≈ 0.3 assumed). The POM/dust ratio would have to be higher than 0.3, as only a fraction of the POM dissociates into CO. This is not only in disagreement with the estimate of the mass ratio POM to dust, but also means that a very substantial fraction of the grains would have to evaporate, although the evaporated grains are not necessarily included in the dust production rate, as this rate is partly based on dust densities measured at distances $> 8\,000$ km. Therefore, despite the fact that POM offers a qualitative explanation, it seems not to account quantitatively for the amount of CO observed in the extended source. Furthermore, it must be pointed out that no direct evidence for the presence of POM in comet Halley exists.

Irrespective of the chemical nature of the CO-containing compounds in the dust grains, the problem of understanding how such a large amount of CO is released in a relatively short time after the dust left the nucleus remains. Therefore, as an alternative, anisotropic outgassing from the nucleus should be considered. Images from the Vega and Giotto missions have shown that the emission of dust and (presumably) gas, from the nuclear surface was highly anisotropic. Although it can be expected that an anisotropic localized outgassing will become isotropized at larger distances, owing to the lateral pressure gradients close to the nucleus, as well as due to the increasing gas temperature from chemical heating further out, numerical calculations by Kitamura (1986) and Kömle & Ip (1986) have shown that recognizable anisotropies in the outflow still persist at a few thousand kilometres distance. However, no noticeable structures seem to be present in the radial profile of H_2O along the Giotto trajectory (see Fig. 7) which would be expected from bulk anisotropies. Another, perhaps more likely, alternative, is a locally enhanced CO/H_2O ratio in one or several of the active regions on the nucleus. Combi (1986) has shown that such trace gas jets would remain fairly well focused even at large cometocentric distances. The localized injection of a trace gas does not disturb the expansion of the bulk and other trace gases. Also, temporal variations in the CO/H_2O ratio at the nucleus would result in radial changes of the coma gas composition without necessarily influencing the bulk gas properties. It is obvious that such variability in the composition of the volatiles has implications for the formation of comet Halley as well as for its evolution during earlier passages through the inner Solar System.

5 DISCUSSION AND SUMMARY

The inventory of parent molecules other than H_2O detected in the coma of comet Halley is summarized in Table 1. According to the mass spectra of gases measured by the NMS experiment on Giotto, water constitutes $\approx 80\%$ by volume of the gases in the coma (Krankowsky et al. 1986). This is excluding the amount of CO shown to be probably released from an extended source (Eberhardt et al. 1986a, 1987a). Thus one could argue, by simply adding the appropriate numbers in Table 1, that the species listed account approximately for the remaining 20% of the gaseous emanation from the nucleus, considering the uncertainties in some of the relative gas production rates quoted. But from the NMS mass spectra it is obvious that a few more species not listed in Table 1, and not yet all identified completely, have been present in the coma. Furthermore, heavier species outside the mass range of the M-analyzer of the NMS, except CO_2, were not included in the estimate of water abundance made by Krankowsky et al. (1986).

There is one caveat in interpreting the numbers in Table 1 as the nuclear gas production rates and as being indicative of the composition of the ices in the nucleus. Even the in situ mass spectrometric measurements of the radial density profiles of gaseous species with high spatial resolution to as close as $\approx 1\,000$ km distance cannot discern between gas emanation from the nucleus and release from the dust grain halo although such a distinction was possible in the case of CO. It is conceivable that some of the observed species, as in the case of CO and the progenitors of CN and H_2CO, are adsorbed on, or incorporated in,

grains and are released within the first 1 000 km from the nucleus. In this case it is also reasonable to argue that such volatiles are contained in the solid fraction of the nucleus rather than in ices. It is unfortunate, but very likely true, that the observational data from the 1985/86 apparition of comet Halley will not allow distinction between ice and rocks as the source of most of the gas species identified. Because no observations to the contrary are available, we shall assume in further discussion the conventional view incorporating volatiles such as CO, CO_2, CH_4, NH_3, and N_2 in the water ice. The possibility of a dust-related origin for at least a fraction of the volatiles must, however, be kept in mind.

From a mass balance standpoint there is enough water in Halley to include these species, which add up to about 12% by number, as clathrates and hydrates in the water ice. Also, Bar-Nun et al. (1985) have shown experimentally that gases in excess of the clathrate–hydrate limit of 0.17 can be trapped at low temperatures in water ice. Scenarios that are considered for comet formation invoke dust and condensable gases of interstellar origin and the solar nebula condensates. In the thermodynamical equilibrium models (Lewis 1972) and in later disequilibrium models (Lewis & Prinn 1980), clathrate and hydrates were supposed to be the low-temperature condensates which retained in the outer Solar System nitrogen and carbon either as ammonia hydrate $NH_3 \bullet H_2O$ and methane clathrate $CH_4 \bullet 6H_2O$ or as $N_2 \bullet 6H_2O$ and $CO \bullet 6H_2O$ clathrates. In the gas phase of the interstellar medium, carbon and nitrogen occur predominantly in the molecules CO and N_2 (Mitchell et al. 1981). Thus from the relative distribution of these two elements among the molecular forms CO, CH_4, N_2, and NH_3, hints can be obtained as to the chemical origin of the volatiles in comet Halley.

As was pointed out by Allen et al. (1987) the abundance ratios $CH_4/CO \approx 0.1$ and $NH_3/N_2 > 0.5$ in Halley's coma greatly exceed those which are observed in the interstellar medium. This speaks against a pristine interstellar nature of the volatiles in Halley. According to Prinn & Fegley (1987), chemical disequilibrium probably prevailed in the outer regions of the solar nebula, and then kinetic quenching of the $CO \Rightarrow CH_4$ and $N_2 \Rightarrow NH_3$ conversion leads to CH_4/CO and NH_3/N_2 ratios much smaller than the

Halley values. Subnebulae around the giant proto-planets were suggested as the environment where at least part of the volatiles in comet Halley condensed, because there conditions favour CH_4 and NH_3 very much over CO and N_2. Therefore, Prinn & Fegley hypothesized that Halley's CH_4 and NH_3 constitutes a heterogeneous mixture of condensates from inter-stellar cloud, solar nebula, and subnebula material.

The relative amount of nitrogen in the gas phase of Halley is less than expected from its solar abundance (Balsiger et al. 1986). The abundance of nitrogen is also low if gas plus dust are considered (Geiss 1987), whereas the amounts of carbon and oxygen are very close to their relative solar abundances. The apparent deficit of nitrogen in Halley could be explained if nitrogen was mainly present as N_2 when Halley formed. Although the N_2 clathrate is expected as a major low-temperature condensate in the solar nebula (Lewis & Prinn 1980), it is less stable than the CO clathrate. Furthermore, if all the CO observed in Halley's coma resided as clathrate in the water ice, then available cavities could have been occupied.

Prialnik & Bar-Nun (1987) have studied the thermal evolution of an initially very cold object on a Halley-like orbit. The phase transition from amor-phous to crystalline ice is a major internal heat source. The transition occurs episodically, and the energy release during the amorphous to cubic transition at 137 K is sufficient to raise the temperature above the cubic to hexagonal transition temperature of 160 K. Each transformation episode converts a layer of several tens to several hundreds of metres in thick-ness. A large fraction of gases such as CO, CH_4, and N_2 trapped at very low temperatures are released during the warming and phase transition. Prialnik & Bar-Nun (1987) argue that a fraction of the released gas may be trapped in gas pockets. In later apparitions such gas pockets may be vented explosively. Thus, deeper layers of still amorphous, gas-rich ice can be exposed. The implications of this model for the present discussion are twofold. If most of the H_2O in the coma stems from the transformed surface shell of hexagonal ice, then the observed volatiles CO, N_2, CH_4 etc. are probably highly depleted compared to their initial abundance in the pristine ice. The depletion factors can be different for the different molecules. Conclusions on the origin of the cometary

volatiles based on the observed abundance of CO, N_2 CH_4, NH_3, etc. in the coma must thus be viewed with caution. Exposed parts of the deeper, not yet transformed amorphous ice layer, or cracks venting the gas pockets, could be sources for the gas with enhanced CO/H_2O ratios as discussed in the chapter on extended gas sources.

The large amount of CO ($\approx 50\%$ of the total CO in the coma) seemingly being released from dust grains, poses a problem by itself. As no indication for a comparable source of water was found, CO must be stored in the dust in a form other than ice. Geiss (1987) remarked that CO, in contrast to N_2, bonds weakly to solids, metals in particular. However, in view of the very large amount of CO that needs to be adsorbed this position is hardly tenable. From the total mass of dust observed in the coma $\approx 30\%$ must be attributed to CO. Considering that the bulk of the mass of dust is in the larger particles, the situation is even less favourable. The same reservation holds for refractory organics such as the CHON component in dust grains, or the proposed POM fraction as CO carrier. The alternative to a dust origin discussed before (see section on extended sources of gases), namely aniso-tropic outgassing of CO from the nucleus, would require localized areas that contain or outgas water containing CO enhanced by a factor of two. Evidently such a presumption would have implications for the formation of comets and for the sub-surface evolution of the nucleus after the comet has left the Oort cloud and experienced cyclic temperature variations on its eccentric orbit around the Sun.

Isotopic abundances of oxygen and sulphur in the gas phase of comet Halley agree with the terrestrial values. Despite the relatively wide limits on the deuterium abundance found in Halley's water, some conclusions with respect to the origin of the cometary water are evident (Eberhardt et al. 1986b, 1987b). The deuterium abundance in Halley is comparable to the D/H ratio in the Solar System objects poor in hydrogen such as the Earth, the silicate fraction in meteorites, and Titan. But Halley's water is distinctly enriched in deuterium compared to the protosolar hydrogen and to hydrogen accreted in gaseous form from the solar nebula in bodies such as Jupiter and Saturn. In contrast, some molecules in interstellar molecular clouds show a much larger enrichment due to deuteration by ion–molecule reactions. As noted by Geiss (1987), this does not preclude IMC's from having contributed to cometary water, because condensation could have interrupted deuteration of H_2O. Wyckoff *et al.* (1989) have argued that the approximately 35% enrichment in ^{13}C observed in CN relative to the bulk Solar System value is not compatible with the formation of cometary nuclei in the vicinity of Uranus and Neptune. Measurements of the $^{12}C/^{13}C$ ratio in the outer Solar System presently agree with the solar value (de Bergh 1988). However, CN in Halley is only a minor species, comprising less than 1% of the volatile carbon inventory of the comet (cf. Table 1). Meteorites were most likely formed within the orbit of Jupiter. Nevertheless, they contain minor phases showing isotopic anomalies including ^{13}C enrichment due to incomplete mixing within the Solar System of components from different nucleosynthesis sites. The highest observed ^{13}C enrichment of about 3000% corresponds to $^{12}C/^{13}C = 3$ (Wopenka *et al.* 1989). There is strong evidence that a sizeable fraction (50% to 90%) of the cometary CN stems from CHON particles (A'Hearn *et al.* 1986, A'Hearn 1988). If only 2% by number of the CN originated from CHON particles with $^{12}C/^{13}C \approx 4$, corresponding to the equilibrium value for CNO burning in stars (Iben 1975), then the $^{13}C/^{12}C$ of CN is lowered from 89 to 66*. As the isotopic composition of the bulk carbon in Halley (CO, CO_2) is at present unknown, any conclusions based on bulk carbon isotopic composition seem premature.

ACKNOWLEDGEMENTS

This work was supported by the German Bundesminister für Forschung und Technologie (Grant 01OF8512) and by the Swiss National Science Foundation (Grant 25545.88).

---

* Based on an erroneous calculation Wyckoff *et al.* (1989) discussed and discarded the possibility that ^{13}C rich CHON particles could substantially lower the $^{12}C/^{13}C$ ratio in CN. The assumptions of Wyckoff *et al.* (10% of the CHON particles have $^{12}C/^{13}C \approx 4$ and 50% of the CN stems from CHON particles) would give $^{12}C/^{13}C = 48$, considerably lower than the measured value of 65.

REFERENCES

A'Hearn, M.F., & Cowan, J.J. (1980). Vaporization in comets: the icy grain halo of Comet West. *Moon and Planets* **22** 41–52

A'Hearn, M.F., Feldman, P.D., & Schleicher, D.G. (1983) The discovery of S_2 in the coma of comet IRAS-Araki-Alcock. *Astrophys. J.* **274** L99-L103

A'Hearn, M.F., Dwek, E., & Tokunage, A.T. (1984) Infrared photometry of Comet Bowell and other comets. *Astrophys. J.* **282** 803–806

A'Hearn, M.F., Hoban, S., Birch, P.V., Bowers, C., Martin, R., & Klinglesmith III, D.A. (1986a) Cyanogen jets in Comet Halley. *Nature* **324** 649–651

A'Hearn, M.F., Hoban, S., Birch, P.V., Bowers, C., Martin, R., & Klinglesmith III, D.A. (1986b) Gaseous jets in Comet P/Halley. *Proc. 20th ESLAB Symposium on the Exploration of Halley's comet* ESA SP-250 **I** 438–486

A'Hearn, M.F. (1988) private communication; quoted in Wyckoff *et al.* (1989)

Allen, M., Delitsky, M., Huntress, W., Yung, Y., Ip, W.-H., Schwenn, R., Rosenbauer, H., Shelley, E., Balsiger, H., & Geiss, J. (1987) Evidence for methane and ammonia in the coma of Comet Halley. *Astron. Astrophys.* **187** 502–512

Altenhoff, W.J., Batrla, W., Huchtmeier, W.K., Schmidt, J., Stumpf, P., & Walmsley, M. (1983) *Astron. Astrophys.* **125** L19

Balsiger, H., Altwegg, K., Bühler, F., Geiss, J., Ghielmetti, A.G., Goldstein, B.E., Huntress, W.T., Ip, W.-H., Lazarus, A.J., Meier, A., Neugebauer, M., Rettemund, U., Rosenbauer, H., Schwenn, R., Sharp, R.D., Shelley, E.G., Ungstrup, E., & Young, D.T. (1986) Ion composition and dynamics at Comet Halley. *Nature* **321** 330–334

Bar-Nun, A., Herman, G., Laufer, D., & Rappaport, M.L. (1985) Trapping and release of gases by water ice and implications for icy bodies. *Icarus* **63** 317–332

Biermann, L., Giguere, P.T., & Huebner, W.F. (1982) A model of a comet coma with interstellar molecules in the nucleus. *Astron. Astrophys.* **108** 221–226

Bockelée-Morvan, D., Crovisier, J., Despois, D., Forveille, T., Gérard, E., Schraml, J., & Thum, C. (1986) A search for parent molecules at millimetre wavelengths in Comets Giacobini-Zinner 1984e and P/Halley 1982i. *Proc. 20th ESLAB Symposium on the Exploration of Halley's Comet*, ESA SP-250 **I** 365–367

Campins, H., Rieke, G.H., & Lebfsky, M.J. (1983) Ice in Comet Bowell. *Nature* **301** 405

Clark, B., Mason, L.W., & Kissel, J. (1986) Systematics of the CHON and other light-element particle populations in Comet Halley. *Proc. 20th ESLAB Symposium on the Exploration of Halley's Comet*, ESA SP-250 **III** 353–358

Combes, M., Moroz, V., Crifo, J.P., Bibring, J.P., Coron, N., Crovisier, J., Encrenaz, T., Sanko, N., Grigoriev, A., Bockelée-Morvan, D., Gispert, R., Emerich, C., Lamarre, J.M., Rocard, F., Krasnopolsky, V., & Owen, T. (1986) Detection of parent molecules in Comet Halley from the IKS-experiment. *Proc. 20th ESLAB Symposium on the Exploration of Halley's Comet*, ESA SP-250 **I** 353–358

Combes, M., Moroz, V.I, Crrovisier, J., Encranez, T., Bibring, J.P., Grigoriev, A.V., Sanko, N.E., Coron, N., Crifo, J.F., Gispert, R., Bockelée-Morvan, D., Nikolsky, Yu.V., Kransnopolsky, V.A., Owen, T., Emerich, C., Lamarre, J.M., & Rocard, F. (1988) The 2.5–12 μm spectrum of comet Halley from the IKS-Vega experiment. *Icarus* **76** 404–436

Combi, M.R. (1987) Sources of cometary radicals and their jets: gases or grains. *Icarus* **71** 178–191

Combi. J.R., & Delsemme, A.H. (1980) Neutral cometary atmospheres 1. An average random walk model of photodissociation in comets. *Astrophys. J.* **237** 633–640

Crifo J.F. (1983) Visible and infrared emissions from volatile and refractory cometary dust. A new interpretation of Comet Kohoutek observations. In: *Cometary exploration*, ed. T.I. Gombosi, Hungarian Acad. Sci. **II** 167–176

Crifo, J.F. (1989) Water clusters in the coma of Comet Halley and their effect on gas density, temperature and velocity. *Icarus* (in press)

Crovisier, J., Despois, D., Gerard, E., Irvine, W.M., Kazes, I., Robinson, S., & Schloerb, F.P. (1981) A search for the 1.35 cm line of H_2O in comets Kohler (1977 XIV) and Meier (1978 XXI). *Astron. Astrophys.* **97** 195–198

Crovisier, J. (1989) On the photodissociation of water in cometary atmospheres. *Astron. Astrophys.* **213** 459–464

de Bergh, C. (1988) private communication; quoted in Wyckoff *et al.* (1989)

Delsemme, A.H. (1972) Vaporization theory and non-gravitational forces in comets. In *Origin of the solar system*, ed. H. Reeves (Paris: C.E.R.N.) 305–310. Also in *Asteroids, Comets, Meteoritic Matter*, ed. Cristescu, W. Klepczynski, & B. Milet (Romania: Publ. Acad. Soc. Republ. of Romania, 1974), 315–322

Delsemme, A.H. (1975) Physical Interpretation of the brightness law of Comet Kohoutek. In: *Comet Kahoutek* ed. G.A. Gary, NASA SP-355, Washington, 195

Delsemme, A.H. (1985) The sublimation temperature of the cometary nucleus: Observational evidence for H_2O snow. In: *Ices in the solar system*, ed. J. Klinger, Benest, D., Dollfus, A., & Smoluchowski, R., NATO ASI Series, Series C: Mathematical and Physical Sciences **156** D. Reidel Publishing Company, 367–387

Delsemme A.H., & Swings, P. (1952) Hydrates de gaz dans les noyaux cométaires et les grains interstellaires. *Ann. d'Astrophys.* **15** 1–6

Despois, D., Crovisier, J., Bockelée-Morvan, D., Schraml, J., Forveille, T., & Gérard, E. (1986) Observations of hydrogen cyanide in Comet Halley. *Astron. Astrophys.* **160** 11–12

Drapatz, S., Larson, H.P., & Davis, D.S. (1986) Search for methane in Comet Halley. *Proc. 20th ESLAB Symposium on the Exploration of Halley's Comet*, ESA SP-250 **I** 347–352

Eberhardt, P., Krankowsky, D., Schulte, W., Dolder, U., Lämmerzahl, P., Berthelier, J.J., Woweries, J., Stubbemannn, U., Hodges, R.R., Hoffman, J.H., & Illiano, J.M. (1986a) On the CO and N_2 abundance in Comet Halley. *Proc. 20th ESLAB Symposium on the Exploration of Halley's Comet*, ESA SP-250 **I** 383–386

Eberhardt, P., Dolder, U., Schulte, W., Krankowsky, D., Lämmerzahl, P., Hoffman, J.H., Hodges, R.R., Berthelier, J.J., & Illiano, J.M. (1986b) The D/H ratio in water from Halley. *Proc. 20th ESLAB Symposium on the Exploration of Halley's Comet*, ESA SP-250 **I** 539–541

Eberhardt, P., Krankowsky, D., Schulte, W., Dolder, U., Lämmerzahl, P., Berthelier, J.J., Woweries, J., Stubbemann, U., Hodges, R.R., Hoffman, J.H., & Illiano, J.M. (1987a) On the CO and N_2 abundance in Comet P/Halley. *Astron. Astrophys.* **187** 481–484

Eberhardt, P., Dolder, U., Schulte, W., Krankowsky, D., Lämmerzahl, P., Hoffman, J.H., Hodges, R.R., Berthelier, R.R., &

Illiano, J.M. (1987b.) The D/H ratio in water from Comet P/Halley. *Astron. Astrophy.* **187** 435–437

Feldman, P.D., Festou, M.C., A'Hearn, M.F., Arpigny, C., Butterworth, P.S., Cosmovici, C.B., Danks, A.C., Gilmozzi, R., Jackson, W.M., McFadden, L.A., Patriarchi, P., Schleicher, D.G., Tozzi, G.P., Wallis, M.K., Weaver, H.A., & Woods, T.N. (1986) IUE observations of Comet Halley: evolution of the UV spectrum between September 1985 and July 1986. *Proc. 20th ESLAB Symposium on the Exploration of Halley's Comet*, ESA SP-250 **1** 325–328

Festou, M.C. (1981) The density distribution of neutral compounds in cometary atmospheres. *Astron. Astrophys.* **95** 69–79

Festou, M.C., Feldman, P.D., A'Hearn, M.F., Arpigny, C., Cosmovici, C.B., Danks, A.C., McFadden, L.A., Gilmozzi, R., Patriarchi, P., Tozzi, G.P., Wallis, M.K., & Weaver, H.A. (1986) IUE observations of Comet Halley during the VeGa and Giotto encounters. *Nature* **321** 361–363

Fowler, A. (1910) Investigations relating to the spectra of comets. *Mon. Not. Roy. Astron. Soc.* **70** 484–496

Geiss, J. (1987) Composition measurements and the history of cometary matter. *Astron. Astrophys.* **187** 859–866

Greenberg, J.M. (1982) Laboratory dust experiments — tracing the composition of cometary dust. In: *Cometary exploration*, ed. T.I. Gombosi, Hungarian Acad. Sci. **II** 23–54

Hanner, M. (1984) Comet Cernis: Icy grains at last? *Astrophys. J.* **277** L75

Hartmann, W.K., & Cruikshank, D.P. (1983) Systematics of ices among remote comets, asteroids, and satellites. *Bull. Amer. Astron. Soc.* **15** 808 (abstract)

Haser, L. (1957) Distribution d'intensité dans la tête d'une comète. *Bull. Acad. Roy. Belgique*, Classe de Sciences **43** 740–750

Haser, L. (1966) Calcul de distribution d'intensité relatif dans une tête cométaire. *Mém. Soc. Roy. Liège*, Ser. 5 **12** 233–241

Herzberg, G., & Lew, H. (1974) Tentative identification of the H_2O^+ ion in Comet Kohoutek. *Astron. Astrophys.* **31** 123–124

Hirn, G.A. (1889) *Constitution de l'espace céleste.* Gauthier-Villars, Paris

Hodges Jr., R.R. (1989) Monte Carlo Simulation of Nonadiabatic Expansion in Cometary Atmospheres. *Icarus* (in press)

Hollis, J.M., Brandt, J.C., Hobbs, R.W., Maran, S., & Feldman, P.D. (1981) Radio observations of Comet Bradfield (1979l) *Astrophys. J.* **244** 355–357

Huebner, W.F., & Carpenter, C.W. (1979) Solar Photo Rate Coefficients. Los Alamos Scientific Report. No. LA-8085-MS

Huebner, W.F. (1987) First polymer in space identified in Comet Halley. *Science* **237** 628–630

Iben Jr., I. (1975). Thermal pulses; p-capture, s-process nucleosynthesis; and convective mixing in a star of intermediate mass. *Astrophys. J.* **196** 525–547

Ip, H.-W. (1989) Ion Composition and Chemistry. *Adv. Space Res.* **9**(3) 141–150

Magee-Saur, K., Scherb, F., Roessler, F.L., & Harlander, J. (1988) Fabry-Perot Observations of NH_2 Emissions from Comet Halley. *Bull. Am. Astron. Soc.* **20**(3) 827 (abstract)

Jackson, W.M., Clark, T., & Donn, B. (1976) Radio detection of H_2O in Comet Bradfield (1974b) In: *The study of comets*, eds. Donn, B., Mumma, M., Jackson, W., A'Hearn, M., & Harrington, R., (Washington: NASA SP-393), 272–280

Kissel, J., Brownlee, D.E., Büchler, K., Clark, B.C., Fechtig, H., Grün, E., Hornung, K., Igenbergs, E.B., Jessberger, E.K., Krueger, F.R., Kuczera, H., McDonnell, J.A.M., Morfill, G.M., Rahe, J., Schwehm, G.H., Sekanina, Z., Utterback,

N.G., Völk, H.J., & Zook, H.A. (1986) Composition of Comet Halley dust particles from Giotto observations. *Nature* **321** 336–337

Kitamura, Y. (1986) Axisymmetric dusty gas jets in the inner coma of a comet. *Icarus* **66** 241–257

Kömle, N.I., & Ip, W.-H. (1986) Anisotropic non-stationary gas flow dynamics in the coma of Comet Halley. *Proc. 20th ESLAB Symposium on the Exploration of Halley's Comet*, ESA SP-250 **I** 523–527

Korth, A., Richter, A.K., Loidl, A., Anderson, K.A., Carlson, C.W., Curtis, D.W., Lin, R.P., Réme, H., Sauvaud, J.A., d'Uston, C., Cotin, F., Cros, A., & Mendis, D.A. (1986) Mass spectra of heavy ions near Comet Halley. *Nature* **321** 335–336

Krankowsky, D., Lämmerzahl, P., Herrwerth, I., Woweries, J., Eberhardt., P., Dolder, U., Herrmann, U., Schulte, W., Berthelier, J.J., Illiano, J.M., Hodges, R.R., & Hoffman, J.H. (1986) *In situ* gas and ion composition measurements at Comet Halley. *Nature* **321** 326–329

Krasnopolsky, V.A., Gogoshev, M., Moreels, G., Moroz, V.I., Krysko, A.A., Gogosheva, Ts., Palazov, K., Sargoichev, S., Clairmidi, J., Vincent, M., Bertaux, J.L., Blamont, J.E., Troshin, V.S., & Valniček, B. (1986a) Spectroscopic study of Comet Halley by the VeGa-2 three-channel spectrometer. *Náture* **321** 269–271

Krasnopolsky, V.A., Moroz, V.I., Gogoshev, M., Gogosheva, Ts., Moreels, G., Clairemidi, J., Krysko, A.A., Parisot, V.S., & Tkachuk, A.Yu. (1986b) Near infrared spectroscopy of Comet Halley by the VeGa-2 three channel spectrometer. *Proc. 20th ESLAB Symposium on the Exploration of Halley's Comet*, ESA SP-250 **I** 459–463.

Lacy, J.H., Baas, F., Allamandola, L.J., Persson, S.E., McGregor, P.J., Lonsdale, C.J., Geballe, T.R., & van de Bult, C.E.P. (1984) 4.6 micron absorption features due to solid phase CO and cyano group molecules toward compact infra-red source. *Astrophys. J.* **276** 533–543

Léger, A. (1983) Does CO condense on dust in molecular clouds? *Astron. Astrophys.* **123** 271–278

Lewis, J.S. (1972) Low temperature condensation in the solar nebula. *Icarus* **16** 241–252

Lewis, J.S., & Prinn, R. (1980) Kinetic inhibition of CO and N_2 reduction in the solar nebula. *Astrophys. J.* **238** 357–364

Lämmerzahl, P., Krankowsky, D., Hodges, R.R., Stubbemann, U., Woweries, J., Herrwerth, I., Berthelier, J.J., Illiano, J.M., Eberhardt, P., Dolder, U., Schulte, W., & Hoffman, J.H. (1986) Expansion velocity and temperature of gas and ions measured in the coma of Comet Halley. *Proc. 20th ESLAB Symposium on the Exploration of Halley's Comet*, ESA SP-250 **I** 179–182

Lämmerzahl, P., Krankowsky, D., Hodges, R.R., D., Stubbemann, U., Woweries, J., Herrwerth, I., Berthelier, J.J., Illiano, J.M., Eberhardt, P., Dolder, U., Schulte, W., & Hoffman, J.H. (1987) Expansion velocity and temperature of gas and ions measured in the coma of Comet P/Halley. *Astron. Astrophys.* **187** 169–173

Laplace, P.S. (1813). *Exposition du système du Monde*. 4th edition, Paris, 130

Magee-Saur, K., Scherb, F., Roessler, F.L., & Harlander, J. (1988) Fabry-Perot Observations of NH_2 emissions from comet Halley. *Bull. Am. Astron. Soc.* **20** (3) 827

Mazets, E.P., Aptekar, R.L., Golenetskii, S.V., Guryan, Yu.A., Dyachkov, A.V., Ilyinskii, V.N., Panov, V.N., Petrov, G.G., Savvin, A.V., Sagdeev, R.Z., Sokolov, I.A., Khavenson, N.G.,

Shapiro, V.D., & Shevchenko, V.I. (1986) Comet Halley dust environment from SP-2 detector measurements. *Nature* **321** 276–278

McDonnell, J.A.M., Alexander, W.M., Burton, W.M., Bussoletti, E., Clark, D.H., Grard, R.J.L., Grün, E., Hanner, M.S., Hughes, D.W., Igenbergs, E., Kuczera, H., Lindblad, B.A., Mandeville, J.C., Schwehm, G.H., Sekanina, Z., Wallis, M.K., Zarnecki, J.C., Chakaveh, S.C., Evans, G.C., Evans, S.T., Firth, J.G., Littler, A.N., Massonne, L., Olearczyk, R.E., Pankiewicz, G.S., Stevenson, T.J., & Turner, R.F. (1986) Dust density and mass distribution near Comet Halley from Giotto observations. *Nature* **321** 338–341

Miller, F.D. (1980) H_2O^+ in the tails of 13 comets. *Astron. J.* **85** 468–473

Millis, R.L., & Schleicher, D.G. (1986) Rotational period of Comet Halley. *Nature* **324** 646–649

Mitchell, G.F., Prasad, S.S., & Huntress, W.T. (1981) Chemical model calculations of C_2, C_3, CH, CN, OH, and NH_2 abundances in cometary comae. *Astrophys. J.* **244** 1087–1093

Mitchell, D.L., Lin, R.P., Anderson, K.A., Carlson, C.W., Curtis, D.W., Korth, A., Richter, A.K., Rème, H., Sauvaud, J.A., d'Uston, C., & Mendis, D.A. (1986) Derivation of heavy (10–210 AMU) ion composition and flow parameters for the Giotto PICCA instrument. *Proc. 20th ESLAB Symposium on the Exploration of Halley's Comet*, ESA SP-250 **I** 203–205

Mitchell, D.L., Lin, R.P., Anderson, K.A., Carlson, C.W., Curtis, D.W., Korth, A., Réme, H., Sauvaud, J.A., d'Uston, C., & Mendis, D.A. (1987) Evidence for chain molecules enriched in carbon, hydrogen, and oxygen in Comet Halley. *Science* **237** 626–628

Mumma, M.J., Weaver, H.A., Larson, H.P., Davis, D.S., & Wiliams, M. (1986) Detection of water vapor in Halley's Comet *Science* **232** 1523–1528

Ogilvie K.W., Coplan, M.A., Bochsler, P., & Geiss, J. (1986) Ion composition results during the International Cometary Explorer encounter with Giacobini–Zinner. *Science* **232** 374–377

Prialnik, D., & Bar-Nun, A. (1987) On the evolution and activity of cometary nuclei. *Astrophys. J.* **313** 893–905

Prinn, R.G., & Fegley, Jr. B. (1987) Solar nebula chemistry: origin of planetary, satellite, and cometary volatiles. In: *Planetary and satellite atmospheres: origin and evolution*, University of Arizona Press, in preparation

Schleicher, D.G., A'Hearn, M.F., and the NASA and ESA IUE teams for Comet Halley, (1986) Comets P/Halley and P/Giacobini–Zinner at high dispersion. *Proc. IUE conference* (*London*)

Schloerb, F.P., Kinzel, W.M., Swade, D.A., & Irvine, W.M. (1986) HCN production from Comet Halley. *Astrophys. J.* **310** L55–L60

Snyder, L.E., Palmer, & De Pater, I. (1989) Radio Detection of Formaldehyde Emission from Comet Halley. *Astron. J.* **97**(1) 246–253

Swings, P., & Page, T.L. (1948) The spectrum of Comet 1947n. *Astrophys. J.* **108** 526–536

Vaisberg, O., Smirnov, V., & Omelchenko, A. (1986) Spatial distribution of low-mass dust particles ($m < 10^{-10}$g) in Comet Halley coma. *Proc. 20th ESLAB Symposium on the Exploration of Halley's Comet*, ESA SP-250 **II** 17–23

Weaver, H.A., Feldman, P.D., Festou, M.C., & A'Hearn, M.F. (1981a) Water production models for Comet Bradfield (1979 X). *Astrophys. J.* **251** 809

Weaver, H.A., P.D., Feldman, M.C., Festou, M.F., A'Hearn, &

Keller, H.U. (1981b.) IUE observations of faint comets. *Icarus* **47** 449

Weaver, H.A., Mumma, M.J., Larson, H.P., & Davis, D.S. (1986) Airborne infrared investigation of water in the coma of Halley's Comet. *Proc. 20th ESLAB Symposium on the Exploration of Halley's Comet*, ESA SP-250 **I** 329–334

Whipple, F.L. (1950) A comet model. I. The acceleration of Comet Encke. *Astrophys. J.* **111** 375–394

Whipple, F.L. (1951) A comet model. II. Physical relations for comets and meteors. *Astrophys. J.* **113** 464–474

Woods, T.N., Feldman, P.D., Dymond, K.F., & Sahnow, D.J. (1986) Rocket ultraviolet spectroscopy of Comet Halley and abundance of carbon monoxide and carbon. *Nature* **324** 436–438

Wopenka, B., Virag, A., Zinner, E., Amari, S., Lewis, R.S., & Anders, E. (1989). Isotopic and optical properties of large individual SiC crystals from the Murchison chondrite. *Meteoritics* **24** (in press)

Wurm, K. (1943) Die Natur der Kometen. *Mitt. Ham. Sternwarte* **8** No. 51

Wyckoff, S., & Wehinger, P.A. (1976) Molecular ions in comet tails. *Astrophys. J.* **204** 604–615

Wyckoff, S., & Lindholm, E. (1989) On the Carbon and Nitrogen Isotope Abundance Ratios in Comet Halley. *Adv. Space Res.* **9**(3) 151–155

Wyckoff, S., & Theobald, J. (1989) Molecular Ions in Comets. *Adv. Space Res.* **9**(3) 157–161

Wyckoff, S., Lindholm, E., Wehlnger, P.A., Peterson, B.A., Zucconi, J.-M., & Festou, M.C. (1989). The $^{12}C/^{13}C$ abundance ratio in Comet Halley *Astrophys. J.* **339** 488–500

Wyckoff, S., Tegler, S., & Engel, L. (1989) Ammonia Abundances in Comets. *Adv. Space Res.* **9**(3) 169–176

Yamamoto, T., Nakagawa, N., & Fukui, Y. (1983) The chemical composition and thermal history of the ices of a cometary nucleus. *Astron, Astrophys.* **122** 171–176

Index